Corinna Hager

Towards the efficient numerical simulation of a car tire

Corinna Hager

Towards the efficient numerical simulation of a car tire

Robust numerical algorithms for dynamic frictional contact problems with different time and space scales

Südwestdeutscher Verlag für Hochschulschriften

Imprint

Publisher:
Südwestdeutscher Verlag für Hochschulschriften
is a trademark of
Dodo Books Indian Ocean Ltd., member of the OmniScriptum S.R.L Publishing group
str. A.Russo 15, of. 61, Chisinau-2068, Republic of Moldova Europe
Printed at: see last page
ISBN: 978-3-8381-2442-1

Zugl. / Approved by: Stuttgart, Universität Stuttgart, Diss., 2010

Acknowledgments

This thesis summarizes the main results of my research activities during the last four years at the chair "Numerische Mathematik für Höchstleistungsrechner" of the Institut für Angewandte Analysis und Numerische Simulation at the Universität Stuttgart.

First of all, I would like to express my sincere gratitude to my supervisor Prof. Dr. Barbara Wohlmuth for her surpassing guidance and mentoring during this time. Besides her numerous activities and engagements, she has always been willing to provide advice and assistance for all kinds of questions. Her outstanding dedication to the research activities of the work group has been truly inspiring to me. In particular, I would like to thank her for giving me the possibility to present my work on some international conferences and for putting me into touch with several designated experts, which has lead to fruitful and interesting discussions.

Special thanks go to Prof. Dr. Tod Laursen and Prof. Dr. Axel Klawonn for their willingness and effort to write the referee reports for this thesis. Further, I would like to thank Prof. Dr. Helmut Harbrecht for his participation in the oral defense. I am much obliged to Prof. Dr. Patrick Le Tallec for inviting me to Ecole Polytechnique and to him and Dr. Patrice Hauret for many stimulating and inspiring discussions. Many thanks also to Prof. Dr. Luca Pavarino and Dr. Marilena Munteanu for their kind welcome at Università di Milano, as well as to Prof. Dr. Rainer Helmig and Andreas Lauser for the good cooperation.

During my employment at the IANS, I greatly enjoyed the cooperative and friendly atmosphere in our work group. I would like to thank Dr. Stephan Brunßen, Dr. Yufei Cao, Dr. Bernd Flemisch, Dr. Arpiruk Hokpunna, Dr. Andreas Klimke, Dr. Bishnu Lamichhane, Julia Niemeyer, Dr. Iryna Rybak, Dr. Evgeny Savenkov, Marc Schlienger, Dr. Igor Shevchenko, Brit Steiner, Alexander Weiß and especially Dr. Stefan Hüeber for their help, patience and advice in various topics.

I am deeply indepted to my parents for their continuous support and understanding, as well as for providing me with the possibility to pursuit an academic career. Thanks also to my friends for their moral backup and their welcome distractions.

Finally, I would like to express my deepest appreciation to Johannes for his constant encouragement and support. Without his love and patience, it would not have been possible to complete this work.

Stuttgart, July 2010 Corinna Hager

Contents

Abstract

In many technical and engineering applications, numerical simulation is becoming more and more important for the design of products or the optimization of industrial production lines. However, the simulation of complex processes like the forming of sheet metal or the rolling of a car tire is still a very challenging task, as nonlinear elastic or elastoplastic material behaviour needs to be combined with frictional contact and dynamic effects. In addition, these processes often feature a small mobile contact zone which needs to be resolved very accurately to get a good picture of the evolution of the contact stress. In order to be able to perform an accurate simulation of such intricate systems, there is a huge demand for a robust numerical scheme that combines a suitable multiscale discretization of the geometry with an efficient solution algorithm capable of dealing with the material and contact nonlinearities. The aim of this thesis is to design such an algorithm by combining several different methods which are described in the following.

Our main field of application is structural mechanics. Here, we base the implementation on finite element methods in space and implicit finite difference schemes in time. The conditions for both plasticity and frictional contact are given in terms of a set of local inequality constraints which are formulated by introducing additional inner or dual degrees of freedom. For the plastic contributions, the dual variables are defined with respect to the elements, whereas the contact multipliers are associated with the potential contact nodes. As the meshes are generally non-matching at the contact interface, we employ mortar techniques to incorporate the contact constraints in a variationally consistent way. By using biorthogonal basis functions for the discrete multiplier space, the contact conditions can be enforced node-wise, and a two-body contact problem can be solved in the same way as a one-body problem.

The next step in the construction of an efficient solution algorithm is to reformulate the local inequality conditions for plasticity and contact in terms of nondifferentiable equalities. These nonlinear complementarity (NCP) functions can be combined with the equations for the bulk material to form a set of nonlinear semismooth equations which are then solved by means of a generalized form of the Newton method for semismooth systems. Due to the local structure of the inequality constraints, this iterative scheme can be implemented as an active set strategy where the active sets are updated in each Newton iteration. Further, the additional dual degrees of freedom can easily be eliminated using local static condensation, such that only a system of the size of the displacement needs to be solved in each step. We remark that the well-known radial return method is a special case of this general framework if the plastic hardening laws are linear.

However, the convergence properties of the Newton iteration strongly depend on the choice of the NCP function. In this context, we show that the function corresponding to the radial return method is not optimal, and we present a family of modified NCP functions which allow for better convergence results.

Another important issue for the robust simulation of dynamic contact problems is related to the inertia terms. If standard time discretization schemes like the trapezoidal rule are used, the contact stress often shows spurious oscillations in time that become worse when the time step is refined. In order to avoid this effect, we employ a modified mass matrix where no mass is associated with the contact nodes. By this, the original semi-discrete system decouples into an algebraic equation in time for the contact nodes and an ordinary differential equation (ODE) in time for the other nodes. This in turn leads to much smoother results for the contact stress. We present an efficient way of obtaining the modified mass matrix by means of non-standard quadrature formulas used only for the elements near the contact boundary. Furthermore, we prove optimal a priori error estimates for the modified semi-discrete as well as for the fully discrete system, provided that the contact stress is given and that the solution is sufficiently regular.

In the last part of the thesis, we deal with the situation that the body features fine local structures near the contact zone by incorporating the multiscale aspect into the discretization. For this, the domain is decomposed into several overlapping subdomains which have different grid spacing; one global mesh that does not resolve the details and overlapping local patches with a fine triangulation. Based on a surface coupling by means of the mortar method, we construct an iterative solution scheme for the coupled problem whose convergence rate is bounded independently of the mesh size or the Lamé parameters and which can also be applied to the nearly incompressible case. Finally, we employ the subdomain decomposition for introducing a finer time step size on the patch. We present suitable interface conditions with no numerical dissipation and prove a priori error estimates with respect to time for the resulting coupled energy-conserving system. The latter can efficiently be solved by the iterative procedure presented before.

The thesis is organized as follows. The first introductory part contains the governing equations for the dynamics of an elastoplastic body subject to frictional contact in Chapter 1, followed by the derivation of the corresponding discrete nonlinear system in Chapter 2. In the second part, we present robust nonlinear solvers for the elastoplastic contact problem. After a short review of semismooth Newton methods in Chapter 3, we apply them to the cases of friction and plasticity in Chapter 4. The third part deals with the stabilization of dynamic contact problems using a modified mass matrix. In Chapter 5, the use of the mass matrix is motivated, and its efficient construction by means of non-standard quadrature rules is described. Optimal a priori error estimates are shown in Chapter 6. The final part of the thesis is concerned with the multiscale aspect using overlapping domain decomposition. In Chapter 7, we derive and analyse an iterative solution algorithm for the linear case, whereas Chapter 8 covers the nonlinear effects. Finally, Chapter 9 analyses the use of a finer time step size on the patch. We refer to the introductions at the beginning of each chapter for a detailed description of its content.

The implementation for all the numerical examples presented in this thesis is based on the finite element toolbox UG [9] and the fast optimized direct solver package PARDISO [150, 151]. The visualization has been accomplished using ParaView [1].

Part I

Introduction and problem formulation

1 Continuum mechanics

In this chapter, we present the governing equations for the dynamic frictional contact of an elastoplastic body with a given obstacle. We start with a short description of dynamic (hyper)elastic behaviour in Section 1.1, where we introduce the corresponding strain, stress and energy measures. For the derivation of the elasticity equations and further examples of elastic materials, we refer to the books [40, 123, 135, 167]. The Hamilton–Lagrange formalism and the corresponding derivation of the equilibrium equations is explained in [5, 65, 124] and the references therein. In Section 1.2, the extension of elastic media to elastoplastic materials is sketched; the latter notion allows for incorporating irreversible effects by means of additional internal variables. The concept of plasticity is explained in more detail in [77, 78, 157] and the references therein. Finally, in Section 1.3, we present the conditions for unilateral frictional contact of a deformable body with a fixed obstacle. For the theoretical background of contact problems, we refer to the textbooks [53, 104, 116, 168] and the works cited therein.

1.1 Dynamic elasticity

In the following, we consider a hyperelastic body $\Omega \subset \mathbb{R}^d$, where $d \in \{2,3\}$ denotes the number of spatial dimensions. A schematic plot of a two-dimensional setting is shown in Figure 1.1. The deformation of this body within the time interval $(0,T)$ can be described using the displacement function $\mathbf{u} : (0,T) \times \Omega \to \mathbb{R}^d$. The boundary $\partial\Omega$ is assumed to be polyhedral such that the unit outer normal $\mathbf{n}$ is defined almost everywhere. Further, we consider a partition of $\partial\Omega$ into three nonoverlapping parts Γ_D, Γ_N, Γ_C and assume $\bar{\Gamma}_D \cap \bar{\Gamma}_C = \emptyset$ for simplicity. On Γ_D, the value of the displacement is fixed to $\mathbf{u}_D$, whereas a surface load denoted by $\mathbf{g}_N$ acts on the boundary Γ_N. The boundary conditions on Γ_C are determined by the possibility of frictional contact with a fixed obstacle Ω_{obs} and are described in Section 1.3. Furthermore, the volume load acting on Ω is denoted by the vector $\mathbf{l}$.

Denoting the gradient of the deformation of Ω by $\mathbf{F} = \mathbf{F}(\mathbf{u}) = \mathrm{Id} + \nabla\mathbf{u}$, we define the right Cauchy–Green tensor $\mathbf{C} = \mathbf{C}(\mathbf{u})$ as well as the Green–Saint Venant strain tensor $\mathbf{E} = \mathbf{E}(\mathbf{u})$ by

$$\mathbf{C} := \mathbf{F}^T\mathbf{F}, \quad \mathbf{E} := \frac{1}{2}(\mathbf{C} - \mathrm{Id}). \tag{1.1}$$

With these strain tensors, we can characterize the internal energy of Ω in terms of a strain–energy density function $W(\mathbf{E}) = \widehat{W}(\mathbf{C})$. Together with the external and the kinematic energy contributions, the total energy of Ω at time $t \in (0,T)$ is then given by

$$\mathbb{E} = \int_\Omega \left(\frac{1}{2}\varrho\|\dot{\mathbf{u}}\|^2 + W(\mathbf{E}) - \mathbf{l}\cdot\mathbf{u}\right) d\mathbf{x} - \int_{\Gamma_N} \mathbf{g}_N \cdot \mathbf{u}\, ds, \tag{1.2}$$

with $\varrho = \varrho(\mathbf{x})$ denoting the density of Ω and $\dot{\mathbf{u}} := \frac{\partial}{\partial t}\mathbf{u}$ standing for the time derivative of $\mathbf{u}$.

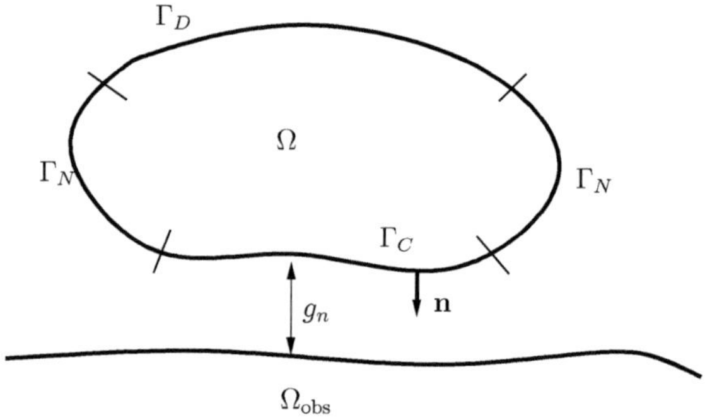

Figure 1.1: Geometry for one-body contact problem

Further, we make use of the energy density W to define the second Piola–Kirchhoff tensor $\mathbf{S} = \mathbf{S}(\mathbf{u})$ and the first Piola–Kirchhoff stress tensor $\mathbf{P} = \mathbf{P}(\mathbf{u})$ by

$$\mathbf{S} = \frac{\partial W(\mathbf{E})}{\partial \mathbf{E}} = 2\frac{\partial \widehat{W}(\mathbf{C})}{\partial \mathbf{C}}, \qquad \mathbf{P} = \mathbf{FS}. \tag{1.3}$$

The latter tensor is used in the equations of motion describing the balance of linear momentum

$$\begin{aligned} \varrho\ddot{\mathbf{u}} - \operatorname{div}\mathbf{P} &= \mathbf{l}, && \text{in } \Omega, \\ \mathbf{u} &= \mathbf{u}_D, && \text{on } \Gamma_D, \\ \mathbf{Pn} &= \mathbf{g}_N, && \text{on } \Gamma_N. \end{aligned} \tag{1.4}$$

As one can see from (1.2), (1.3), the behaviour of an elastic material is characterized by its energy density W. There are many different choices for this function (see, e.g., [95, 135]); in this work, we focus on three different cases. To begin with, we consider the Saint Venant–Kirchhoff material which has a quadratic energy function W and a linear strain–stress relationship (1.3). The latter can be expressed using the fourth order Hooke tensor $\mathbb{C}^{\text{el}} = \mathbb{C}^{\text{el}}(\mathbf{x})$ whose components are given in terms of the Lamé parameters λ, μ by

$$\mathbb{C}^{\text{el}}_{ijkl} = \lambda\delta_{ij}\delta_{kl} + \mu(\delta_{ik}\delta_{jl} + \delta_{il}\delta_{jk}), \tag{1.5}$$

with the Kronecker delta function

$$\delta_{ij} = \begin{cases} 1, & i = j, \\ 0, & \text{else}. \end{cases} \tag{1.6}$$

With this, the energy and stress definitions for the Saint Venant–Kirchhoff material read

$$W(\mathbf{E}) = \frac{1}{2}\mathbf{E} : \mathbb{C}^{\text{el}}\mathbf{E} = \frac{\lambda}{2}(\operatorname{tr}\mathbf{E})^2 + \mu\operatorname{tr}(\mathbf{E}^2), \tag{1.7a}$$

$$\mathbf{S} = \mathbb{C}^{\text{el}}\mathbf{E} = \lambda(\operatorname{tr}\mathbf{E})\mathrm{Id} + 2\mu\mathbf{E}. \tag{1.7b}$$

We remark that (1.5) can also be described in terms of the elastic modulus E and the Poisson ratio ν by using the relations

$$\lambda = \frac{E\nu}{(1+\nu)(1-2\nu)}, \quad \mu = \frac{E}{2(1+\nu)}, \quad \nu = \frac{\lambda}{2(\lambda+\mu)}, \quad E = \frac{\mu(3\lambda+2\mu)}{\lambda+\mu}. \tag{1.8}$$

The linear relationship (1.7b) can only be used if the norm of the strains remains small; otherwise, more general energy functions have to be employed. As an example, we present the Mooney–Rivlin law [43, 135] which is often used to model rubber–like materials; its energy density and second Piola–Kirchhoff stress are defined according to

$$\widehat{W}(\mathbf{C}) = \frac{\lambda}{4}\left(J^2 - 1 - 2\ln J\right) - \mu \ln J + \mu c_m(2-d)\ln J \tag{1.9a}$$
$$+ \frac{\mu}{2}\left((1-c_m)(\operatorname{tr}\mathbf{C} - d) + \frac{c_m}{2}\left((\operatorname{tr}\mathbf{C})^2 - \operatorname{tr}(\mathbf{C}^2) - d(d-1)\right)\right),$$
$$\mathbf{S} = \frac{\lambda}{2}(J^2-1)\mathbf{C}^{-1} + \mu\Big((1-c_m)(\mathrm{Id} - \mathbf{C}^{-1}) + c_m\big((\operatorname{tr}\mathbf{C})\mathrm{Id} - \mathbf{C} - (d-1)\mathbf{C}^{-1}\big)\Big), \tag{1.9b}$$

with the notation $J := \det \mathbf{F} > 0$.

The third material we look at is the linearized Saint Venant–Kirchhoff law which can be employed if the displacements are small. Here, the Green–Saint Venant strain tensor $\mathbf{E}$ used in (1.7a) is replaced by its linearized version

$$\boldsymbol{\varepsilon} = \boldsymbol{\varepsilon}(\mathbf{u}) := \frac{1}{2}\left(\nabla\mathbf{u} + \nabla\mathbf{u}^T\right), \tag{1.10}$$

and the corresponding stress tensor (1.7b) is denoted by the Cauchy stress $\boldsymbol{\sigma} = \mathbb{C}^{\mathrm{el}}\boldsymbol{\varepsilon}$. In this case, Equations (1.4) become

$$\begin{aligned} \varrho\ddot{\mathbf{u}} - \operatorname{div}\boldsymbol{\sigma} &= \mathbf{l}, && \text{in } \Omega, \\ \mathbf{u} &= \mathbf{u}_D, && \text{on } \Gamma_D, \\ \boldsymbol{\sigma}\mathbf{n} &= \mathbf{g}_N, && \text{on } \Gamma_N. \end{aligned} \tag{1.11}$$

In the rest of this work, we mainly consider the problem (1.11) due to the fact that the stress $\boldsymbol{\sigma}$ linearly depends on the displacement $\mathbf{u}$.

Remark 1.1. An important aspect of all materials introduced above concerns their compressibility. The Lamé parameters of compressible elastic media are of the same order of magnitude, whereas nearly incompressible materials are characterized by a large ratio of the Lamé parameters, i.e., $\lambda \gg \mu$ or equivalently $\nu \to 0.5$. Such materials are more complicated to handle numerically, as standard displacement-based formulations often lead to volume locking (see, e.g., [6, 22, 26, 94, 146]). This topic is considered in more detail in [69] and in Section 7.3.7.

1.2 Plasticity

In spite of the generality of the concept of hyperelastic media, there are many physical phenomena that cannot be captured by these equations, e.g., damage or plastification. The characteristic feature of elastoplastic materials is that they behave as elastic media within a certain stress range, termed the elastic region; if the stress reaches the boundary of this region, plastification takes place. The left picture of Figure 1.2 illustrates a typical strain-stress diagram for an one-dimensional plastic problem. As long as the stress is below a given yield stress σ_0, the relationship between strain and stress is linear. But if the strain is increased further, the stress value is bounded from above by σ_0 for the case of perfect plasticity or by the value of a

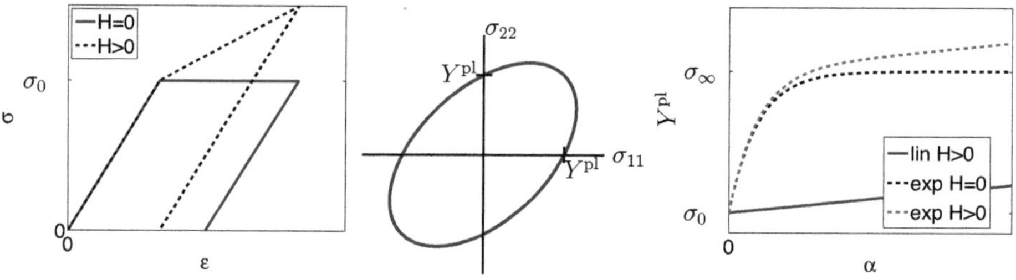

Figure 1.2: Left: strain-stress diagram for one-dimensional plasticity with no hardening ($H = 0$) and linear hardening ($H > 0$); middle: von Mises yield curve for plane stress; right: hardening functions Y^{pl} and $Y^{\mathrm{pl}}_{\mathrm{exp}}$.

so-called yield function if hardening is present. The irreversibility of this process can be seen from the fact that there is some remaining plastic strain after unloading to the stress-free state.

To incorporate this effect, the model presented in Section 1.1 is extended by a set of additional inner variables such that the stored energy and the stress tensor are functions not only of the strain but also of these internal values. In this section, we describe the equations for elastoplastic media with hardening but restrict ourselves to small displacements and strains, i.e., to the formulation (1.11). For finite plasticity, we refer to, e.g., [131, 132, 133, 157].

From thermodynamic considerations [77], one can deduce that the total strain $\boldsymbol{\varepsilon}$ as well as the internal energy W can additively be decomposed into an elastic part and a plastic part by

$$\boldsymbol{\varepsilon} = \boldsymbol{\varepsilon}^{\mathrm{el}} + \boldsymbol{\varepsilon}^{\mathrm{pl}}, \quad W = W^{\mathrm{el}} + W^{\mathrm{pl}}. \tag{1.12}$$

Using the quadratic function (1.7a) for the elastic energy $W^{\mathrm{el}} = W^{\mathrm{el}}(\boldsymbol{\varepsilon}^{\mathrm{el}})$, one obtains that the Cauchy stress depends only on the elastic part of the strain:

$$\sigma = \frac{\partial W}{\partial \boldsymbol{\varepsilon}^{\mathrm{el}}} = \mathbb{C}^{\mathrm{el}}\boldsymbol{\varepsilon}^{\mathrm{el}} = \mathbb{C}^{\mathrm{el}}\left(\boldsymbol{\varepsilon} - \boldsymbol{\varepsilon}^{\mathrm{pl}}\right). \tag{1.13}$$

The plastic part W^{pl} of the energy describes the hardening behaviour of the material; as a typical example, we consider the case of linear kinematic and isotropic hardening [157], where the inner variables are given by the plastic strain tensor $\boldsymbol{\varepsilon}^{\mathrm{pl}}$ and the equivalent plastic strain α:

$$W^{\mathrm{pl}} = W^{\mathrm{pl}}(\alpha, \boldsymbol{\varepsilon}^{\mathrm{pl}}) = \frac{1}{2}a_0^{-2}K\|\boldsymbol{\varepsilon}^{\mathrm{pl}}\|^2 + \frac{1}{2}H\alpha^2. \tag{1.14}$$

The two constants K, H for the kinematic and isotropic hardening are assumed to be nonnegative, and the scaling factor $a_0^2 := \frac{d}{d-1}$ is used in order to have a consistent notation for both the two- and the three-dimensional case.

Similarly to (1.13), one can define the inner variable χ conjugate to α by

$$\chi := -\frac{\partial W}{\partial \alpha} = -H\alpha.$$

This quantity is used to describe the plastification process by means of the yield function $\widehat{f}^{\mathrm{pl}}(\chi, \boldsymbol{\sigma})$ which defines the elastic region consisting of those sets of generalized stresses $(\chi, \boldsymbol{\sigma})$

with $\widehat{f}^{\mathrm{pl}}(\chi, \boldsymbol{\sigma}) < 0$. Together with its boundary, the elastic region forms the set of admissible generalized stresses $\mathcal{E}_{\mathrm{pl}}$ (see (2.3a) in the next chapter for a proper definition). As an example, the middle picture of Figure 1.2 shows the boundary of a typical elastic region $\mathcal{E}_{\mathrm{pl}}$ within a plane stress setting.

Given that the principle of maximum plastic dissipation [77, 157] is valid, it follows that $\mathcal{E}_{\mathrm{pl}}$ is convex as well as that the evolution of the plastic inner variables satisfies an associative flow rule, i.e., their derivatives with respect to time are determined by the yield function:

$$\dot{\boldsymbol{\varepsilon}}^{\mathrm{pl}} = \gamma^{\mathrm{pl}} \frac{\partial \widehat{f}^{\mathrm{pl}}(\chi, \boldsymbol{\sigma})}{\partial \boldsymbol{\sigma}}, \qquad \dot{\alpha} = \gamma^{\mathrm{pl}} \frac{\partial \widehat{f}^{\mathrm{pl}}(\chi, \boldsymbol{\sigma})}{\partial \chi}. \tag{1.15}$$

Above, γ^{pl} is a nonnegative variable called the plastic multiplier which can only be nonzero if the generalized stresses $(\chi, \boldsymbol{\sigma})$ are on the boundary $\partial\mathcal{E}_{\mathrm{pl}}$ of the elastic region. This fact leads to the complementarity or Karush–Kuhn–Tucker (KKT) conditions

$$\widehat{f}^{\mathrm{pl}}(\chi, \boldsymbol{\sigma}) \leq 0, \quad \gamma^{\mathrm{pl}} \geq 0, \quad \gamma^{\mathrm{pl}} \widehat{f}^{\mathrm{pl}}(\chi, \boldsymbol{\sigma}) = 0. \tag{1.16}$$

Remark 1.2. There are also models of elastoplastic materials which do not rely on the principle of maximum plastic dissipation and can thus have non-associative flow rules; however, they pose some further difficulties from the analytical point of view. We refer to [17, 89, 157, 162] for details.

For cases like metal plasticity, one observes that the plastic deformation barely inflicts a change in the volume. In this case, which is also called von Mises- or J_2-plasticity [157], it is assumed that the plastic strain $\boldsymbol{\varepsilon}^{\mathrm{pl}}$ has no volumetric part, i.e.

$$\boldsymbol{\varepsilon}^{\mathrm{pl}} = \operatorname{dev} \boldsymbol{\varepsilon}^{\mathrm{pl}} := \boldsymbol{\varepsilon}^{\mathrm{pl}} - \frac{1}{d}(\operatorname{tr} \boldsymbol{\varepsilon}^{\mathrm{pl}})\mathrm{Id}, \quad \operatorname{tr} \boldsymbol{\varepsilon}^{\mathrm{pl}} = 0. \tag{1.17}$$

Introducing the relative deviatoric stress

$$\boldsymbol{\eta} := \operatorname{dev} \boldsymbol{\sigma} - a_0^{-2} K \boldsymbol{\varepsilon}^{\mathrm{pl}} \tag{1.18}$$

whose second term follows from the definition of the plastic energy function (1.14), we obtain the yield function for combined linear kinematic and isotropic hardening:

$$\widehat{f}^{\mathrm{pl}}(\chi, \boldsymbol{\sigma}) := \|\boldsymbol{\eta}\| - \widehat{Y}^{\mathrm{pl}}(\chi), \quad \widehat{Y}^{\mathrm{pl}}(\chi) := a_0^{-1}(\sigma_0 - \chi) = a_0^{-1}(\sigma_0 + H\alpha). \tag{1.19}$$

In (1.19), the yield stress σ_0 is assumed to be positive. The middle picture of Figure 1.2 illustrates the yield function $\widehat{f}^{\mathrm{pl}}$ by showing a cut of the yield surface $\widehat{f}^{\mathrm{pl}} = 0$ for the case of plane stress and $K = 0$.

Substituting (1.19) into (1.15), we get the evolution equations

$$\dot{\boldsymbol{\varepsilon}}^{\mathrm{pl}} = \gamma^{\mathrm{pl}} \frac{\partial \|\boldsymbol{\eta}\|}{\partial \boldsymbol{\sigma}} = \begin{cases} \gamma^{\mathrm{pl}} \frac{\boldsymbol{\eta}}{\|\boldsymbol{\eta}\|}, & \|\boldsymbol{\eta}\| > 0, \\ \mathbf{0}, & \|\boldsymbol{\eta}\| = 0, \end{cases} \tag{1.20a}$$

$$\dot{\alpha} = -\gamma^{\mathrm{pl}} (\widehat{Y}^{\mathrm{pl}})'(\chi) = a_0^{-1} \gamma^{\mathrm{pl}}. \tag{1.20b}$$

For some numerical simulations, we will extend the linear isotropic hardening in (1.14), (1.19) to a nonlinear exponential one that can for example be found in [157]:

$$W^{\mathrm{pl}}_{\exp}(\alpha, \boldsymbol{\varepsilon}^{\mathrm{pl}}) := \frac{1}{2} a_0^{-2} K \|\boldsymbol{\varepsilon}^{\mathrm{pl}}\|^2 + \frac{1}{2} H \alpha^2 + \frac{\sigma_\infty - \sigma_0}{k_e} e^{-k_e \alpha}, \tag{1.21a}$$

$$\chi = -\frac{\partial W^{\mathrm{pl}}_{\exp}}{\partial \alpha} = -H\alpha + (\sigma_\infty - \sigma_0) e^{-k_e \alpha}, \tag{1.21b}$$

$$\widehat{Y}^{\mathrm{pl}}_{\exp}(\chi) = a_0^{-1}(\sigma_\infty - \chi) = a_0^{-1}\left((\sigma_0 + H\alpha) + (\sigma_\infty - \sigma_0)\left(1 - e^{-k_e \alpha}\right)\right). \tag{1.21c}$$

Above, the material parameters σ_0, σ_∞ denote the initial and the final yield stress, and k_e is the hardening exponent. On the right of Figure 1.2, examples for the hardening functions $\widehat{Y}^{\mathrm{pl}}$ and $\widehat{Y}^{\mathrm{pl}}_{\exp}$ are depicted.

As we work only with the variable α and not with the conjugate force χ in the subsequent chapters, we use the notation $Y^{\mathrm{pl}}(\alpha) = \widehat{Y}^{\mathrm{pl}}(\chi)$ and $f^{\mathrm{pl}}(\alpha, \boldsymbol{\eta}) = \widehat{f}^{\mathrm{pl}}(\chi, \boldsymbol{\sigma})$ from now on.

Equations (1.10), (1.11), (1.13), (1.15), (1.16), (1.17), combined with the definition of the yield function as in (1.19), fully describe the dynamic deformation of an elastoplastic body, if the volume and boundary forces are given. In the next section, we further extend this model by allowing for unilateral contact along the boundary Γ_C.

1.3 Frictional contact

In this subsection, we state the contact constraints on the potential contact boundary Γ_C. For ease of notation, we assume that the contact obstacle Ω_{obs} is rigid and fixed (see Figure 1.1); if contact of several elastic bodies is considered, we reformulate it on the algebraic level as a one-body contact problem, as described in [93, 166] and Remark 2.3. Furthermore, we restrict ourselves to the linearized version of the contact conditions, i.e., we neglect the dependence of the normal vector $\mathbf{n}$ on the actual displacement. A rigorous derivation of the linearized contact conditions can be found in [104], the case of frictional contact between several bodies undergoing large deformations is treated in, e.g., [140, 142, 172].

The contact conditions are given in terms of the displacement $\mathbf{u}$ and the contact stress $\boldsymbol{\lambda} := -\boldsymbol{\sigma}\mathbf{n}$ on Γ_C. Decomposing these vectors into their normal parts $u_n := \mathbf{u} \cdot \mathbf{n}$, $\lambda_n := \boldsymbol{\lambda} \cdot \mathbf{n}$ and their tangential components $\mathbf{u}_t := \mathbf{u} - u_n \mathbf{n}$, $\boldsymbol{\lambda}_t := \boldsymbol{\lambda} - \lambda_n \mathbf{n}$, we can formulate the conditions for normal contact on Γ_C [81, 120, 147]:

$$u_n \leq g_n, \quad \lambda_n \geq 0, \quad (u_n - g_n)\lambda_n = 0. \tag{1.22}$$

In (1.22), the scalar gap function g_n measures the distance of a point on Γ_C and its projection onto the obstacle Ω_{obs} in the direction of the normal $\mathbf{n}$; we refer to [116, 167] for a detailed definition. The first inequality in (1.22), usually termed non-penetration condition, describes the fact that the body Ω is not allowed to penetrate into Ω_{obs}, whereas the second inequality ensures that the contact forces are compressive. The last complementarity condition expresses the fact that nonzero contact forces can only arise at those points where the gap is closed.

The contact conditions with respect to the tangential direction have a similar structure as the plastic evolution laws presented in the previous section. Combining the cases of Tresca and

Coulomb friction [34, 116, 120, 147], the tangential contact conditions are given in terms of the frictional yield function

$$f^{\mathrm{co}}(\lambda_n, \boldsymbol{\lambda}_t) := \|\boldsymbol{\lambda}_t\| - Y^{\mathrm{co}}(\lambda_n), \quad Y^{\mathrm{co}}(\lambda_n) := g_t + \mathfrak{F}\lambda_n, \tag{1.23}$$

the flow rule

$$\dot{\mathbf{u}}_t = \begin{cases} \gamma^{\mathrm{co}} \frac{\boldsymbol{\lambda}_t}{\|\boldsymbol{\lambda}_t\|}, & \|\boldsymbol{\lambda}_t\| > 0, \\ \mathbf{0}, & Y^{\mathrm{co}}(\lambda_n) > \|\boldsymbol{\lambda}_t\| = 0, \\ \text{free}, & Y^{\mathrm{co}}(\lambda_n) = \|\boldsymbol{\lambda}_t\| = 0, \end{cases} \tag{1.24}$$

and the KKT conditions

$$f^{\mathrm{co}}(\lambda_n, \boldsymbol{\lambda}_t) \leq 0, \quad \gamma^{\mathrm{co}} \geq 0, \quad \gamma^{\mathrm{co}} f^{\mathrm{co}}(\lambda_n, \boldsymbol{\lambda}_t) = 0. \tag{1.25}$$

The case of Tresca friction with the fixed friction bound $g_t \geq 0$ can be obtained by setting $\mathfrak{F} = 0$, whereas Coulomb friction with the friction coefficient $\mathfrak{F} \geq 0$ is given by $g_t = 0$. The locations where $\gamma^{\mathrm{co}} = 0$ and thus $\dot{\mathbf{u}}_t = \mathbf{0}$ holds are called sticky, whereas frictional sliding occurs at the slippy points where $\gamma^{\mathrm{co}} > 0$ and $f^{\mathrm{co}}(\lambda_n, \boldsymbol{\lambda}_t) = 0$.

Comparing the equations for plasticity and friction, many parallels can be seen, especially between the yield functions (1.19) (1.23), the flow rules (1.20a), (1.24) and the complementarity conditions (1.16), (1.25). However, one of the main differences between the conditions for contact and plasticity is that the latter law is associative due to the evolution law (1.20b). For the contact case, we have a set of complementarity conditions (1.22) for λ_n instead of an differential equation in time, such that the contact law cannot be associative. Another important difference is the higher regularity of the plastic inner stress compared to the contact boundary stress, as discussed in more detail in Section 2.1. Last but not least, the pure elastoplastic problem without contact is still an ordinary differential equation (ODE) with respect to time, whereas the normal contact problem without plastic effects has the structure of a differential-algebraic equation (DAE), making its numerical solution more challenging. This will be the topic of Chapter 5.

2 Discretization techniques

The system of differential–algebraic equations and inequalities derived in the previous chapter usually has to be solved on a complex geometry. As this cannot be done analytically, the problem needs to be approximated by a finite dimensional formulation which is then solved numerically. In this chapter, we sketch the discretization techniques used to transform the problem into a system of nonlinear equations. First, in Section 2.1, the problem is converted into its weak form, resulting in a quasi-variational inequality. Section 2.2 contains the description of the spatial discretization, including some a priori error estimates for the quasi-static contact problem. The following Section 2.3 describes the time stepping method used for the approximation of the resulting ODE and gives some remarks about energy conservation. Finally, in Section 2.4, we describe how the inequality constraints for contact and plasticity can be reformulated as a set of nondifferentiable equations. The solution of the resulting nonlinear system by means of a generalization of Newton's method is the topic of Chapter 3.

2.1 Weak formulation

To derive the weak formulation of the strong relations given in Chapter 1, we introduce the following (affine) function spaces for the displacement $\mathbf{u}$:

$$\mathbf{V} := \mathbf{H}^1(\Omega) := [H^1(\Omega)]^d, \tag{2.1a}$$

$$\mathbf{V}_D := \{\mathbf{v} \in \mathbf{V} : \mathbf{v} = \mathbf{u}_D \text{ on } \Gamma_D\}, \quad \mathbf{V}_0 := \{\mathbf{v} \in \mathbf{V} : \mathbf{v} = \mathbf{0} \text{ on } \Gamma_D\}. \tag{2.1b}$$

The inner variables $(\alpha, \boldsymbol{\varepsilon}^{\text{pl}})$ for the plastic effects are elements of the space

$$\mathbf{M}_{\text{pl}} := L_2(\Omega) \times \left\{\boldsymbol{\tau} \in [L_2(\Omega)]^{d\times d} : \boldsymbol{\tau} = \boldsymbol{\tau}^T,\ \text{tr}\ \boldsymbol{\tau} = 0\right\}. \tag{2.2}$$

For the contact contributions, we define the trace space $\mathbf{W} := [H^{1/2}(\Gamma_C)]^d$ and its dual space $\mathbf{M}_{\text{co}} := \mathbf{W}'$ on Γ_C. The notations $\boldsymbol{\mu} = \mu_n \mathbf{n} + \boldsymbol{\mu}_t \in \mathbf{M}_{\text{co}}$ and $(\mu_n, \boldsymbol{\mu}_t) \in \mathbf{M}_{\text{co}}$ are used equivalently in the following.

The sets of inner and dual variables that are admissible, i.e., satisfy the KKT conditions (1.16), (1.22) and (1.25), are subsets of the spaces $\mathbf{M}_{\text{pl}}$, $\mathbf{M}_{\text{co}}$, respectively:

$$\mathcal{E}_{\text{pl}} := \{(\beta, \boldsymbol{\tau}) \in \mathbf{M}_{\text{pl}} : f^{\text{pl}}(\beta, \boldsymbol{\tau}) \le 0\}, \tag{2.3a}$$

$$\mathcal{E}_{\text{co}}(\lambda_n) := \{(\mu_n, \boldsymbol{\mu}_t) \in \mathbf{M}_{\text{co}} : \mu_n \ge 0, f^{\text{co}}(\lambda_n, \boldsymbol{\mu}_t) \le 0\}. \tag{2.3b}$$

For the yield functions f^{pl}, f^{co} defined in (1.19), (1.23), the admissible sets (2.3) are convex, as can be seen from the examples sketched in Figure 2.1. We remark that the inequality in (2.3a) holds for almost all $\mathbf{x} \in \Omega$, and that the inequalities in (2.3b) need to be interpreted in a weak sense, i.e. for a function $\mu \in H^{1/2}(\Gamma_C)'$, we denote

$$\mu \ge 0 \quad \Leftrightarrow \quad \langle \mu, v\rangle_{\Gamma_C} \ge 0, \quad v \in H^{1/2}(\Gamma_C),\ v \ge 0,$$

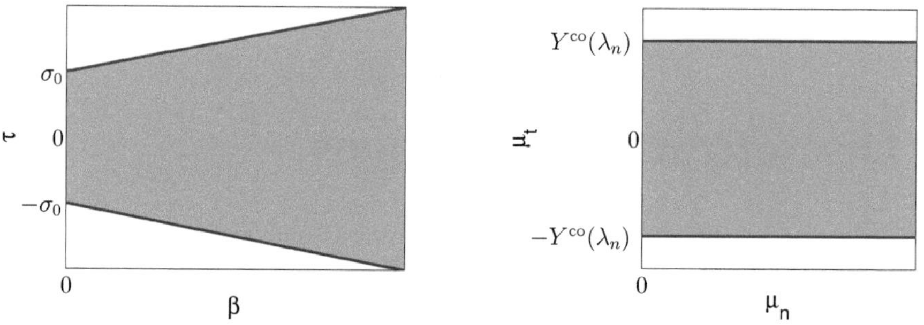

Figure 2.1: Examples of elastic regions in 1D; left: $\mathcal{E}_{\mathrm{pl}}$; right: $\mathcal{E}_{\mathrm{co}}(\lambda_n)$ for fixed λ_n.

where $\langle\cdot,\cdot\rangle_{\Gamma_C}$ stands for the dual pairing on Γ_C.

If only plastic effects and normal contact with Tresca friction (i.e., $\mathfrak{F}=0$) are considered, the dependence of $\mathcal{E}_{\mathrm{co}}$ on λ_n vanishes and one can rewrite the plastic contact problem in terms of a variational inequality [30, 52, 77, 79, 80, 88, 105]. But for $\mathfrak{F}>0$, the weak formulation has the structure of a quasi-variational inequality (see, e.g., [63, 78, 104] for the derivation): find $\mathbf{u}\in\mathbf{V}_D$, $(\alpha,\boldsymbol{\eta})\in\mathcal{E}_{\mathrm{pl}}$, $\boldsymbol{\lambda}=(\lambda_n,\boldsymbol{\lambda}_t)\in\mathcal{E}_{\mathrm{co}}(\lambda_n)$ with $\boldsymbol{\eta}$ given by (1.18) such that for each time $t\in(0,T]$

$$m(\ddot{\mathbf{u}},\mathbf{w})+a(\mathbf{u},\mathbf{w})+b_{\mathrm{pl}}(\boldsymbol{\varepsilon}^{\mathrm{pl}},\boldsymbol{\varepsilon}(\mathbf{w}))+b_{\mathrm{co}}(\boldsymbol{\lambda},\mathbf{w})-\mathbf{f}(\mathbf{w})=0,\quad \mathbf{w}\in\mathbf{V}_0, \tag{2.4a}$$

$$-b_{\mathrm{pl}}\left((C^{\mathrm{el}})^{-1}(\boldsymbol{\tau}-\boldsymbol{\eta}),\dot{\boldsymbol{\varepsilon}}^{\mathrm{pl}}\right)-\int_\Omega H\dot{\alpha}(\beta-\alpha)\,\mathrm{d}\mathbf{x}\le 0,\quad (\beta,\boldsymbol{\tau})\in\mathcal{E}_{\mathrm{pl}}, \tag{2.4b}$$

$$b_{\mathrm{co}}(\boldsymbol{\mu}-\boldsymbol{\lambda},\dot{\mathbf{u}}_t)\le 0,\quad \boldsymbol{\mu}\in\mathcal{E}_{\mathrm{co}}(\lambda_n), \tag{2.4c}$$

$$b_{\mathrm{co}}(\boldsymbol{\mu}-\boldsymbol{\lambda},u_n\mathbf{n})-\int_{\Gamma_C} g_n(\mu_n-\lambda_n)\,\mathrm{d}s\le 0,\quad \boldsymbol{\mu}\in\mathcal{E}_{\mathrm{co}}(\lambda_n). \tag{2.4d}$$

with appropriate initial conditions for $\mathbf{u}$, $(\alpha,\boldsymbol{\eta})$ and the definitions

$$m(\ddot{\mathbf{u}},\mathbf{w}):=\int_\Omega \varrho\,\ddot{\mathbf{u}}\cdot\mathbf{w}\;\mathrm{d}\mathbf{x}, \tag{2.5a}$$

$$a(\mathbf{u},\mathbf{w}):=\int_\Omega C^{\mathrm{el}}\boldsymbol{\varepsilon}(\mathbf{u}):\boldsymbol{\varepsilon}(\mathbf{w})\,\mathrm{d}\mathbf{x}, \tag{2.5b}$$

$$b_{\mathrm{pl}}(\boldsymbol{\varepsilon},\boldsymbol{\tau}):=-\int_\Omega C^{\mathrm{el}}\boldsymbol{\varepsilon}:\boldsymbol{\tau}\,\mathrm{d}\mathbf{x}, \tag{2.5c}$$

$$b_{\mathrm{co}}(\boldsymbol{\mu},\mathbf{w}):=\int_{\Gamma_C}\boldsymbol{\mu}\cdot\mathbf{w}\,\mathrm{d}s, \tag{2.5d}$$

$$\mathbf{f}(\mathbf{w}):=\int_\Omega \mathbf{l}\cdot\mathbf{w}\,\mathrm{d}\mathbf{x}+\int_{\Gamma_N}\mathbf{t}\cdot\mathbf{w}\,\mathrm{d}s. \tag{2.5e}$$

2.2 Spatial discretization

For the spatial discretization of (2.4), we define a finite dimensional approximation of the function spaces $\mathbf{V}$, $\mathbf{M}_{\mathrm{pl}}$, $\mathbf{M}_{\mathrm{co}}$ on Ω by means of a shape-regular triangulation $\mathcal{T}^h$ consisting of simplices or quadrilaterals/hexahedrals. We assume that the discretization resolves the Dirichlet boundary Γ_D and denote all vertices of $\mathcal{T}^h$ by the index set $\mathcal{N}^h$. With $\mathcal{N}_D^h$ ($\mathcal{N}_0^h$), we summarize all nodes (not) belonging to Γ_D. The nodes lying on the potential contact boundary Γ_C are pooled in the set $\mathcal{N}_{\text{co}}^h \subset \mathcal{N}^h$, whereas the elements are numbered according to the set $\mathcal{N}_{\text{pl}}^h$. An example of these sets is sketched in Figure 2.2.

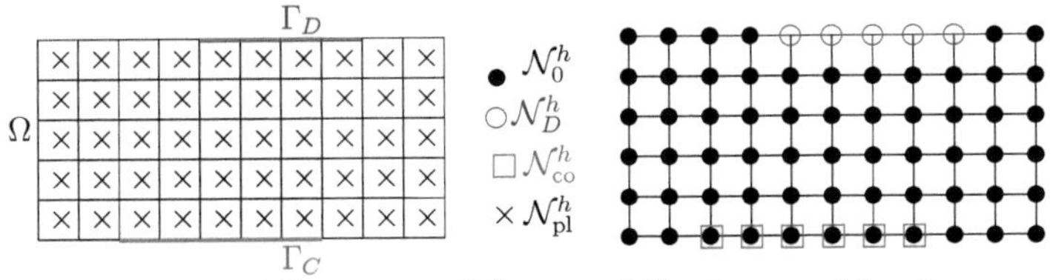

Figure 2.2: Illustration of the sets of the degrees of freedom

For $\omega \subset \mathbb{R}^d$ and $k \in \mathbb{N}_0$, we define $\mathcal{P}_k(\omega)$ as the space of polynomials with the basis functions $\mathbf{x}^{\boldsymbol{\alpha}}$ with $\mathbf{x} \in \omega$, $\boldsymbol{\alpha} \in \mathbb{N}_0^d$ and $0 \le \sum_{i=1}^d \alpha_i \le k$. Furthermore, $\mathcal{Q}_k(\omega)$ is the polynomial space spanned by the functions $\mathbf{x}^{\boldsymbol{\alpha}}$ with $\max_{1\le i \le d}(\alpha_i) \le k$.

Because a dynamic contact problem usually does not yield a high regularity of the solution [53], we only consider first order finite element spaces; we refer to [26, 152] for ansatz spaces with higher order polynomials. Let $\{\phi_p\}_{p\in\mathcal{N}^h}$ denote the lowest order ($k = 1$) conforming nodal finite element basis functions associated with the nodes in $\mathcal{N}^h$, spanning the finite dimensional test space $\mathbf{V}^h \subset \mathbf{V}$ with $\dim \mathbf{V}^h = d \cdot |\mathcal{N}^h|$. The discrete displacement is denoted by $\mathbf{u}^h = \sum_{p\in\mathcal{N}_0^h} \phi_p \mathbf{u}_p + \mathbf{u}_D^h \in \mathbf{V}_D^h \subset \mathbf{V}^h$, with $\mathbf{u}_D^h = \sum_{p\in\mathcal{N}_D^h} \phi_p \mathbf{u}_{D,p}$ being a suitable finite element interpolation of the Dirichlet data. The discrete test space $\mathbf{V}_0^h$ is given by $\mathbf{V}_0^h := \mathbf{V}^h \cap \mathbf{V}_0$.

For the plastic inner variables, we define the indicator function χ_p of the element $K_p \in \mathcal{T}^h$ and the mean value $\overline{\boldsymbol{\varepsilon}(\mathbf{u}^h)}_p$ of $\boldsymbol{\varepsilon}(\mathbf{u}^h)$ on K_p as follows:

$$\chi_p(\mathbf{x}) = \delta_{pq} \text{ for } \mathbf{x} \in K_q \in \mathcal{T}^h, \qquad \overline{\boldsymbol{\varepsilon}(\mathbf{u}^h)}_p := \frac{\int_{K_p} \boldsymbol{\varepsilon}(\mathbf{u}^h)\,\mathrm{dx}}{\int_{K_p} 1\,\mathrm{dx}}, \tag{2.6}$$

with δ_{pq} already defined in (1.6). The finite dimensional space $\mathbf{M}_{\text{pl}}^h \subset \mathbf{M}_{\text{pl}}$ is now constructed as the span of the piecewise constant functions $\{\chi_p\}_{p\in\mathcal{N}_{\text{pl}}^h}$, leading to the discrete inner variables

$$\alpha^h = \sum_{p\in\mathcal{N}_{\text{pl}}^h} \chi_p \alpha_p, \quad \boldsymbol{\varepsilon}^{\text{pl},h} = \sum_{p\in\mathcal{N}_{\text{pl}}^h} \chi_p \boldsymbol{\varepsilon}_p^{\text{pl}}, \quad \boldsymbol{\eta}^h = \sum_{p\in\mathcal{N}_{\text{pl}}^h} \chi_p \left(2\mu \operatorname{dev}\left(\overline{\boldsymbol{\varepsilon}(\mathbf{u}^h)}_p\right) - a_0^{-2} K \boldsymbol{\varepsilon}_p^{\text{pl}}\right). \tag{2.7}$$

Remark 2.1. In the literature, the degrees of freedom of the plastic variables are often associated with the Gauß integration points (see, e.g., [157, 163]). This approach can be interpreted as partitioning each element $K_p \in \mathcal{T}^h$ containing n_p^{GP} Gauß points into n_p^{GP} "subelements" and defining the corresponding basis functions as the indicator functions of these subelements. Equation (2.7) is a special case of this procedure for $n_p^{\text{GP}} = 1$. We remark that for a simplex K_p, $\boldsymbol{\varepsilon}(\mathbf{u}^h)$ is constant on K_p such that the above definition of $\boldsymbol{\eta}^h$ yields the correct discrete value.

In the rest of this work, we employ the elementwise definition of $\mathbf{M}_{\text{pl}}^h$ as given in (2.6), (2.7), but the results can be transferred onto the Gauß point case as well.

For the finite dimensional version of the contact conditions, we introduce the discrete trace space $\mathbf{W}^h \subset \mathbf{W}$ spanned by the basis functions $\varphi_p := \phi_p|_{\Gamma_C}$, $p \in \mathcal{N}^h_{\text{co}}$. In addition, we need a discrete dual space $\mathbf{M}^h_{\text{co}} \subset \mathbf{M}_{\text{co}}$ for the approximation of the contact stress; it is defined as the span of a set of basis functions $\{\psi_p\}_{p\in\mathcal{N}^h_{\text{co}}}$ associated with $\mathcal{N}^h_{\text{co}}$, implying $\dim(\mathbf{W}^h) = \dim(\mathbf{M}^h_{\text{co}})$. There are several suitable possibilities to define these basis functions (see [129, 164, 165] and the references therein for more details); in the rest of this work, we use the so-called dual basis functions $\{\psi_p\}_{p\in\mathcal{N}^h_{\text{co}}}$ which are piecewise (bi)linear and satisfy the requirement $\text{supp}(\psi_p) = \text{supp}(\varphi_p)$ as well as the biorthogonality property

$$\int_{\Gamma_C} \varphi_q \psi_p \, \mathrm{d}s = \delta_{qp} \int_{\Gamma_C} \varphi_q \, \mathrm{d}s, \quad p, q \in \mathcal{N}^h_{\text{co}}. \tag{2.8}$$

Examples of these functions for a planar interface Γ_C and locally uniform triangulations are sketched in Figure 2.3. For the construction of appropriate dual basis functions for non-planar interfaces, we refer to [59].

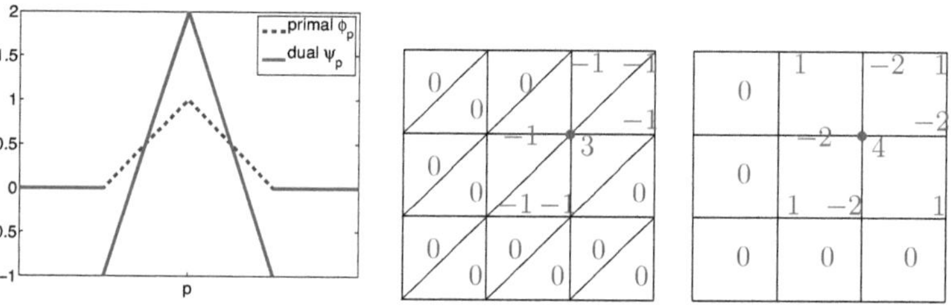

Figure 2.3: Examples of primal and dual basis functions; left: values of φ_p and ψ_p in 1D; middle and right: values of ψ_p for triangles and quadrilaterals in 2D.

Using the spaces $\mathbf{M}^h_{\text{pl}}$, $\mathbf{M}^h_{\text{co}}$, we define the discrete counterparts of the admissible sets (2.3) as

$$\mathcal{E}^h_{\text{pl}} := \Big\{(\beta^h, \boldsymbol{\tau}^h) \in \mathbf{M}^h_{\text{pl}} : f^{\text{pl}}(\beta_p, \boldsymbol{\tau}_p) \le 0,\, p \in \mathcal{N}^h_{\text{pl}}\Big\}, \tag{2.9a}$$

$$\mathcal{E}^h_{\text{co}}(\lambda^h_n) := \Big\{(\mu^h_n, \boldsymbol{\mu}^h_t) \in \mathbf{M}^h_{\text{co}} : \mu_{p,n} \ge 0,\, f^{\text{co}}(\lambda_{p,n}, \boldsymbol{\mu}_{p,t}) \le 0,\, p \in \mathcal{N}^h_{\text{co}}\Big\}. \tag{2.9b}$$

With these definitions, we can formulate the semi-discrete version of (2.4): find $\mathbf{u}^h \in \mathbf{V}^h_D$, $(\alpha^h, \boldsymbol{\eta}^h) \in \mathcal{E}^h_{\text{pl}}$, $\boldsymbol{\lambda}^h = (\lambda^h_n, \boldsymbol{\lambda}^h_t) \in \mathcal{E}^h_{\text{co}}(\lambda^h_n)$ such that for each time $t \in (0, T]$

$$m(\ddot{\mathbf{u}}^h, \mathbf{w}^h) + a(\mathbf{u}^h, \mathbf{w}^h) + b_{\text{pl}}(\boldsymbol{\varepsilon}^{\text{pl},h}, \boldsymbol{\varepsilon}(\mathbf{w}^h)) + b_{\text{co}}(\boldsymbol{\lambda}^h, \mathbf{w}^h) - \mathbf{f}(\mathbf{w}^h) = 0,\ \ \mathbf{w}^h \in \mathbf{V}^h_0, \tag{2.10a}$$

$$-b_{\text{pl}}\left((C^{\text{el}})^{-1}(\boldsymbol{\tau}^h - \boldsymbol{\eta}^h), \dot{\boldsymbol{\varepsilon}}^{\text{pl},h}\right) - \int_\Omega H\dot{\alpha}^h(\beta^h - \alpha^h)\,\mathrm{d}\mathbf{x} \le 0,\ (\beta^h, \boldsymbol{\tau}^h) \in \mathcal{E}^h_{\text{pl}}, \tag{2.10b}$$

$$b_{\text{co}}(\boldsymbol{\mu}^h - \boldsymbol{\lambda}^h, \dot{\mathbf{u}}^h_t) \le 0,\ \ \boldsymbol{\mu}^h \in \mathcal{E}^h_{\text{co}}(\lambda^h_n), \tag{2.10c}$$

$$b_{\text{co}}(\boldsymbol{\mu}^h - \boldsymbol{\lambda}^h, u^h_n\mathbf{n}) - \int_{\Gamma_C} g_n(\mu^h_n - \lambda^h_n)\,\mathrm{d}s \le 0,\ \ \boldsymbol{\mu}^h \in \mathcal{E}^h_{\text{co}}(\lambda^h_n). \tag{2.10d}$$

The reason for choosing the basis functions of the spaces $\mathbf{M}^h_{\text{pl}}$, $\mathbf{M}^h_{\text{co}}$ as stated in (2.6), (2.8), respectively, is that the inequality conditions (2.10bcd) decouple into separate inequalities for each of the inner/dual degrees of freedom. Hence, the conditions for plasticity and contact can be enforced locally, leading to the nodal flow rules and complementarity conditions

$$\left.\begin{aligned} \dot{\alpha}_p &= a_0^{-1}\gamma_p^{\mathrm{pl}}, \\ \dot{\boldsymbol{\varepsilon}}_p^{\mathrm{pl}} &= \begin{cases} \gamma_p^{\mathrm{pl}}\widehat{\boldsymbol{\eta}}_p, & \|\boldsymbol{\eta}_p\| > 0, \\ \mathbf{0}, & \|\boldsymbol{\eta}_p\| = 0, \end{cases} \\ f^{\mathrm{pl}}(\alpha_p, \boldsymbol{\eta}_p) \leq 0, \quad & \gamma_p^{\mathrm{pl}} \geq 0, \quad \gamma_p^{\mathrm{pl}} f^{\mathrm{pl}}(\alpha_p, \boldsymbol{\eta}_p) = 0, \end{aligned}\right\} \quad p \in \mathcal{N}_{\mathrm{pl}}^h, \tag{2.11a}$$

$$\left.\begin{aligned} u_{p,n} \leq g_{p,n}, \quad \lambda_{p,n} \geq 0, \quad & (u_{p,n} - g_{p,n})\lambda_{p,n} = 0, \\ \dot{\mathbf{u}}_{p,t} &= \begin{cases} \gamma_p^{\mathrm{co}}\widehat{\boldsymbol{\lambda}}_{p,t}, & \|\boldsymbol{\lambda}_{p,t}\| > 0, \\ \mathbf{0}, & Y^{\mathrm{co}}(\lambda_{p,n}) > \|\boldsymbol{\lambda}_{p,t}\| = 0, \\ \text{free}, & Y^{\mathrm{co}}(\lambda_{p,n}) = \|\boldsymbol{\lambda}_{p,t}\| = 0, \end{cases} \\ f^{\mathrm{co}}(\lambda_{p,n}, \boldsymbol{\lambda}_{p,t}) \leq 0, \quad & \gamma_p^{\mathrm{co}} \geq 0, \quad \gamma_p^{\mathrm{co}} f^{\mathrm{co}}(\lambda_{p,n}, \boldsymbol{\lambda}_{p,t}) = 0 \end{aligned}\right\} \quad p \in \mathcal{N}_{\mathrm{co}}^h, \tag{2.11b}$$

where we have used the notation $\widehat{\boldsymbol{\eta}} := \frac{\boldsymbol{\eta}}{\|\boldsymbol{\eta}\|}$ for a normed vector or tensor.

In [90, 92], the following a priori error estimates for the quasi-static linear elastic contact problem without plasticity can be found:

Theorem 2.2. *Let $(\mathbf{u}, \boldsymbol{\lambda}) \in \mathbf{V}_D \times \mathbf{M}_{co}$ with $\mathbf{u} \in [H^{3/2+\tau}(\Omega)]^d$, $0 < \tau \leq \frac{1}{2}$, be the solution of (2.4), and let $(\mathbf{u}^h, \boldsymbol{\lambda}^h) \in \mathbf{V}_D^h \times \mathbf{M}_{co}^h$ be the solution of (2.10) for $\boldsymbol{\varepsilon}^{pl,h} = \mathbf{0}$, $\boldsymbol{\varepsilon}^{pl} = \mathbf{0}$ and $\varrho = 0$. Consider one of the following cases:*

a) $d \in \{2, 3\}$ and there is no friction, i.e., $g_t = \mathfrak{F} = 0$,

b) $d = 2$ and only Tresca friction, i.e., $\mathfrak{F} = 0$.

Then, under some technical assumptions on the shape of the actual contact boundary, there exists a constant C independent of the mesh size h such that

$$\|\mathbf{u} - \mathbf{u}^h\|_{\mathbf{H}^1(\Omega)} + \|\boldsymbol{\lambda} - \boldsymbol{\lambda}^h\|_{\mathbf{H}^{-1/2}(\Gamma_C)} \leq Ch^{1/2+\tau}|\mathbf{u}|_{\mathbf{H}^{3/2+\tau}(\Omega)}.$$

For an a priori error analysis of the pure elastoplastic problem without contact, we refer to, e.g., [3, 30, 77].

Remark 2.3. In our numerical examples, we also consider the more general case of contact between two elastic bodies, with one master body Ω^{ma} and one slave body Ω^{sl}. Here, the displacement $\mathbf{u}(t, \mathbf{x})$, $\mathbf{x} \in \Gamma_C$, in the non-penetration condition (1.22) and the flow rule (1.24) has to be replaced by the relative contact displacement

$$[\mathbf{u}(t, \mathbf{x})] := \left(\mathbf{u}^{\mathrm{sl}}(t, \mathbf{x}) - \mathbf{u}^{\mathrm{ma}}(t, R\mathbf{x})\right), \quad \mathbf{x} \in \Gamma_C^{\mathrm{sl}},$$

with a mapping $R : \Gamma_C^{\mathrm{sl}} \to \Gamma_C^{\mathrm{ma}}$ that is assumed to be smooth and injective (we refer to [116] for details). In this case, we proceed as described in [93, 166] and perform a local basis transformation on the master side, such that the slave coefficients with respect to the transformed basis describe the relative displacement $[\mathbf{u}]$ of the two bodies. Hence, the non-penetration and friction conditions, formulated in the transformed basis, have the same local structure as the conditions (2.11b) for one-body contact. Details about the implementation can be found in [93, 166].

Finally, we state the algebraic form of the semidiscrete balance equation (2.10a):

$$M^h \ddot{\mathbf{u}}^h + A^h \mathbf{u}^h + B^h_{\mathrm{pl}} \boldsymbol{\varepsilon}^{\mathrm{pl},h} + B^h_{\mathrm{co}} \boldsymbol{\lambda}^h - \mathbf{f}^h = \mathbf{0}, \tag{2.12}$$

where the matrices M^h A^h, B^h_{pl}, B^h_{co} and the vector $\mathbf{f}^h$ are the discrete approximations of the continuous (bi)linear forms defined in (2.5) and given by $\langle A^h \mathbf{u}^h, \mathbf{w}^h \rangle := a(\mathbf{u}^h, \mathbf{w}^h)$, $\langle M^h \mathbf{u}^h, \mathbf{w}^h \rangle := m(\mathbf{u}^h, \mathbf{w}^h)$, $\langle B^h_{\mathrm{pl}} \boldsymbol{\varepsilon}^{\mathrm{pl},h}, \mathbf{w}^h \rangle := b_{\mathrm{pl}}(\boldsymbol{\varepsilon}^{\mathrm{pl},h}, \boldsymbol{\varepsilon}(\mathbf{w}^h))$, $\langle B^h_{\mathrm{co}} \boldsymbol{\lambda}^h, \mathbf{w}^h \rangle := b_{\mathrm{co}}(\boldsymbol{\lambda}^h, \mathbf{w}^h)$ and $\langle \mathbf{f}^h, \mathbf{w}^h \rangle := \mathbf{f}(\mathbf{w}^h)$ for $\mathbf{w}^h \in \mathbf{V}^h_0$. In (2.12) and later on, we denote the coefficient vector with respect to the discrete basis with the same symbol as the corresponding function.

To solve the resulting time-dependent problem (2.11), (2.12) numerically, we need to employ a time discretization method, which is the topic of the next section.

2.3 Time stepping

In order to discretize (2.11), (2.12) in time, we partition the time interval $(0, T)$ into equidistant time steps $t_j = j\Delta t$ of step size Δt and denote the increments of the displacement $\mathbf{u}$ from t_j to t_{j+1} by $\Delta \mathbf{u}_{j+1} := (\mathbf{u}_{j+1} - \mathbf{u}_j)$. The increments of the other physical quantities are defined analogeously.

In the following, we consider a general implicit time discretization scheme of the system (2.11), (2.12) which is given by the combination of the θ-scheme with $\theta \in \left[\frac{1}{2}, 1\right]$ applied to the first order ODEs (2.11) and the Newmark scheme [94] with $\gamma = \theta$, $\beta = \frac{\theta}{2}$ for the second order equation (2.12). The latter method yields the system

$$\begin{aligned} \frac{1}{\Delta t} M^h \Delta \mathbf{v}^h_{j+1} + A^h \mathbf{u}^h_{j+\theta} + B^h_{\mathrm{pl}} \boldsymbol{\varepsilon}^{\mathrm{pl},h}_{j+\theta} + B^h_{\mathrm{co}} \boldsymbol{\lambda}^h_{j+\theta} - \mathbf{f}_{j+\theta} &= \mathbf{0}, \\ \mathbf{v}^h_{j+1/2} - \frac{1}{\Delta t} \Delta \mathbf{u}^h_{j+1} &= \mathbf{0}, \end{aligned} \tag{2.13a}$$

with the definition $\cdot_{j+\theta} := (1-\theta) \cdot_j + \theta \cdot_{j+1}$, whereas the inequality constraints and first order ODEs become

$$\left.\begin{aligned} \Delta \alpha_{p,j+1} &= a_0^{-1} \gamma^{\mathrm{pl}}_{p,j+\theta}, \\ \Delta \boldsymbol{\varepsilon}^{\mathrm{pl}}_{p,j+1} &= \begin{cases} \gamma^{\mathrm{pl}}_{p,j+\theta} \widehat{\boldsymbol{\eta}}_{p,j+\theta}, & \|\boldsymbol{\eta}_{p,j+\theta}\| > 0, \\ \mathbf{0}, & \|\boldsymbol{\eta}_{p,j+\theta}\| = 0, \end{cases} \\ f^{\mathrm{pl}}(\alpha_{p,j+\theta}, \boldsymbol{\eta}_{p,j+\theta}) \le 0, \quad & \gamma^{\mathrm{pl}}_{p,j+\theta} \ge 0, \quad \gamma^{\mathrm{pl}}_{p,j+\theta} f^{\mathrm{pl}}(\alpha_{p,j+\theta}, \boldsymbol{\eta}_{p,j+\theta}) = 0, \end{aligned}\right\} \quad p \in \mathcal{N}^h_{\mathrm{pl}}, \tag{2.13b}$$

$$\left.\begin{aligned} u_{p,j+1,n} - g_{p,n} \le 0, \quad & \lambda_{p,j+\theta,n} \ge 0, \quad (u_{p,j+1,n} - g_{p,n}) \lambda_{p,j+\theta,n} = 0, \\ \Delta \mathbf{u}_{p,j+1,t} &= \begin{cases} \gamma^{\mathrm{co}}_{p,j+\theta} \widehat{\boldsymbol{\lambda}}_{p,j+\theta,t}, & \|\boldsymbol{\lambda}_{p,j+\theta,t}\| > 0, \\ \mathbf{0}, & Y^{\mathrm{co}}(\lambda_{p,j+\theta,n}) > \|\boldsymbol{\lambda}_{p,j+\theta,t}\| = 0, \\ \text{free}, & Y^{\mathrm{co}}(\lambda_{p,j+\theta,n}) = \|\boldsymbol{\lambda}_{p,j+\theta,t}\| = 0, \end{cases} \\ f^{\mathrm{co}}(\lambda_{p,j+\theta,n}, \boldsymbol{\lambda}_{p,j+\theta,t}) \le 0, \quad & \gamma^{\mathrm{co}}_{p,j+\theta} \ge 0, \quad \gamma^{\mathrm{co}}_{p,j+\theta} f^{\mathrm{co}}(\lambda_{p,j+\theta,n}, \boldsymbol{\lambda}_{p,j+\theta,t}) = 0, \end{aligned}\right\} \quad p \in \mathcal{N}^h_{\mathrm{co}}. \tag{2.13c}$$

We remark that the discrete consistency parameters $\gamma^{\mathrm{pl}}_{p,j+\theta}$, $\gamma^{\mathrm{co}}_{p,j+\theta}$ are implicitely scaled with the time step size Δt. Furthermore, one can see that the contact stress $\boldsymbol{\lambda}$ only appears at time

$t_{j+\theta}$ in (2.13). Hence, in contrast to the plastic variables $(\alpha, \boldsymbol{\varepsilon}^{\text{pl}})$, the past values of the contact stress do not contribute to the results of the current time interval.

To avoid overloaded indices, we omit the time index $\cdot_{j+\theta}$ from now on.

Remark 2.4. Due to (2.13b), the variables $\Delta\alpha$ and λ_n are supposed to be nonnegative. But if the problem is solved iteratively and the intermediate variables are not necessarily admissible, also negative values can occur (cf. [35, 91]). For this reason, we slightly modify the definitions of the functions Y^{pl}, Y^{co} to

$$Y^{\text{pl}}(\alpha) := a_0^{-1}(\sigma_0 + H\alpha_j + H\max(0, \Delta\alpha)), \qquad f^{\text{pl}}(\alpha, \boldsymbol{\eta}) = \|\boldsymbol{\eta}\| - Y^{\text{pl}}(\alpha) \tag{2.14a}$$

$$Y^{\text{co}}(\lambda_n) := g_t + \mathfrak{F}\max(0, \lambda_n), \qquad f^{\text{co}}(\lambda_n, \boldsymbol{\lambda}_t) = \|\boldsymbol{\lambda}_t\| - Y^{\text{co}}(\lambda_n). \tag{2.14b}$$

still denoted by the same symbol for ease of notation. If the quantities λ_n, $\Delta\alpha$ are nonnegative, these definitions give the same result as (1.19), (1.23).

In the rest of this section, we list some statements about the stability, the convergence and the energy consistency of the time stepping scheme used in (2.13). First, we note that if both plastic and contact contributions are omitted, i.e., $\boldsymbol{\varepsilon}^{\text{pl},h} = \mathbf{0}$, $\boldsymbol{\lambda}^h = \mathbf{0}$, the problem becomes linear and the Newmark time stepping scheme (2.13) is unconditionally stable for $\theta \in \left[\frac{1}{2}, 1\right]$. Further, it is first order convergent with respect to Δt for $\theta > \frac{1}{2}$ and second order convergent for $\theta = \frac{1}{2}$. The latter scheme is well-known as the trapezoidal rule which conserves the discrete version of the total energy (1.2) given by

$$\mathbb{E}_j^h := \mathbb{E}_j^{h,\text{kin}} + \mathbb{E}_j^{h,\text{el}} := \frac{1}{2}(\mathbf{v}_j^h)^T M^h \mathbf{v}_j^h + \left(\frac{1}{2}(\mathbf{u}_j^h)^T A^h \mathbf{u}_j^h - (\mathbf{f}_j^h)^T \mathbf{u}_j^h\right). \tag{2.15}$$

However, if the elastic part of (2.13a) is extended to a nonlinear hyperelastic material like in (1.3), the energy conservation property of the trapezoidal rule is no longer valid and needs some modifications – we refer to [14, 67, 113, 119, 156] for several possibilities. Other second order time stepping schemes are given in [39, 86] but are not considered in this work because of their numerical dissipativity. An alternative time stepping approach has recently been presented in [66].

The plastic part of the time stepping scheme (2.13) has been analysed in [136], proving the same stability and convergence results as for the linear system without plastification. The contact part of the discretization is investigated in [116] where it is shown that the scheme can be numerically unstable for $\theta = \frac{1}{2}$. This can be avoided if the time discretization of the normal contact conditions $(2.13\text{c})_1$ is changed like, e.g., described in [4, 34, 85, 98, 117, 118, 137]. As an example, we sketch the approach given in [117] which restores the energy consistency of the time stepping scheme. It is based on the enforcement of the persistency condition $\lambda_n \frac{\mathrm{d}}{\mathrm{d}t}(u_n - g_n) = 0$ instead of the complementarity condition $\lambda_n(u_n - g_n) = 0$. Assuming that the gap function g_n does not explicitly depend on time, the discrete version of the contact constraints at a given node $p \in \mathcal{N}_{\text{co}}^h$ using the persistency condition reads

$$\begin{aligned} u_{p,j,n} - g_{p,n} < 0 \quad &\Rightarrow \quad \lambda_{p,j+\theta,n} = 0, \\ u_{p,j,n} - g_{p,n} \geq 0 \quad &\Rightarrow \quad \begin{cases} \lambda_{p,j+\theta,n} \geq 0, \\ \Delta u_{p,j+1,n} \leq 0, \\ \Delta u_{p,j+1,n}\lambda_{p,j+\theta,n} = 0. \end{cases} \end{aligned} \tag{2.16}$$

We remark that the persistency condition is equivalent to the non-penetration condition in the time-continuous setting if we have $u_n - g_n = 0$ at the first time the body comes into contact; if this is not the case, small penetrations may occur. Further, (2.16) tends to the original non-penetration condition as the time step size Δt goes to zero.

In the next section, we present a possibility how the inequality constraints (2.13bc), (2.16) can be transformed into a set of nonlinear equations.

2.4 Reformulation of the inequality constraints

The discretization procedure sketched in the previous subsections leads to the problem (2.13) containing global nonlinear equations as well as local complementarity conditions. There exist many different approaches to treat the inequality constraints, e.g., interior point methods [37, 169], SQP algorithms [139, 163], radial return mapping or cutting plane methods [157], as well as penalty or augmented Lagrangian approaches [62, 116]. For an overview of most of these methods, we refer to [60].

Another approach is to rewrite the inequality constraints as a set of nondifferentiable equations, termed nonlinear complementarity (NCP) functions; some examples of such functions can be found in [2, 35, 36, 47, 111, 155]. The latter approach has the advantage that the complementarity conditions can be treated with the same nonlinear solver as the discrete balance equations, avoiding an additional nested iteration. Thus, in this section, we are going to reformulate the complementarity conditions (2.13b), (2.13c) as a set of equalities. For this purpose, we introduce so-called trial values which are defined by choosing positive constants c^{pl}, c_t^{co}, c_n^{co} and setting

$$\boldsymbol{\eta}_p^{\mathrm{tr}} := \boldsymbol{\eta}_p + c^{\mathrm{pl}} \Delta \boldsymbol{\varepsilon}_p^{\mathrm{pl}}, \tag{2.17a}$$

$$h_p := \left(\int_{\Gamma_C} \phi_p \,\mathrm{d}s \right)^{1/(d-1)}, \qquad \begin{aligned} \boldsymbol{\lambda}_{p,t}^{\mathrm{tr}} &:= \boldsymbol{\lambda}_{p,t} + c_t^{\mathrm{co}} h_p^{-1} \Delta \mathbf{u}_{p,j+1,t}, \\ \lambda_{p,n}^{\mathrm{tr}} &:= \lambda_{p,n} + c_n^{\mathrm{co}} h_p^{-1} (u_{p,j+1,n} - g_{p,n}), \end{aligned} \tag{2.17b}$$

where all variables without time index are taken at $\cdot_{j+\theta}$. With these definitions, we define the NCP functions locally by

$$\mathbf{C}^{\mathrm{pl}}(\mathbf{u}^h, (\alpha^h, \boldsymbol{\varepsilon}^{\mathrm{pl},h})) = (C_p^{\mathrm{pl}}(\mathbf{u}^h, (\alpha_p, \boldsymbol{\varepsilon}_p^{\mathrm{pl}})))_{p \in \mathcal{N}_{\mathrm{pl}}^h}, \quad \mathbf{C}^{\mathrm{co}}(\mathbf{u}^h, \boldsymbol{\lambda}^h) = (C_p^{\mathrm{co}}(\mathbf{u}_p, \boldsymbol{\lambda}_p))_{p \in \mathcal{N}_{\mathrm{co}}^h}$$

with

$$C_p^{\mathrm{pl}}(\mathbf{u}^h, (\alpha_p, \boldsymbol{\varepsilon}_p^{\mathrm{pl}})) := \begin{pmatrix} \max(0, f^{\mathrm{pl}}(\alpha_p, \boldsymbol{\eta}_p^{\mathrm{tr}})) - a_0 c^{\mathrm{pl}} \Delta \alpha_p \\ \boldsymbol{\eta}_p - \min\left(1, \frac{Y^{\mathrm{pl}}(\alpha_p)}{\|\boldsymbol{\eta}_p^{\mathrm{tr}}\|}\right) \boldsymbol{\eta}_p^{\mathrm{tr}} \end{pmatrix}, \quad p \in \mathcal{N}_{\mathrm{pl}}^h, \tag{2.18a}$$

$$C_p^{\mathrm{co}}(\mathbf{u}_p, \boldsymbol{\lambda}_p) := \begin{pmatrix} C_{p,n}^{\mathrm{co}}(\mathbf{u}_p, \boldsymbol{\lambda}_p) \\ C_{p,t}^{\mathrm{co}}(\mathbf{u}_p, \boldsymbol{\lambda}_p) \end{pmatrix} := \begin{pmatrix} \lambda_{p,n} - \max(0, \lambda_{p,n}^{\mathrm{tr}}) \\ \boldsymbol{\lambda}_{p,t} - \min\left(1, \frac{Y^{\mathrm{co}}(\lambda_{p,n}^{\mathrm{tr}})}{\|\boldsymbol{\lambda}_{p,t}^{\mathrm{tr}}\|}\right) \boldsymbol{\lambda}_{p,t}^{\mathrm{tr}} \end{pmatrix}, \quad p \in \mathcal{N}_{\mathrm{co}}^h. \tag{2.18b}$$

For the persistency condition (2.16), we set

$$C_{p,n}^{\mathrm{en}}(\mathbf{u}_p, \boldsymbol{\lambda}_p) := \lambda_{p,n} - \max\left(0, \lambda_{p,n}^{\mathrm{en}}\right), \tag{2.19}$$

where

$$\lambda_{p,n}^{\text{en}} := \begin{cases} c_n^{\text{co}} h_p^{-1}(u_{p,j,n} - g_{p,n}) & \text{if } u_{p,j,n} - g_{p,n} < 0, \\ \lambda_{p,n} + c_n^{\text{co}} h_p^{-1} \Delta u_{p,j+1,n} & \text{if } u_{p,j,n} - g_{p,n} \geq 0. \end{cases}$$

The NCP functions above are constructed such that the following lemma holds:

Lemma 2.5. *The inequality constraints (2.13b), (2.13c) can be rewritten as*

$$\mathbf{C}^{pl}(\mathbf{u}_{j+\theta}^h, (\alpha, \boldsymbol{\varepsilon}^{pl})_{j+\theta}^h) = \mathbf{0}, \qquad \mathbf{C}^{co}(\mathbf{u}_{j+1}^h, \boldsymbol{\lambda}_{j+\theta}^h) = \mathbf{0}.$$

Furthermore, the contact constraints (2.16) are equivalent to

$$\mathbf{C}_n^{en}(\mathbf{u}_{j+1}^h, \boldsymbol{\lambda}_{j+\theta}^h) = 0.$$

Proof. We only treat the case of plasticity – the contact conditions can be proven similarly (see also [2, 35]). Using the relation

$$\Delta\alpha_{j+\theta} = (1-\theta)\alpha_j + \theta\alpha_{j+1} - \alpha_j = \theta\Delta\alpha_{j+1}$$

and omitting the time index $\cdot_{j+\theta}$ as well as the node index $\cdot_p$ for sake of clarity, the following equivalence needs to be verified:

$$\left.\begin{aligned} \theta^{-1}\Delta\alpha &= a_0^{-1}\gamma^{\text{pl}}, \\ \theta^{-1}\Delta\boldsymbol{\varepsilon}^{\text{pl}} &= \begin{cases} \gamma^{\text{pl}}\hat{\boldsymbol{\eta}}, & \|\boldsymbol{\eta}\| > 0, \\ \mathbf{0}, & \boldsymbol{\eta} = \mathbf{0}, \end{cases} \\ f^{\text{pl}}(\alpha, \boldsymbol{\eta}) &\leq 0, \\ \gamma^{\text{pl}} &\geq 0, \\ \gamma^{\text{pl}} f^{\text{pl}}(\alpha, \boldsymbol{\eta}) &= 0, \end{aligned}\right\} \Leftrightarrow \left\{ \begin{aligned} a_0\, c^{\text{pl}}\Delta\alpha &= \max(0, f^{\text{pl}}(\alpha, \boldsymbol{\eta}^{\text{tr}})), \\ \boldsymbol{\eta} &= \min\left(1, \tfrac{Y^{\text{pl}}(\alpha)}{\|\boldsymbol{\eta}^{\text{tr}}\|}\right)\boldsymbol{\eta}^{\text{tr}}. \end{aligned}\right. \tag{2.20}$$

In a first step, we assume that the left side of (2.20) holds. We start with the case $\gamma^{\text{pl}} = 0$, which leads to $\Delta\alpha = 0$ and $\Delta\boldsymbol{\varepsilon}^{\text{pl}} = \mathbf{0}$. Thus, definitions (2.14a) and (2.17) yield $\boldsymbol{\eta}^{\text{tr}} = \boldsymbol{\eta}$, $f^{\text{pl}}(\alpha, \boldsymbol{\eta}^{\text{tr}}) \leq 0$, $Y^{\text{pl}}(\alpha) \geq \|\boldsymbol{\eta}^{\text{tr}}\|$, and the right side of (2.20) is satisfied. Next, we consider the case $f^{\text{pl}}(\alpha, \boldsymbol{\eta}) = 0$ and $\gamma^{\text{pl}} > 0$ and find $\|\boldsymbol{\eta}\| > 0$. The definition of $\boldsymbol{\eta}^{\text{tr}}$ gives $\boldsymbol{\eta}^{\text{tr}} = \left(1 + \frac{c^{\text{pl}}\theta\gamma^{\text{pl}}}{\|\boldsymbol{\eta}\|}\right)\boldsymbol{\eta}$ and $f^{\text{pl}}(\alpha, \boldsymbol{\eta}^{\text{tr}}) = \|\boldsymbol{\eta}\| - Y^{\text{pl}}(\alpha) + c^{\text{pl}}\theta\gamma^{\text{pl}} = c^{\text{pl}}\theta\gamma^{\text{pl}}$. Thus, the right side of (2.20) also holds.

In the second step, we assume that the right side of (2.20) is fulfilled. Let $Y^{\text{pl}}(\alpha) \geq \|\boldsymbol{\eta}^{\text{tr}}\|$, then $\boldsymbol{\eta} = \boldsymbol{\eta}^{\text{tr}}$, $f^{\text{pl}}(\alpha, \boldsymbol{\eta}) \leq 0$ and $\Delta\alpha = 0$, $\Delta\boldsymbol{\varepsilon}^{\text{pl}} = \mathbf{0}$. Setting $\gamma^{\text{pl}} = 0$, we find that the left side of (2.20) is satisfied. Next, we assume $Y^{\text{pl}}(\alpha) < \|\boldsymbol{\eta}^{\text{tr}}\|$, then $0 < \|\boldsymbol{\eta}\| = Y^{\text{pl}}(\alpha)$ and $f^{\text{pl}}(\alpha, \boldsymbol{\eta}) = 0$. Setting $\gamma^{\text{pl}} = (c^{\text{pl}})^{-1}\theta^{-1} f^{\text{pl}}(\alpha, \boldsymbol{\eta}^{\text{tr}}) > 0$, we get

$$\Delta\boldsymbol{\varepsilon}^{\text{pl}} = (c^{\text{pl}})^{-1}(\boldsymbol{\eta}^{\text{tr}} - \boldsymbol{\eta}) = (c^{\text{pl}})^{-1}\left(\frac{\|\boldsymbol{\eta}^{\text{tr}}\|}{\|\boldsymbol{\eta}\|} - 1\right)\boldsymbol{\eta} = (c^{\text{pl}})^{-1} f^{\text{pl}}(\alpha, \boldsymbol{\eta}^{\text{tr}})\hat{\boldsymbol{\eta}} = \theta\gamma^{\text{pl}}\hat{\boldsymbol{\eta}}$$

from (2.17), and the left side of (2.20) also holds. □

Remark 2.6. For good numerical performance, the trial values (2.17) should be a linear combination of quantities that show the same dependence on the time step size Δt, the local mesh size and the material parameters λ, μ. Thus, it is advisable to choose the plastic constant c^{pl} in (2.17a) independently of the mesh size or of Δt but of the order $\mathcal{O}(\mu)$ to appropriately balance strains and stresses. For the contact case, we need to equilibrate displacements and stresses which yields a dependence not only on the material parameters λ, μ but also on the mesh size. This is respected by the additional scaling factor h_p^{-1} in (2.17b), with h_p being of the order of the local mesh size.

Remark 2.7. For $\theta = 1$ and the specific choice $c^{\text{pl}} = 2\mu + a_0^{-2}K$, we obtain the trial relative stress $\boldsymbol{\eta}_{j+1}^{\text{tr}} = 2\mu \operatorname{dev} \boldsymbol{\varepsilon}(\mathbf{u}_{j+1}) - (2\mu + a_0^{-2}K)\boldsymbol{\varepsilon}_j^{\text{pl}}$ which does not depend on the new value $\boldsymbol{\varepsilon}_{j+1}^{\text{pl}}$ of the plastic strain. For the linear hardening laws (1.14), (1.19), this trial value corresponds to the trial elastic state of the radial return method [157]. There, the following integration algorithm is used for $p \in \mathcal{N}_{\text{pl}}^h$:

$$(\alpha_p, \boldsymbol{\varepsilon}_p^{\text{pl}})_{j+1} = \begin{cases} (\alpha_p, \boldsymbol{\varepsilon}_p^{\text{pl}})_j, & \text{if } f^{\text{pl}}(\alpha_{p,j}, \boldsymbol{\eta}_{p,j+1}^{\text{tr}}) \le 0, \\ \left(\alpha_{p,j} + a_0^{-1}\gamma_p^{\text{pl}}, \boldsymbol{\varepsilon}_{p,j}^{\text{pl}} + \gamma_p^{\text{pl}}\widehat{\boldsymbol{\eta}}_{p,j+1}^{\text{tr}}\right), & \text{if } f^{\text{pl}}(\alpha_{p,j}, \boldsymbol{\eta}_{p,j+1}^{\text{tr}}) > 0, \end{cases}$$

with the consistency parameter

$$\gamma_p^{\text{pl}} = \frac{f^{\text{pl}}(\alpha_{p,j}, \boldsymbol{\eta}_{p,j+1}^{\text{tr}})}{2\mu + a_0^{-2}(H + K)}.$$

One can easily verify that the plastic parameter γ_p^{pl} above is the same as the one in the proof of Lemma 2.5. Hence, the formulation (2.18a) can be interpreted as a generalization of the radial return method.

With Lemma 2.5 and static condensation of the velocity $\mathbf{v}_{j+1}^h$ in (2.13a)$_1$, we obtain that (2.13) is equivalent to the following problem: given $(\mathbf{u}, \mathbf{v}, (\alpha, \boldsymbol{\varepsilon}^{\text{pl}}))_j \in \mathbf{V}_D^h \times \mathbf{V}^h \times \mathbf{M}_{\text{pl}}^h$, find $(\mathbf{u}, \mathbf{v}, (\alpha, \boldsymbol{\varepsilon}^{\text{pl}}))_{j+1} \in \mathbf{V}_D^h \times \mathbf{V}^h \times \mathbf{M}_{\text{pl}}^h$, $\boldsymbol{\lambda}_{j+\theta} \in \mathbf{M}_{\text{co}}^h$ satisfying

$$K^h \mathbf{u}_{j+1}^h + B_{\text{pl}}^h \boldsymbol{\varepsilon}_{j+\theta}^{\text{pl},h} + B_{\text{co}}^h \boldsymbol{\lambda}_{j+\theta}^h - \boldsymbol{\varrho}_j^h = \mathbf{0}, \tag{2.21a}$$

$$\mathbf{v}_{j+1/2}^h - \frac{1}{\Delta t}\Delta \mathbf{u}_{j+1}^h = \mathbf{0}, \tag{2.21b}$$

$$\mathbf{C}^{\text{pl}}(\mathbf{u}_{j+\theta}^h, (\alpha, \boldsymbol{\varepsilon}^{\text{pl}})_{j+\theta}^h) = \mathbf{0}, \tag{2.21c}$$

$$\mathbf{C}^{\text{co}}(\mathbf{u}_{j+1}^h, \boldsymbol{\lambda}_{j+\theta}^h) = \mathbf{0}. \tag{2.21d}$$

In (2.21a)$_1$, we have used the matrix

$$K^h := \frac{2}{\Delta t^2} M^h + \theta A^h \tag{2.22}$$

and the residual vector

$$\boldsymbol{\varrho}_j^h := \mathbf{f}_{j+\theta}^h + \frac{2}{\Delta t} M_\omega^h \mathbf{v}_j^h - A_\omega^h \mathbf{u}_j^h + K_\omega^h \mathbf{u}_j^h. \tag{2.23}$$

As (2.21) is a system of nonlinear equations, it can be solved by means of a variant of the Newton method [57, 87] which will be explained in more detail in the next chapter. In anticipation of the results obtained there, we present a numerical example illustrating that the convergence rate of the Newton iteration strongly depends on the form of the nonlinear complementarity functions C_p^{co}, C_p^{pl}.

We consider the quasi-static plastic benchmark problem of a plate with a circular hole which is pulled from above with a given force f, applied in several load steps. (The geometry and the material parameters are described in more detail in Subsection 4.1.2.) The problem is treated by the nondamped radial return method corresponding to the case $c^{\mathrm{pl}} = 2\mu + a_0^{-2}K$ as explained in Remark 2.7. In Figure 2.4, the decay of the relative error is displayed for different values of f. One can see that if the surface load f exceeds a certain value, the method does not converge

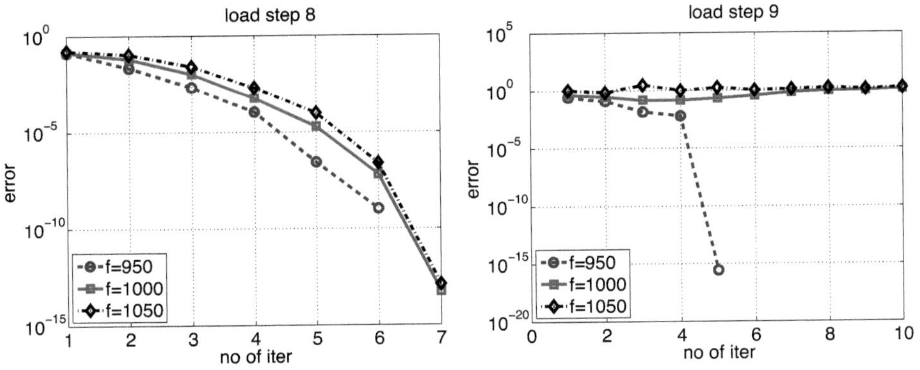

Figure 2.4: Relative error of Newton iteration with radial return at load steps t_8 and t_9 with respect to maximal boundary traction f.

any more at load step t_9. The numerical results in Section 4.2.2 show that a similar effect arises in the contact case. Hence, one of the objectives of the next chapter is the investigation of how modified complementarity functions $C_p^{\mathrm{pl},s}$, $C_p^{\mathrm{co},s}$ with better stability properties can be defined.

Part II

Nonlinear solvers for frictional contact and plasticity

3 Semismooth Newton methods

In the previous part, we have seen that the approximate solution of the frictional contact of elastoplastic bodies leads to a system of nonlinear nondifferentiable equations. The second part of this work is concerned with the construction of a suitable solver for this system by means of a generalized version of the Newton method for semismooth equations. We refer to [49] and the references therein for an overview of Newton methods and to [57, 87, 130, 144] for more details on the subject of semismoothness.

This chapter is structured as follows: First, we introduce the notation by stating the definition of a semismooth function in Section 3.1 and present the most important examples of this class of functions. In Section 3.2, we quote a local convergence result for the semismooth Newton method. Afterwards, Section 3.3 contains the application of the semismooth Newton scheme to an abstract system of equations that covers the cases of both frictional contact and plasticity. With regard to the numerical example at the end of Chapter 2, we introduce a modified version of the NCP function that contains an additional factor featuring a scalar parameter s. We compute the directional derivatives of the new NCP function, investigate its properties and illustrate why the additional factor is able to improve the robustness of the Newton iteration. The specification of this abstract setting to the cases of plasticity and contact is given later in Chapter 4.

We remark that the semismooth Newton method based on suitable NCP functions is a very general approach and by no means restricted to the continuum mechanical context considered here; it has successfully been applied to problems with inequality constraints in mathematical finance and porous media in [28, 71, 72, 121].

3.1 Semismooth functions

In this section, we quote the definitions of different types of smoothness for a given vector-valued function $\mathbf{f}$. For convenience, we start with (directionally) differentiable functions that are defined in, e.g., [57, 144]:

Definition 3.1. A function $\mathbf{f} : \mathbb{R}^n \to \mathbb{R}^m$ is directionally differentiable at $\mathbf{z} \in \mathbb{R}^n$ if for every $\mathbf{v} \in \mathbb{R}^n$ the limit

$$\mathbf{f}'(\mathbf{z};\mathbf{v}) := \lim_{h\to 0,\, h>0} \frac{\mathbf{f}(\mathbf{z}+h\mathbf{v}) - \mathbf{f}(\mathbf{z})}{h}$$

exists. The value $\mathbf{f}'(\mathbf{z};\mathbf{v})$ is called the directional derivative of $\mathbf{f}$ at $\mathbf{z}$ in the direction $\mathbf{v}$.

Definition 3.2. A function $\mathbf{f} : \mathbb{R}^n \to \mathbb{R}^m$ is (Fréchet–)differentiable at $\mathbf{z} \in \mathbb{R}^n$ if there exists a bounded linear operator $\nabla\mathbf{f}(\mathbf{z}) : \mathbb{R}^n \to \mathbb{R}^m$ such that

$$\lim_{\mathbf{h}\to\mathbf{0}} \frac{\|\mathbf{f}(\mathbf{z}+\mathbf{h}) - \mathbf{f}(\mathbf{z}) - \nabla\mathbf{f}(\mathbf{z})\mathbf{h}\|}{\|\mathbf{h}\|} = 0.$$

The matrix $\nabla \mathbf{f}(\mathbf{z})$ is called the Jacobian of $\mathbf{f}$ at $\mathbf{z}$.

Clearly, if $\mathbf{f}$ is differentiable at $\mathbf{z}$, then $\mathbf{f}$ is directionally differentiable at $\mathbf{z}$ with $\mathbf{f}'(\mathbf{z};\mathbf{v}) = \nabla \mathbf{f}(\mathbf{z})\mathbf{v}$ for every $\mathbf{v} \in \mathbb{R}^n$.

Another concept used later on is the following:

Definition 3.3. A function $\mathbf{f} : \mathbb{R}^n \to \mathbb{R}^m$ is piecewise smooth at $\mathbf{z} \in \mathbb{R}^n$ if there exist an open neighbourhood $\mathcal{U}$ of $\mathbf{z}$ and a finite number of continuously differentiable functions $\mathbf{f}_i : \mathcal{U} \to \mathbb{R}^m$, $i = 1, \ldots, p$, such that for any $\widetilde{\mathbf{z}} \in \mathcal{U}$ we have $\mathbf{f}(\widetilde{\mathbf{z}}) = \mathbf{f}_i(\widetilde{\mathbf{z}})$ for at least one index $i \in \{1, \ldots, p\}$.

The notion of a semismooth function is a generalization of the concept of Fréchet–differentiability; roughly speaking, it postulates that the directional derivatives converge uniformly in all directions, but without the linearity requirement stated in Definition 3.2.

There are several different characterizations of a semismooth function $\mathbf{f} : \mathbb{R}^n \to \mathbb{R}^m$, but all of them need $\mathbf{f}$ to be a locally Lipschitz continuous function and thus differentiable almost everywhere. With $U_{\mathbf{f}}$, we denote the set of points where $\mathbf{f}$ is differentiable. In [35, 57], the following definition of semismoothness is given:

Definition 3.4. A function $\mathbf{f} : \mathbb{R}^n \to \mathbb{R}^m$ is semismooth at $\mathbf{z}$ if it is locally Lipschitz continuous in a neighbourhood of $\mathbf{z}$, directionally differentiable at $\mathbf{z}$ and the following limit holds:

$$\lim_{\mathbf{h} \to \mathbf{0}; (\mathbf{z}+\mathbf{h}) \in U_{\mathbf{f}}} \frac{\|\mathbf{f}'(\mathbf{z}+\mathbf{h};\mathbf{h}) - \mathbf{f}'(\mathbf{z};\mathbf{h})\|}{\|\mathbf{h}\|} = 0.$$

If in addition the stronger requirement

$$\limsup_{\mathbf{h} \to \mathbf{0}; (\mathbf{z}+\mathbf{h}) \in U_{\mathbf{f}}} \frac{\|\mathbf{f}'(\mathbf{z}+\mathbf{h};\mathbf{h}) - \mathbf{f}'(\mathbf{z};\mathbf{h})\|}{\|\mathbf{h}\|^2} < \infty,$$

holds, we say that $\mathbf{f}$ is strongly semismooth at $\mathbf{z}$.

Example 3.5. A characteristic example for a scalar function which is not differentiable but semismooth is given by $m := \max(0, \cdot) : \mathbb{R} \to \mathbb{R}_0^+ : x \mapsto \max(0, x)$, where $\mathbb{R}_0^+$ denotes the set of nonnegative real numbers. This function is continuously differentiable for all $x \neq 0$ and directionally differentiable at 0 with

$$m'(0;h) = \begin{cases} 0, & h < 0, \\ 1, & h > 0. \end{cases} \tag{3.1}$$

Because of $m'(h;h) = m'(0;h)$ for all $h \neq 0$, we obtain from Definition 3.4 that m is strongly semismooth.

In a similar fashion, one can show that the norm function $\|\cdot\|_p : \mathbb{R}^n \to \mathbb{R}_0^+$ is strongly semismooth for every $p \in [1, \infty]$ [57, Prop. 7.4.8].

Another characterization of a semismooth function can be formulated using the so-called limiting Jacobian or B(ouligand)-subdifferential $\boldsymbol{\partial}_B \mathbf{f}(\mathbf{z})$ defined by [57, 144]

$$\boldsymbol{\partial}_B \mathbf{f}(\mathbf{z}) := \left\{ \lim_{U_{\mathbf{f}} \ni \widetilde{\mathbf{z}} \to \mathbf{z}} \nabla \mathbf{f}(\widetilde{\mathbf{z}}) \right\}. \tag{3.2}$$

With this, the generalized Jacobian $\partial_C \mathbf{f}(\mathbf{z})$ according to Clarke [41] is given by the convex hull of (3.2), i.e.,

$$\partial_C \mathbf{f}(\mathbf{z}) := \operatorname{co}\big(\partial_B \mathbf{f}(\mathbf{z})\big). \tag{3.3}$$

Example 3.6. In the situation of Example 3.5, the B-subdifferential and the generalized Jacobian of $m = \max(0, \cdot)$ are given by

$$\partial_B m(x) = \begin{cases} \{0\}, & x < 0, \\ \{0,1\}, & x = 0, \\ \{1\}, & x > 0, \end{cases} \qquad \partial_C m(x) = \begin{cases} \{0\}, & x < 0, \\ [0,1], & x = 0, \\ \{1\}, & x > 0. \end{cases} \tag{3.4}$$

With (3.3), we obtain the following alternative characterizations of semismoothness (for the proof, see [57, 144]):

Lemma 3.7. *Let $\mathbf{f} : \mathbb{R}^n \to \mathbb{R}^m$ be locally Lipschitz continuous. The following statements are equivalent:*

(i) *$\mathbf{f}$ is semismooth at $\mathbf{z}$.*

(ii) *The limit*

$$\lim_{\tilde{\mathbf{v}} \to \mathbf{v},\, h \to 0,\, V \in \partial_C \mathbf{f}(\mathbf{z} + h\tilde{\mathbf{v}})} V\tilde{\mathbf{v}}$$

exists for every $\mathbf{v} \in \mathbb{R}^n$.

(iii)

$$\lim_{\mathbf{h} \to 0,\, V \in \partial_C \mathbf{f}(\mathbf{z}+\mathbf{h})} \frac{\|V\mathbf{h} - \mathbf{f}'(\mathbf{z}; \mathbf{h})\|}{\|\mathbf{h}\|} = 0.$$

The class of semismooth functions is quite large, as the following results from [32, 57, 58, 84] show:

Lemma 3.8. *(i) A vector-valued function is (strongly) semismooth if each of its components is (strongly) semismooth.*

(ii) The composition of two (strongly) semismooth functions is (strongly) semismooth.

(iii) If $f : \mathbb{R}^n \to \mathbb{R}$ is continuously differentiable (with a Lipschitz continuous gradient) in a neighbourhood of $\mathbf{z}$, then f is (strongly) semismooth at $\mathbf{z}$. The same result holds if f is only piecewise smooth near $\mathbf{z}$ [84].

(iv) If $f : \mathbb{R}^n \to \mathbb{R}$ is convex in a neighbourhood of $\mathbf{z}$, then f is semismooth at $\mathbf{z}$.

In the next section, we treat the iterative solution of systems containing semismooth functions; the investigation of the degree of semismoothness of the NCP functions (2.18) is done in Section 3.3.

3.2 Newton method for semismooth functions

The classical Newton method is an iterative scheme for computing a solution to $\mathbf{f}(\mathbf{z}) = \mathbf{0}$ for a given differentiable function $\mathbf{f} : \mathbb{R}^n \to \mathbb{R}^n$. In the k-th iteration step, we solve the equation

$$\nabla \mathbf{f}(\mathbf{z}^{(k)})\delta\mathbf{z}^{(k)} = -\mathbf{f}(\mathbf{z}^{(k)}) \tag{3.5}$$

for a search direction $\delta\mathbf{z}^{(k)}$ and obtain the new iterate $\mathbf{z}^{(k+1)}$ via

$$\mathbf{z}^{(k+1)} = \mathbf{z}^{(k)} + \delta\mathbf{z}^{(k)}. \tag{3.6}$$

A possible generalization of this method to the case that $\mathbf{f}$ is only semismooth is obtained if (3.5) is replaced by

$$V_k\delta\mathbf{z}^{(k)} = -\mathbf{f}(\mathbf{z}^{(k)}), \quad V_k \in \boldsymbol{\partial}_B\mathbf{f}(\mathbf{z}^{(k)}). \tag{3.7}$$

In [143, Theorem 3.1], the following convergence result is proved for the iterative scheme (3.7), (3.6):

Theorem 3.9. *Suppose that $\mathbf{z}^*$ satisfies $\mathbf{f}(\mathbf{z}^*) = \mathbf{0}$, $\mathbf{f}$ is semismooth in a neighbourhood of $\mathbf{z}^*$, and all $V \in \boldsymbol{\partial}_B\mathbf{f}(\mathbf{z}^*)$ are nonsingular. Then, the iterative method (3.7), (3.6) is well-defined and superlinearly convergent to $\mathbf{z}^*$ in a neighbourhood of $\mathbf{z}^*$. If in addition $\mathbf{f}$ is strongly semismooth at $\mathbf{z}^*$, then the convergence rate of (3.7), (3.6) is locally quadratic.*

Like in the differentiable case, one can define an inexact version of the above Newton scheme, as for example proposed in [56]: Given a sequence of scalar tolerances $\eta_k \geq 0$ and a starting vector $\mathbf{z}^{(0)}$, one solves the equation

$$V_k\delta\mathbf{z}^{(k)} = -\mathbf{f}(\mathbf{z}^{(k)}) + \mathbf{r}^{(k)}, \quad V_k \in \boldsymbol{\partial}_B\mathbf{f}(\mathbf{z}^{(k)}) \tag{3.8}$$

instead of (3.7), where the residual vector $\mathbf{r}^{(k)}$ satisfies the condition

$$\|\mathbf{r}^{(k)}\| \leq \eta_k\|\mathbf{f}(\mathbf{z}^{(k)})\|.$$

The convergence of this inexact scheme is summarized in [56, Theorem 3.2]:

Theorem 3.10. *Under the assumptions of Theorem 3.9, there are numbers $\eta > 0$ and $\varepsilon > 0$ such that, if $\|\mathbf{z}^{(0)} - \mathbf{z}^*\| \leq \varepsilon$ and $\eta_k \leq \eta$ for all k, the sequence $\{\mathbf{z}^{(k)}\}$ generated by (3.8), (3.6) is well-defined and converges linearly to $\mathbf{z}^*$. Furthermore, if $\eta_k \to 0$, the convergence rate is superlinear. Finally, if $\mathbf{f}$ is strongly semismooth at $\mathbf{z}^*$ and for some $\bar{\eta} > 0$, $\eta_k \leq \bar{\eta}\|\mathbf{f}(\mathbf{z}^{(k)})\|$ for all k, then the rate of convergence is quadratic.*

Similar theorems for differentiable functions are given in, e.g., [48, 49, 54]. For possible globalization strategies of the scheme (3.7) based on an Armijo line search, we refer to [35, 56, 138].

3.3 Abstract framework

As we have seen in Section 2.4, the complementarity conditions for plasticity and friction as well as their reformulations in terms of (2.18) have a very similar structure. In this section, we are going to consider an abstract setting that covers both cases. The aim is to derive a more general form of the complementarity functions (2.18) featuring a scalar parameter s that can be tuned in order to increase the robustness of the corresponding semismooth Newton scheme (3.7). The results have already been published in [74].

Let $(\mathbf{u},\mathbf{r}) \in \mathbb{R}^{m_u} \times \mathbb{R}^{m_r}$, $m_u, m_r \in \mathbb{N}$, be a discrete coefficient vector, where the second vector is splitted into $\mathbf{r}^T = \left(r_n, \mathbf{r}_t^T\right)$ with $r_n \in \mathbb{R}$, $\mathbf{r}_t \in \mathbb{R}^{m_r-1}$. With regard to the form of (2.18), we consider a NCP function $C : \mathbb{R}^{m_u} \times \mathbb{R}^{m_r} \to \mathbb{R}^{m_r}$ of the form

$$C(\mathbf{u},\mathbf{r}) := (C_n(\mathbf{u},\mathbf{r}),\ C_t(\mathbf{u},\mathbf{r}))$$

with an arbitrary but fixed semismooth scalar function C_n and a possibly vector-valued function C_t defined by

$$C_t(\mathbf{u},\mathbf{r}) := \begin{cases} \mathbf{g}(\mathbf{u},\mathbf{r}_t), & Y(\mathbf{u},r_n) = 0, \\ \mathbf{g}(\mathbf{u},\mathbf{r}_t) - \min\left(1, \dfrac{Y(\mathbf{u},r_n)}{\|\mathbf{g}^{\mathrm{tr}}(\mathbf{u},\mathbf{r}_t)\|}\right)\mathbf{g}^{\mathrm{tr}}(\mathbf{u},\mathbf{r}_t), & Y(\mathbf{u},r_n) > 0. \end{cases} \tag{3.9}$$

In (3.9), we have used the functions $Y : \mathbb{R}^{m_u} \times \mathbb{R} \to \mathbb{R}_0^+$ and $\mathbf{g}, \mathbf{g}^{\mathrm{tr}} : \mathbb{R}^{m_u} \times \mathbb{R}^{m_r-1} \to \mathbb{R}^{m_r-1}$ which are assumed to be semismooth.

Example 3.11. The local plastic NCP function (2.18a) is a special case of the above general framework with

$$\mathbf{r} = (\alpha_p, \boldsymbol{\eta}_p), \quad m_r = \frac{1}{2}d(d+1), \quad Y(\mathbf{u},r_n) = Y^{\mathrm{pl}}(\alpha_p), \quad \mathbf{g}(\mathbf{u},\mathbf{r}_t) = \boldsymbol{\eta}_p.$$

The contact case (2.18b) can be obtained with

$$\mathbf{r} = (\lambda_{p,n}, \boldsymbol{\lambda}_{p,t}), \quad m_r = d, \quad Y(\mathbf{u},r_n) = Y^{\mathrm{co}}(\lambda_{p,n}^{\mathrm{tr}}), \quad \mathbf{g}(\mathbf{u},\mathbf{r}_t) = \boldsymbol{\lambda}_{p,t}.$$

In the following, we denote the partial derivatives of Y, $\mathbf{g}$ and $\mathbf{g}^{\mathrm{tr}}$ with $\partial_{\mathbf{u}}$, ∂_{r_n} and $\partial_{\mathbf{r}_t}$, respectively, and omit the arguments $(\mathbf{u},\mathbf{r})$ for ease of notation. Using the rule

$$\partial_* \hat{\mathbf{g}}^{\mathrm{tr}} = \frac{1}{\|\mathbf{g}^{\mathrm{tr}}\|}\left(\mathrm{Id} - \hat{\mathbf{g}}^{\mathrm{tr}} \otimes \hat{\mathbf{g}}^{\mathrm{tr}}\right)\partial_* \mathbf{g}^{\mathrm{tr}}, \quad * \in \{\mathbf{u}, r_n, \mathbf{r}_t\},$$

and the B-subdifferential of $\max(0,\cdot)$ given in (3.4), we compute the partial derivatives of the function C_t in (3.9):

$$\partial_{\mathbf{u}} C_t = \begin{cases} (\partial_{\mathbf{u}}\mathbf{g}), & 0 = Y, \\ (\partial_{\mathbf{u}}\mathbf{g}) - (\partial_{\mathbf{u}}\mathbf{g}^{\mathrm{tr}}), & 0 < Y \geq \|\mathbf{g}^{\mathrm{tr}}\|, \\ (\partial_{\mathbf{u}}\mathbf{g}) - \hat{\mathbf{g}}^{\mathrm{tr}}(\partial_{\mathbf{u}}Y) - \frac{Y}{\|\mathbf{g}^{\mathrm{tr}}\|}\left(\mathrm{Id} - \hat{\mathbf{g}}^{\mathrm{tr}} \otimes \hat{\mathbf{g}}^{\mathrm{tr}}\right)(\partial_{\mathbf{u}}\mathbf{g}^{\mathrm{tr}}), & 0 < Y < \|\mathbf{g}^{\mathrm{tr}}\|, \end{cases} \tag{3.10a}$$

$$\partial_{r_n} C_t = \begin{cases} 0, & 0 = Y, \\ 0, & 0 < Y \geq \|\mathbf{g}^{\text{tr}}\|, \\ -\widehat{\mathbf{g}}^{\text{tr}}(\partial_{r_n} Y), & 0 < Y < \|\mathbf{g}^{\text{tr}}\|, \end{cases} \tag{3.10b}$$

$$\partial_{\mathbf{r}_t} C_t = \begin{cases} (\partial_{\mathbf{r}_t}\mathbf{g}), & 0 = Y, \\ (\partial_{\mathbf{r}_t}\mathbf{g}) - (\partial_{\mathbf{r}_t}\mathbf{g}^{\text{tr}}), & 0 < Y \geq \|\mathbf{g}^{\text{tr}}\|, \\ (\partial_{\mathbf{r}_t}\mathbf{g}) - \frac{Y}{\|\mathbf{g}^{\text{tr}}\|}\left(\text{Id} - \widehat{\mathbf{g}}^{\text{tr}} \otimes \widehat{\mathbf{g}}^{\text{tr}}\right)(\partial_{\mathbf{r}_t}\mathbf{g}^{\text{tr}}), & 0 < Y < \|\mathbf{g}^{\text{tr}}\|. \end{cases} \tag{3.10c}$$

As illustrated by the numerical example at the end of Section 2.4, the domain of convergence of the semismooth Newton method (3.7) applied to (3.9) can be rather small for this choice of $C(\cdot,\cdot)$. Hence, we consider a modified complementarity function

$$C^s(\mathbf{u},\mathbf{r}) := (S(\mathbf{u},\mathbf{r}))^s\, C(\mathbf{u},\mathbf{r}) \tag{3.11}$$

for some fixed value of $s \geq 0$, where the original function is multiplied with the nonsingular matrix $S^s \in \mathbb{R}^{m_r \times m_r}$ given by

$$S^s := \begin{cases} \text{Id}, & 0 = Y, \\ \begin{pmatrix} 1 & \mathbf{0} \\ \mathbf{0} & \left(\max\left(Y, \|\mathbf{g}^{\text{tr}}\|\right)\right)^s \text{Id} \end{pmatrix}, & 0 < Y. \end{cases} \tag{3.12}$$

Clearly, the above multiplication does not alter the set of solutions of (3.9) but yields a different tangential system. We remark that the choice $s = 0$ implies $S^s = \text{Id}$ and thus $C_t^0 = C_t$, whereas for $s \neq 0$, the derivatives of C_t and C_t^s differ:

$$\partial_{\mathbf{u}} C_t^s = \begin{cases} (\partial_{\mathbf{u}}\mathbf{g}), & 0 = Y, \\ Y^s\Big(sY^{-1}(\mathbf{g} - \mathbf{g}^{\text{tr}})(\partial_{\mathbf{u}}Y) + (\partial_{\mathbf{u}}\mathbf{g}) - (\partial_{\mathbf{u}}\mathbf{g}^{\text{tr}})\Big), & 0 < Y \geq \|\mathbf{g}^{\text{tr}}\|, \\ \|\mathbf{g}^{\text{tr}}\|^s \Big((\partial_{\mathbf{u}}\mathbf{g}) - \widehat{\mathbf{g}}^{\text{tr}}(\partial_{\mathbf{u}}Y) - \frac{Y}{\|\mathbf{g}^{\text{tr}}\|}\left(\text{Id} - \mathbf{g}^s \otimes \widehat{\mathbf{g}}^{\text{tr}}\right)(\partial_{\mathbf{u}}\mathbf{g}^{\text{tr}})\Big), & 0 < Y < \|\mathbf{g}^{\text{tr}}\|, \end{cases} \tag{3.13a}$$

$$\partial_{r_n} C_t^s = \begin{cases} 0, & 0 = Y, \\ Y^s\Big(sY^{-1}(\mathbf{g} - \mathbf{g}^{\text{tr}})(\partial_{r_n}Y)\Big), & 0 < Y \geq \|\mathbf{g}^{\text{tr}}\|, \\ -\|\mathbf{g}^{\text{tr}}\|^s \left(\widehat{\mathbf{g}}^{\text{tr}}(\partial_{r_n}Y)\right), & 0 < Y < \|\mathbf{g}^{\text{tr}}\|, \end{cases} \tag{3.13b}$$

$$\partial_{\mathbf{r}_t} C_t^s = \begin{cases} (\partial_{\mathbf{r}_t}\mathbf{g}), & 0 = Y, \\ Y^s\Big((\partial_{\mathbf{r}_t}\mathbf{g}) - (\partial_{\mathbf{r}_t}\mathbf{g}^{\text{tr}})\Big), & 0 < Y \geq \|\mathbf{g}^{\text{tr}}\|, \\ \|\mathbf{g}^{\text{tr}}\|^s \Big((\partial_{\mathbf{r}_t}\mathbf{g}) - \frac{Y}{\|\mathbf{g}^{\text{tr}}\|}\left(\text{Id} - \mathbf{g}^s \otimes \widehat{\mathbf{g}}^{\text{tr}}\right)(\partial_{\mathbf{r}_t}\mathbf{g}^{\text{tr}})\Big), & 0 < Y < \|\mathbf{g}^{\text{tr}}\|. \end{cases} \tag{3.13c}$$

Above, we have used the auxiliary vector

$$\mathbf{g}^s := (1 - s)\widehat{\mathbf{g}}^{\text{tr}} + s\frac{\mathbf{g}}{Y}$$

which is a convex combination of the vectors $\frac{\mathbf{g}}{Y}$ and $\widehat{\mathbf{g}}^{\text{tr}}$.

The main advantage of using the modified complementarity function (3.11) with $s > 0$ is due to the different local matrices in (3.13a), (3.13c) for the case $0 < Y < \|\mathbf{g}^{\text{tr}}\|$. To illustrate the idea, we consider the following model example:

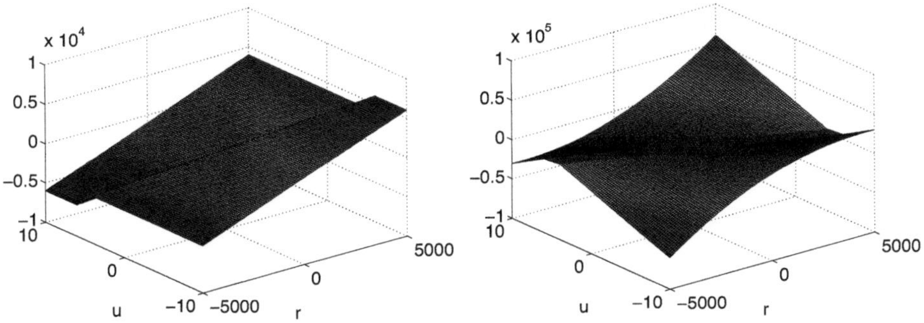

Figure 3.1: Value of modified NCP function C_t^s in 1D for $s = 0$ (left) and $s = 1$ (right).

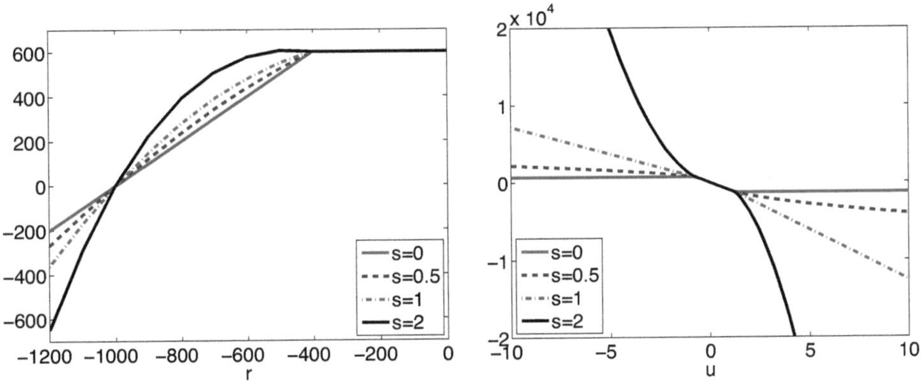

Figure 3.2: Value of modified NCP function C_t^s in 1D for different values of s. Left: cut at $u = -0.6$. Right: cut at $r_2 = -300$.

Example 3.12. We look at scalar values u, r_t and assume that $Y > 0$ is constant and that g, g^{tr} have the form $g(u, r_t) = r_t$, $g^{\text{tr}}(u, r_t) = r_t + cu$ for some constant $c > 0$. In Figures 3.1 and 3.2, the corresponding modified NCP function C_t^s is depicted for $Y = 1000$ and $c = 1000$. Figure 3.1 displays the value of C_t^0 and C_t^1, whereas Figure 3.2 shows cuts of $C_t^s(u, r_t)$, $s \in \{0, 0.5, 1, 2\}$, for $u = -0.6$ and $r_t = -300$, respectively. These pictures illustrate the fact that

$$C_t^0(u, r_t) = \begin{cases} r_t - g^{\text{tr}}(u, r_t) = -cu, & Y \geq |g^{\text{tr}}(u, r_t)|, \\ r_t - Y \operatorname{sign}(g^{\text{tr}}(u, r_t)), & Y < |g^{\text{tr}}(u, r_t)|, \end{cases}$$

is a piecewise linear function which is always constant with respect to one of the variables (u, r_t). Hence, a semismooth Newton scheme based on C^0 is likely to have a small domain of convergence because of the fact that some components of the gradient vanish. In contrast, as one can observe from Figure 3.2, the entries of the tangential matrices of C_t^s generally have a much larger support for $s > 0$ than for $s = 0$. Hence, a Newton scheme applied to C_t^s with $s > 0$ can converge even for a bad starting value of u.

According to Theorem 3.9, the local convergence rate of the semismooth Newton scheme depends on the degree of semismoothness of the target function. In the next lemma, we give sufficient conditions for the function (3.11) to be (strongly) semismooth.

Lemma 3.13. *(i) Assume that $Y : \mathbb{R}^{m_u} \times \mathbb{R} \to \mathbb{R}_0^+$, $\mathbf{g}, \mathbf{g}^{tr} : \mathbb{R}^{m_u} \times \mathbb{R}^{m_r-1} \to \mathbb{R}^{m_r-1}$ are semismooth functions. Then, the NCP function C_t^s, $s \geq 0$, given by (3.9), (3.11), and (3.12) is semismooth at any $(\mathbf{u}, \mathbf{r}) \in \mathbb{R}^{m_u} \times \mathbb{R}^{m_r}$.*

(ii) If, in addition, $Y, \mathbf{g}, \mathbf{g}^{tr}$ are strongly semismooth, then C_t^s, $s \geq 0$, is strongly semismooth at any $(\mathbf{u}, \mathbf{r}) \in \mathbb{R}^{m_u} \times \mathbb{R}^{m_r}$ with $Y(\mathbf{u}, \mathbf{r}) \geq Y_0 > 0$.

Proof. Due to Lemma 3.8 (i) and (ii), it suffices to analyse the (strong) semismoothness of the function $\widetilde{C}_t^s : \mathbb{R}_0^+ \times \mathbb{R}^{m_r-1} \times \mathbb{R}^{m_r-1} \to \mathbb{R}^{m_r-1}$ given by

$$\widetilde{C}_t^s(Y, \mathbf{g}, \mathbf{g}^{\text{tr}}) = \begin{cases} \mathbf{g}, & 0 = Y, \\ Y^s \left(\mathbf{g} - \mathbf{g}^{\text{tr}}\right), & 0 < Y \geq \|\mathbf{g}^{\text{tr}}\|, \\ \|\mathbf{g}^{\text{tr}}\|^s \mathbf{g} - Y \|\mathbf{g}^{\text{tr}}\|^{s-1} \mathbf{g}^{\text{tr}}, & 0 < Y < \|\mathbf{g}^{\text{tr}}\|. \end{cases} \tag{3.14}$$

As (3.14) is piecewise differentiable with the partial derivatives

$$\partial_{(Y, \mathbf{g}, \mathbf{g}^{\text{tr}})} \widetilde{C}_t^s = \begin{cases} \begin{pmatrix} 0 \\ \text{Id} \\ 0 \end{pmatrix}, & 0 = Y, \\ \begin{pmatrix} sY^{s-1}\left(\mathbf{g} - \mathbf{g}^{\text{tr}}\right) \\ Y^s \text{Id} \\ -Y^s \text{Id} \end{pmatrix}, & 0 < Y \geq \|\mathbf{g}^{\text{tr}}\|, \\ \begin{pmatrix} -\|\mathbf{g}^{\text{tr}}\|^{s-1} \mathbf{g}^{\text{tr}} \\ \|\mathbf{g}^{\text{tr}}\|^s \text{Id} \\ -\|\mathbf{g}^{\text{tr}}\|^{s-1} \left(Y \text{Id} - \left((1-s) Y \hat{\mathbf{g}}^{\text{tr}} + s\mathbf{g}\right) \otimes \hat{\mathbf{g}}^{\text{tr}}\right) \end{pmatrix}, & 0 < Y < \|\mathbf{g}^{\text{tr}}\|, \end{cases}$$

the first statement (i) follows with Lemma 3.8 (iii). Further, for $Y \geq Y_0 > 0$, the gradients of (3.14) are locally Lipschitz-continuous, implying the result (ii). □

With Lemma 3.13, we obtain that the NCP functions (2.18) are semismooth and even strongly semismooth if $\sigma_0 > 0$ and $g_t > 0$. Hence, Lemma 3.9 implies that the semismooth Newton scheme applied to (2.21) is locally superlinearly convergent.

The next chapter contains the application of the above abstract considerations to the plastic and frictional yield conditions, respectively. In Section 4.1, we focus on the plastic part and neglect the contact contributions for simplicity, whereas in Section 4.2, we only treat the contact constraints without considering plastic effects.

4 Application to plasticity and frictional contact

In this chapter, we make use of the abstract complementarity function C^s introduced before in order to construct a general iterative algorithm for the solution of the discrete system (2.21). The resulting parameter-dependent scheme contains the radial return method [157] as well as the algorithms described in [35] for plasticity and in [38, 91] for frictional contact as special cases. Investigating the influence of the parameter s, we demonstrate that the parameter combination yielding the radial return algorithm (i.e., $s = 0$) is not optimal and that its robustness can considerably be improved by choosing $s > 0$. Including an additional local stabilization factor a in the first few iterations, a robust quasi-Newton scheme is obtained which converges locally superlinearly in all our numerical examples.

Because of the locality of the NCP functions, the semismooth Newton method can efficiently be implemented by means of a primal-dual active set strategy [29, 87, 91, 96, 97, 114]. This leads to two benefits: one the one hand, the total number of linear systems is reduced due to the unifying treatment of plasticity and frictional contact within the same Newton loop, as all active sets are updated after each Newton iteration. One the other hand, we can easily perform a local static condensation of either the additional inner/dual variables or the corresponding primal degrees of freedom, such that only a system of the size of the displacement has to be solved in each Newton step.

For ease of presentation, we consider the plastic and contact parts separately; the former is dealt with in Section 4.1, whereas the latter is considered in Section 4.2. In each section, we explicitly state the corresponding tangential system as well as its condensed form, followed by a theoretical investigation of the solvability of the system. Here, we make use of a local stabilization factor a that is important during the first few iterations but does not alter the superlinear local convergence of the scheme. Afterwards, several numerical results illustrate the influence of the parameters on the performance of the nonlinear solvers. Finally, in Section 4.3, we combine the results of the two previous sections to a general algorithm which is capable of dealing with complex three-dimensional elastoplastic contact problems, as a numerical example with Coulomb friction and nonlinear isotropic hardening shows.

4.1 Application of abstract framework to plasticity

In this section, we set $\boldsymbol{\lambda}^h = \mathbf{0}$ and apply the semismooth Newton scheme to the plastic part of (2.21), replacing the complementarity function C_p^{pl} in (2.18a) by a modified function $C_p^{\mathrm{pl},s}$ which is constructed as described in (3.11), i.e., we set $C_p^{\mathrm{pl},s} = S_p^{\mathrm{pl},s} C_p^{\mathrm{pl}}$ with the additional factor

$$S_p^{\mathrm{pl},s}(\mathbf{u}^h, (\alpha_p, \boldsymbol{\varepsilon}_p^{\mathrm{pl}})) := \begin{pmatrix} 1 & \mathbf{0} \\ \mathbf{0} & \left(\max\left(Y^{\mathrm{pl}}(\alpha_p), \|\boldsymbol{\eta}_p^{\mathrm{tr}}\|\right)\right)^s \mathrm{Id} \end{pmatrix} \tag{4.1}$$

for some parameter $s \geq 0$.

In Subsection 4.1.1, we derive the tangential system of the k-th Newton iteration and prove its solvability for certain conditions on s and c^{pl}. Subsection 4.1.2 contains numerical tests on the basis of a 2D benchmark example.

4.1.1 Semismooth Newton scheme

Let us recall that the index set $\mathcal{N}_{\mathrm{pl}}^h$ is associated with the degrees of freedom of the plastic variables (2.7), and that all variables without time index are taken at $\cdot_{j+\theta}$, $\theta \in \left[\frac{1}{2}, 1\right]$. The definition of $C_p^{\mathrm{pl},s}$ implies a partition of $\mathcal{N}_{\mathrm{pl}}^h$ into so-called active sets $\mathcal{A}_{\mathrm{pl}}$, $\mathcal{A}_\alpha$ and inactive sets $\mathcal{I}_{\mathrm{pl}}$, $\mathcal{I}_\alpha$ which are defined as

$$\mathcal{A}_{\mathrm{pl}} := \{p \in \mathcal{N}_{\mathrm{pl}}^h : \|\boldsymbol{\eta}_p^{\mathrm{tr}}\| > Y^{\mathrm{pl}}(\alpha_p)\} = \{p \in \mathcal{N}_{\mathrm{pl}}^h : \tau_p < 1\}, \tag{4.2a}$$
$$\mathcal{I}_{\mathrm{pl}} := \{p \in \mathcal{N}_{\mathrm{pl}}^h : \|\boldsymbol{\eta}_p^{\mathrm{tr}}\| \leq Y^{\mathrm{pl}}(\alpha_p)\}, \tag{4.2b}$$
$$\mathcal{A}_\alpha := \{p \in \mathcal{N}_{\mathrm{pl}}^h : \Delta\alpha_p \geq 0\}, \tag{4.2c}$$
$$\mathcal{I}_\alpha := \{p \in \mathcal{N}_{\mathrm{pl}}^h : \Delta\alpha_p < 0\}. \tag{4.2d}$$

The plastic hardening function Y^{pl} employed in (4.2) is defined in (2.14a), and we denote

$$\tau_p := \frac{Y^{\mathrm{pl}}(\alpha_p)}{\|\boldsymbol{\eta}_p^{\mathrm{tr}}\|}, \quad \text{for } \|\boldsymbol{\eta}_p^{\mathrm{tr}}\| > 0. \tag{4.3}$$

For $\|\boldsymbol{\eta}_p^{\mathrm{tr}}\| = 0$, we have $p \in \mathcal{I}_{\mathrm{pl}}$ due to $Y^{\mathrm{pl}}(\alpha_p) > 0$.

The sets (4.2) are closely related to the complementarity conditions (2.13b); for example, the set $\mathcal{A}_{\mathrm{pl}}$ consists of all elements where new plastification occurs, whereas the set $\mathcal{I}_{\mathrm{pl}}$ contains the degrees of freedom in the interior of the elastic region. The second partition into $\mathcal{A}_\alpha$ and $\mathcal{I}_\alpha$ is needed for the correct evaluation of the hardening function (2.14a) during the Newton iteration. However, if the Newton scheme converges to an admissible solution, $\mathcal{I}_\alpha = \emptyset$ is satisfied at convergence.

The characteristic functions $\chi_{\mathcal{A}_{\mathrm{pl}}}, \chi_{\mathcal{I}_{\mathrm{pl}}}, \chi_{\mathcal{A}_\alpha}, \chi_{\mathcal{I}_\alpha}$ of the sets (4.2) are defined according to

$$\chi_{\mathcal{A}_{\mathrm{pl}}} := \chi_{(p \in \mathcal{A}_{\mathrm{pl}})} := \begin{cases} 1, & p \in \mathcal{A}_{\mathrm{pl}}, \\ 0, & p \notin \mathcal{A}_{\mathrm{pl}}. \end{cases} \tag{4.4}$$

The main statement of this section is summarized in the following lemma, where the Newton index $\cdot^{(k)}$ has been omitted for ease of notation:

Lemma 4.1. *For one iteration of the semismooth Newton method on the plastic part of (2.21) with the previous local iterate* $\left(\mathbf{u}_{j+1}^h, (\alpha_p, \boldsymbol{\varepsilon}_p^{pl})_{p \in \mathcal{N}_{pl}^h}\right)$, *the plastic updates* $\left(\delta\alpha_p, \delta\boldsymbol{\varepsilon}_p^{pl}\right)$ *can be computed by*

$$\begin{pmatrix} \delta\alpha_p \\ \delta\boldsymbol{\varepsilon}_p^{pl} \end{pmatrix} = \begin{cases} -\begin{pmatrix} \Delta\alpha_p \\ \Delta\boldsymbol{\varepsilon}_p^{pl} \end{pmatrix}, & p \in \mathcal{I}_{pl} \cap \mathcal{I}_\alpha, \\ -\begin{pmatrix} \Delta\alpha_p \\ \left(1 - \frac{sH\Delta\alpha_p}{a_0 Y^{pl}(\alpha_p)}\right) \Delta\boldsymbol{\varepsilon}_p^{pl} \end{pmatrix}, & p \in \mathcal{I}_{pl} \cap \mathcal{A}_\alpha, \\ \begin{pmatrix} G_p^s \\ M_p^s \end{pmatrix} 2\mu\theta R_p^{dev}(\delta\mathbf{u}_{j+1}^h) + \begin{pmatrix} g_p^s \\ \mathbf{m}_p^s \end{pmatrix}, & p \in \mathcal{A}_{pl}, \end{cases} \tag{4.5}$$

with the abbreviations

$$M_p^s := \frac{1}{\omega_p}\left((1-\tau_p)\, Id + \frac{c^{pl}}{\zeta_p^s} \widetilde{\boldsymbol{\eta}}_p^s \otimes \widehat{\boldsymbol{\eta}}_p^{tr}\right), \tag{4.6a}$$

$$\mathbf{m}_p^s := \begin{cases} \frac{1}{\omega_p}\left(\boldsymbol{\eta}_p - \tau_p \boldsymbol{\eta}_p^{tr}\right) - \frac{\kappa\|\boldsymbol{\eta}_p^{tr}\|\left(\widehat{\boldsymbol{\eta}}_p^{tr}:\bar{\boldsymbol{\eta}}_p - \tau_p\right)}{\omega_p \zeta_p^s} \boldsymbol{\eta}_p^s, & p \in \mathcal{I}_\alpha, \\ \frac{1}{\omega_p}\left(\boldsymbol{\eta}_p - \tau_p \boldsymbol{\eta}_p^{tr}\right) - \frac{\kappa\|\boldsymbol{\eta}_p^{tr}\|\left(\widehat{\boldsymbol{\eta}}_p^{tr}:\bar{\boldsymbol{\eta}}_p - \tau_p\right)}{\omega_p \zeta_p^s} \widetilde{\boldsymbol{\eta}}_p^s & \\ \quad + \frac{H\left(\|\boldsymbol{\eta}_p^{tr}\| - Y^{pl}(\alpha_p) - c^{pl} a_0 \Delta\alpha_p\right)}{\upsilon a_0^2 \omega_p} \left(\frac{\kappa}{\zeta_p^s} \widetilde{\boldsymbol{\eta}}_p^s - \widehat{\boldsymbol{\eta}}_p^{tr}\right), & p \in \mathcal{A}_\alpha. \end{cases} \tag{4.6b}$$

$$G_p^s := \begin{cases} \frac{1}{a_0 \zeta_p^s} \widehat{\boldsymbol{\eta}}_p^{tr} Id, & p \in \mathcal{I}_\alpha, \\ \frac{c^{pl}}{\upsilon a_0 \zeta_p^s} \widehat{\boldsymbol{\eta}}_p^{tr} Id, & p \in \mathcal{A}_\alpha, \end{cases} \tag{4.6c}$$

$$g_p^s := \begin{cases} -\Delta\alpha_p + \frac{\|\boldsymbol{\eta}_p^{tr}\| - Y^{pl}(\alpha_p)}{c^{pl} a_0} - \frac{\|\boldsymbol{\eta}_p^{tr}\| \kappa\left((\widehat{\boldsymbol{\eta}}_p^{tr}:\bar{\boldsymbol{\eta}}_p) - \tau_p\right)}{c^{pl} a_0 \zeta_p^s}, & p \in \mathcal{I}_\alpha, \\ -\Delta\alpha_p + \frac{\|\boldsymbol{\eta}_p^{tr}\| - Y^{pl}(\alpha_{p,old})}{\upsilon a_0} - \frac{\|\boldsymbol{\eta}_p^{tr}\| \kappa\left((\widehat{\boldsymbol{\eta}}_p^{tr}:\bar{\boldsymbol{\eta}}_p) - \tau_p\right)}{\upsilon a_0 \zeta_p^s} & \\ \quad + \frac{\kappa H\left(\|\boldsymbol{\eta}_p^{tr}\| - Y^{pl}(\alpha_p) - c^{pl} a_0 \Delta\alpha_p\right)}{\upsilon^2 a_0^3 \zeta_p^s}, & p \in \mathcal{A}_\alpha, \end{cases} \tag{4.6d}$$

the hardening parameters H, K from (1.14), c^{pl} from (2.17a), Y^{pl} from (1.19), τ_p from (4.3), the fourth order unit tensor Id, the operator

$$R_p^{dev} : \mathbf{u}^h \mapsto \operatorname{dev}\left(\overline{\boldsymbol{\varepsilon}(\mathbf{u}^h)}_p\right), \quad p \in \mathcal{N}_{pl}^h, \tag{4.7}$$

and the further abbreviations

$$\upsilon := c^{pl} + a_0^{-2} H, \qquad \kappa := 2\mu + a_0^{-2} K - c^{pl}, \tag{4.8a}$$

$$\widehat{\boldsymbol{\eta}}_p^{tr} := \frac{\boldsymbol{\eta}_p^{tr}}{\|\boldsymbol{\eta}_p^{tr}\|}, \qquad \bar{\boldsymbol{\eta}}_p := \frac{\boldsymbol{\eta}_p}{\|\boldsymbol{\eta}_p^{tr}\|}, \tag{4.8b}$$

$$\boldsymbol{\eta}_p^s := \left((1 \quad s)\tau_p \widehat{\boldsymbol{\eta}}_p^{tr} + s a_p^{pl} \bar{\boldsymbol{\eta}}_p\right), \qquad \widetilde{\boldsymbol{\eta}}_p^s := \begin{cases} \boldsymbol{\eta}_p^s, & p \in \mathcal{I}_\alpha, \\ \boldsymbol{\eta}_p^s - \frac{H}{\upsilon a_0^2} \widehat{\boldsymbol{\eta}}_p^{tr}, & p \in \mathcal{A}_\alpha, \end{cases} \tag{4.8c}$$

$$\omega_p := \kappa(1-\tau_p) + c^{pl}, \qquad \zeta_p^s := \omega_p + \kappa \widetilde{\boldsymbol{\eta}}_p^s : \widehat{\boldsymbol{\eta}}_p^{tr}. \tag{4.8d}$$

In (4.8c), we have introduced the additional scaling factor $a_p^{pl} := 1$.
Using (4.5), static condensation leads to the following tangential system to be solved for the displacement update $\delta\mathbf{u}_{j+1}^h$ (with $B_{pl}^h \delta\boldsymbol{\varepsilon}^{pl,h} =: \sum_{p\in\mathcal{N}_{pl}^h} B_p^{pl}\delta\boldsymbol{\varepsilon}_p^{pl}$)

$$\Big(K^h + 2\mu\theta \sum_{p\in\mathcal{A}_{pl}} B_p^{pl} M_p^s R_p^{dev}\Big)\delta\mathbf{u}_{j+1}^h \tag{4.9}$$

$$= \boldsymbol{\varrho}_j^h - K^h\mathbf{u}_{j+1}^h - B_{pl}^h\boldsymbol{\varepsilon}^{pl,h} + \sum_{p\in\mathcal{I}_{pl}} B_p^{pl}\Delta\boldsymbol{\varepsilon}_p^{pl} - \sum_{p\in\mathcal{I}_{pl}\cap\mathcal{A}_\alpha} \frac{sH\Delta\alpha_p}{a_0 Y^{pl}(\alpha_p)} B_p^{pl}\Delta\boldsymbol{\varepsilon}_p^{pl} - \sum_{p\in\mathcal{A}_{pl}} B_p^{pl}\mathbf{m}_p^s.$$

Remark 4.2. The variables in (4.6) are only well-defined for $\zeta_p^s \neq 0$. This can be ensured by a proper choice of the parameters c^{pl} and s, as shown in the proof of Lemma 4.3. But if c^{pl}, s are chosen differently and $\zeta_p^s = 0$ holds for some elements $p \in \mathcal{A}_{\mathrm{pl}}$, then the corresponding update $\delta\boldsymbol{\varepsilon}_p^{\mathrm{pl}}$ cannot be computed explicitly as in (4.5), and a mixed problem with both unknowns $\delta\mathbf{u}_{j+1}^h$ and $\delta\boldsymbol{\varepsilon}_p^{\mathrm{pl}}$ remains to be solved.

Proof. Using the relation $\delta\mathbf{u}_{j+\theta}^h = \theta\delta\mathbf{u}_{j+1}^h$, all unknowns are taken at $\cdot_{j+\theta}$ for the rest of this proof. With the definitions (2.14a), (2.17a), (4.1), (4.2) and (4.3), the local NCP function $C_p^{\mathrm{pl},s} := C_p^{\mathrm{pl},s}(\mathbf{u}^h, (\alpha_p, \boldsymbol{\varepsilon}_p^{\mathrm{pl}}))$ reads

$$C_p^{\mathrm{pl},s} = \begin{cases} -c^{\mathrm{pl}}\begin{pmatrix} a_0\Delta\alpha_p \\ Y^{\mathrm{pl}}(\alpha_p)^s\Delta\boldsymbol{\varepsilon}_p^{\mathrm{pl}}\end{pmatrix}, & p\in\mathcal{I}_{\mathrm{pl}}, \\ \begin{pmatrix} \|\boldsymbol{\eta}_p^{\mathrm{tr}}\| - Y^{\mathrm{pl}}(\alpha_p) - c^{\mathrm{pl}}a_0\Delta\alpha_p \\ \|\boldsymbol{\eta}_p^{\mathrm{tr}}\|^s\left(\boldsymbol{\eta}_p - \tau_p\boldsymbol{\eta}_p^{\mathrm{tr}}\right)\end{pmatrix}, & p\in\mathcal{A}_{\mathrm{pl}}. \end{cases} \tag{4.10}$$

We consider one step of the semismooth Newton method applied to (4.10) scaled with the factor $(S_p^{\mathrm{pl},s})^{-1}$:

$$(S_p^{\mathrm{pl},s})^{-1}\left(\partial_{(\alpha_p,\boldsymbol{\varepsilon}_p^{\mathrm{pl}})}C_p^{\mathrm{pl},s}\delta(\alpha_p,\boldsymbol{\varepsilon}_p^{\mathrm{pl}}) + \left(\partial_{\mathbf{u}}C_p^{\mathrm{pl},s}\right)\delta\mathbf{u}^h + C_p^{\mathrm{pl},s}\right) = \mathbf{0}. \tag{4.11}$$

Using the directional derivatives with respect to the variations $(\delta\mathbf{u}^h, \delta\boldsymbol{\varepsilon}^{\mathrm{pl}}, \delta\alpha)$

$$\begin{aligned} (\partial_{\mathbf{u}}\boldsymbol{\eta}_p)\delta\mathbf{u}^h &= 2\mu R_p^{\mathrm{dev}}(\delta\mathbf{u}^h), & (\partial_{\mathbf{u}}\boldsymbol{\eta}_p^{\mathrm{tr}})\delta\mathbf{u}^h &= 2\mu R_p^{\mathrm{dev}}(\delta\mathbf{u}^h), \\ (\partial_{\boldsymbol{\varepsilon}_p^{\mathrm{pl}}}\boldsymbol{\eta}_p)\delta\boldsymbol{\varepsilon}_p^{\mathrm{pl}} &= -(2\mu + a_0^{-2}K)\delta\boldsymbol{\varepsilon}_p^{\mathrm{pl}}, & (\partial_{\boldsymbol{\varepsilon}_p^{\mathrm{pl}}}\boldsymbol{\eta}_p^{\mathrm{tr}})\delta\boldsymbol{\varepsilon}_p^{\mathrm{pl}} &= -(2\mu + a_0^{-2}K - c^{\mathrm{pl}})\delta\boldsymbol{\varepsilon}_p^{\mathrm{pl}}, \\ \partial_*\|\boldsymbol{\eta}_p^{\mathrm{tr}}\| &= \widehat{\boldsymbol{\eta}}_p^{\mathrm{tr}} : \partial_*\boldsymbol{\eta}_p^{\mathrm{tr}}, & \partial_*\widehat{\boldsymbol{\eta}}_p^{\mathrm{tr}} &= \frac{1}{\|\boldsymbol{\eta}_p^{\mathrm{tr}}\|}\left(\mathrm{Id} - \widehat{\boldsymbol{\eta}}_p^{\mathrm{tr}}\otimes\widehat{\boldsymbol{\eta}}_p^{\mathrm{tr}}\right)\partial_*\boldsymbol{\eta}_p^{\mathrm{tr}}, \end{aligned}$$

as well as the definitions (4.8) and the results from (3.13), we obtain

$$\begin{aligned} &(S_p^{\mathrm{pl},s})^{-1}\left(C_p^{\mathrm{pl},s} + \left(\partial_{\mathbf{u}}C_p^{\mathrm{pl},s}\right)\delta\mathbf{u}^h\right) \\ &= \begin{cases} -c^{\mathrm{pl}}\begin{pmatrix} a_0\Delta\alpha_p \\ \Delta\boldsymbol{\varepsilon}_p^{\mathrm{pl}}\end{pmatrix}, & p\in\mathcal{I}_{\mathrm{pl}}, \\ \begin{pmatrix} \|\boldsymbol{\eta}_p^{\mathrm{tr}}\| - Y^{\mathrm{pl}}(\alpha_p) - c^{\mathrm{pl}}a_0\Delta\alpha_p \\ \boldsymbol{\eta}_p - \tau_p\boldsymbol{\eta}_p^{\mathrm{tr}}\end{pmatrix} + 2\mu\begin{pmatrix} \widehat{\boldsymbol{\eta}}_p^{\mathrm{tr}}\mathrm{Id} \\ (1-\tau_p)\mathrm{Id} + \boldsymbol{\eta}_p^s\otimes\widehat{\boldsymbol{\eta}}_p^{\mathrm{tr}}\end{pmatrix} R_p^{\mathrm{dev}}(\delta\mathbf{u}^h), & p\in\mathcal{A}_{\mathrm{pl}}. \end{cases} \end{aligned}$$

The tangential stiffness matrices with respect to the plastic variables read

$$(S_p^{\mathrm{pl},s})^{-1}\partial_{(\alpha_p,\boldsymbol{\varepsilon}_p^{\mathrm{pl}})}C_p^{\mathrm{pl},s} = \begin{cases} -c^{\mathrm{pl}}\begin{pmatrix} a_0 & \mathbf{0} \\ \mathbf{0} & \mathrm{Id}\end{pmatrix}, & p \in \mathcal{I}_{\mathrm{pl}} \cap \mathcal{I}_\alpha, \\ -c^{\mathrm{pl}}\begin{pmatrix} a_0 & \mathbf{0} \\ \frac{Hs}{a_0 Y^{\mathrm{pl}}(\alpha_p)}\Delta\boldsymbol{\varepsilon}_p^{\mathrm{pl}} & \mathrm{Id}\end{pmatrix}, & p \in \mathcal{I}_{\mathrm{pl}} \cap \mathcal{A}_\alpha, \\ -\begin{pmatrix} c^{\mathrm{pl}} a_0 & \kappa\widehat{\boldsymbol{\eta}}_p^{\mathrm{tr}}\mathrm{Id} \\ \mathbf{0} & \omega_p \mathrm{Id} + \kappa\boldsymbol{\eta}_p^s \otimes \widehat{\boldsymbol{\eta}}_p^{\mathrm{tr}}\end{pmatrix}, & p \in \mathcal{A}_{\mathrm{pl}} \cap \mathcal{I}_\alpha, \\ -\begin{pmatrix} v a_0 & \kappa\widehat{\boldsymbol{\eta}}_p^{\mathrm{tr}}\mathrm{Id} \\ a_0^{-1} H \widehat{\boldsymbol{\eta}}_p^{\mathrm{tr}} & \omega_p \mathrm{Id} + \kappa\boldsymbol{\eta}_p^s \otimes \widehat{\boldsymbol{\eta}}_p^{\mathrm{tr}}\end{pmatrix}, & p \in \mathcal{A}_{\mathrm{pl}} \cap \mathcal{A}_\alpha. \end{cases} \tag{4.12}$$

The matrix (4.12) is clearly nonsingular for $p \in \mathcal{I}_{\mathrm{pl}}$; for $p \in \mathcal{A}_{\mathrm{pl}}$, the inverse exists if ζ_p^s as defined in (4.8d) is not zero. If the assumptions of the subsequent Lemma 4.3 are met, this is guaranteed by (4.17); hence, only the case $\zeta_p^s \neq 0$ is considered in the following. Then, the inverse of (4.12) reads

$$\begin{cases} -\frac{1}{c^{\mathrm{pl}}}\begin{pmatrix} a_0^{-1} & \mathbf{0} \\ \mathbf{0} & \mathrm{Id}\end{pmatrix}, & p \in \mathcal{I}_{\mathrm{pl}} \cap \mathcal{I}_\alpha, \\ -\frac{1}{c^{\mathrm{pl}}}\begin{pmatrix} a_0^{-1} & \mathbf{0} \\ -\frac{Hs}{a_0^2 Y^{\mathrm{pl}}(\alpha_p)}\Delta\boldsymbol{\varepsilon}_p^{\mathrm{pl}} & \mathrm{Id}\end{pmatrix}, & p \in \mathcal{I}_{\mathrm{pl}} \cap \mathcal{A}_\alpha, \\ -\begin{pmatrix} \frac{1}{c^{\mathrm{pl}} a_0} & -\frac{\kappa}{c^{\mathrm{pl}} a_0 \zeta_p^s}\widehat{\boldsymbol{\eta}}_p^{\mathrm{tr}}\mathrm{Id} \\ \mathbf{0} & \frac{1}{\omega_p}\left(\mathrm{Id} - \frac{\kappa}{\zeta_p^s}\widetilde{\boldsymbol{\eta}}_p^s \otimes \widehat{\boldsymbol{\eta}}_p^{\mathrm{tr}}\right)\end{pmatrix}, & p \in \mathcal{A}_{\mathrm{pl}} \cap \mathcal{I}_\alpha, \\ -\begin{pmatrix} \frac{\kappa H + v a_0^2 \zeta_p^s}{v^2 a_0^3 \zeta_p^s} & -\frac{\kappa}{v a_0 \zeta_p^s}\widehat{\boldsymbol{\eta}}_p^{\mathrm{tr}}\mathrm{Id} \\ \frac{H}{v a_0^2 \omega_p}\left(\frac{\kappa}{\zeta_p^s}\widetilde{\boldsymbol{\eta}}_p^s - \widehat{\boldsymbol{\eta}}_p^{\mathrm{tr}}\right) & \frac{1}{\omega_p}\left(\mathrm{Id} - \frac{\kappa}{\zeta_p^s}\widetilde{\boldsymbol{\eta}}_p^s \otimes \widehat{\boldsymbol{\eta}}_p^{\mathrm{tr}}\right)\end{pmatrix}, & p \in \mathcal{A}_{\mathrm{pl}} \cap \mathcal{A}_\alpha. \end{cases}$$

Substituting these results into Equation (4.11) and solving for the plastic updates $\delta(\alpha, \boldsymbol{\varepsilon}_p^{\mathrm{pl}})$ leads to the formula given in (4.5). Next, this is inserted into the Newton equation of (2.21a)

$$K^h \delta\mathbf{u}_{j+1}^h + B_{\mathrm{pl}}^h \delta\boldsymbol{\varepsilon}^{\mathrm{pl},h} = \boldsymbol{\varrho}_j^h - K^h \mathbf{u}_{j+1}^h - B_{\mathrm{pl}}^h \boldsymbol{\varepsilon}^{\mathrm{pl},h},$$

which finally gives the tangential system (4.9). □

Next, we investigate the tensor $\boldsymbol{\eta}_p^s$, $p \in \mathcal{A}_{\mathrm{pl}}$, defined in (4.8c) in more detail. For $s \in [0,1]$, it can be regarded as a convex combination of the tensors $\tau_p \widehat{\boldsymbol{\eta}}_p^{\mathrm{tr}}$ and $a_p^{\mathrm{pl}}\bar{\boldsymbol{\eta}}_p$, such that the norm of $\boldsymbol{\eta}_p^s$ is estimated as

$$\|\boldsymbol{\eta}_p^s\| \leq (1-s)\tau_p\|\widehat{\boldsymbol{\eta}}_p^{\mathrm{tr}}\| + s a_p^{\mathrm{pl}}\|\bar{\boldsymbol{\eta}}_p\| \leq \tau_p + s(a_p^{\mathrm{pl}}\|\bar{\boldsymbol{\eta}}_p\| - \tau_p).$$

Especially in the preasymptotic range, we encounter the problem that $\|\bar{\boldsymbol{\eta}}_p\|$ and hence $\|\boldsymbol{\eta}_p^s\|$ cannot be bounded a priori if $a_p^{\mathrm{pl}} = 1$. To account for this fact, we define the stabilization factor a_p^{pl} such that $\|\boldsymbol{\eta}_p^s\| \leq \tau_p$ is always satisfied for $p \in \mathcal{A}_{\mathrm{pl}}$ and $s \in [0,1]$, namely

$$a_p^{\mathrm{pl}} := \min\left(1, \frac{Y^{\mathrm{pl}}(\alpha_p)}{\|\boldsymbol{\eta}_p\|}\right) = \min\left(1, \frac{\tau_p}{\|\bar{\boldsymbol{\eta}}_p\|}\right) \leq 1. \tag{4.13}$$

Due to this modification, the iterative method (4.9) becomes a quasi-Newton scheme. But for $p \in \mathcal{A}_{\text{pl}}$, we have $\|\boldsymbol{\eta}_p\| \to Y^{\text{pl}}(\alpha_p)$ and hence $a_p^{\text{pl}} \to 1$ as $(\mathbf{u}^h, (\alpha_p, \boldsymbol{\varepsilon}_p^{\text{pl}}))$ converges to the solution of (2.21c). Thus, one can expect the superlinear local convergence of the Newton scheme to hold also for the quasi-Newton version.

The stabilization factor (4.13) is important both from the theoretical and the numerical point of view; the former can be seen in Lemma 4.3 analysing the positive definiteness of the tangential matrix, whereas the latter aspect is documented in Subsection 4.1.2.3.

The following lemma gives sufficient criteria for the solvability of the tangential system (4.9).

Lemma 4.3. *Let a_p^{pl} be as defined in (4.13). The symmetric part of the tangential stiffness matrix from (4.9) is positive semidefinite on $\mathbf{V}_0^h$ and positive definite for $\varrho > 0$ or $K > 0$, if either of the following conditions hold:*

$$\begin{array}{lll} (i) & s = 0 & \textit{and} \;\; c^{pl} > 0, \\ (ii) & 0 \le s \le 1 & \textit{and} \;\; c^{pl} \ge 2\mu + a_0^{-2} K. \end{array} \tag{4.14}$$

Hence, if (4.14) and $\varrho > 0$ or $K > 0$ are satisfied, there exists a unique solution $\delta \mathbf{u}_{j+1}^h$ of (4.9).

Remark 4.4. For positive kinematic hardening, the lower bound on c^{pl} in (4.14) *(ii)* is not sharp. A more detailed analysis yields that for $K > 0$, the set of values (s, c^{pl}) implying a positive definite stiffness matrix is larger than the union of the two sets in (4.14) and features a continuous transition between these subsets. But as the bounds in (4.14) are not improved by much for $K \ll \mu$, the details are omitted.

Proof. The case *(i)* can be found in [35]. Here, we consider the case *(ii)*.

As the matrix $K^h = \frac{2}{\Delta t^2} M^h + \theta A^h$ is positive definite for $\varrho > 0$, it suffices to consider the quasi-static case with $\varrho = 0$ and $\theta = 1$. We take the same trial and test function $\mathbf{v} \in \mathbf{V}_0^h$ and consider the symmetric part of the stiffness matrix in (4.9)

$$\begin{aligned} &\left(\lambda + \frac{2\mu}{d}\right) |\operatorname{tr} \boldsymbol{\varepsilon}(\mathbf{v})|^2 + 2\mu \|\operatorname{dev} \boldsymbol{\varepsilon}(\mathbf{v})\|^2 - \sum_{p \in \mathcal{A}_{\text{pl}}} \frac{4\mu^2 (1 - \tau_p)}{\omega_p} \|R_p^{\text{dev}}(\mathbf{v})\|^2 \\ &- \sum_{p \in \mathcal{A}_{\text{pl}}} \frac{4\mu^2 c^{\text{pl}}}{\omega_p \zeta_p^s} \left(\tilde{\boldsymbol{\eta}}_p^s : R_p^{\text{dev}}(\mathbf{v})\right) \left(\hat{\boldsymbol{\eta}}_p^{\text{tr}} : R_p^{\text{dev}}(\mathbf{v})\right), \end{aligned} \tag{4.15}$$

where we have used (2.5) as well as the definition (1.5) of the Hooke tensor $\mathbb{C}^{\text{el}}$. By the Cauchy–Schwarz inequality and Definition (4.7) of R_p^{dev}, we obtain the estimate

$$\|\operatorname{dev} \boldsymbol{\varepsilon}(\mathbf{v})\|^2 \ge \sum_{p \in \mathcal{N}_{\text{pl}}^h} \|R_p^{\text{dev}}(\mathbf{v})\|^2,$$

which becomes an equality if the grid consists of simplicial elements.

The first term of (4.15) containing the volumetric part of $\boldsymbol{\varepsilon}(\mathbf{v})$ is obviously nonnegative, as well as the contribution of all elements $p \in \mathcal{I}_{\text{pl}}$. Hence we focus on the deviatoric part of $\boldsymbol{\varepsilon}(\mathbf{v})$ and consider the following lower bound for the term corresponding to a plastifying element $p \in \mathcal{A}_{\text{pl}}$:

$$\mathfrak{I}_p^s := 2\mu \|R_p^{\text{dev}}(\mathbf{v})\|^2 - \frac{4\mu^2 (1 - \tau_p)}{\omega_p} \|R_p^{\text{dev}}(\mathbf{v})\|^2 - \frac{4\mu^2 c^{\text{pl}}}{\omega_p \zeta_p^s} \left(\tilde{\boldsymbol{\eta}}_p^s : R_p^{\text{dev}}(\mathbf{v})\right) \left(\hat{\boldsymbol{\eta}}_p^{\text{tr}} : R_p^{\text{dev}}(\mathbf{v})\right).$$

Substituting the definitions (4.8d) of ω_p and (4.8c) of $\widetilde{\boldsymbol{\eta}}_p^s$ into the above expression gives

$$\begin{aligned}\mathfrak{I}_p^s &= \frac{2\mu\left(a_0^{-2}K(1-\tau_p)+c^{\mathrm{pl}}\tau_p\right)}{\omega_p}\|R_p^{\mathrm{dev}}(\mathbf{v})\|^2 + \chi_{\mathcal{A}_\alpha}\frac{4\mu^2c^{\mathrm{pl}}H}{\omega_p\zeta_p^s\upsilon a_0^2}\left(\widehat{\boldsymbol{\eta}}_p^{\mathrm{tr}} : R_p^{\mathrm{dev}}(\mathbf{v})\right)^2 \\ &\quad - \frac{4\mu^2c^{\mathrm{pl}}}{\omega_p\zeta_p^s}\left(\boldsymbol{\eta}_p^s : R_p^{\mathrm{dev}}(\mathbf{v})\right)\left(\widehat{\boldsymbol{\eta}}_p^{\mathrm{tr}} : R_p^{\mathrm{dev}}(\mathbf{v})\right). \end{aligned} \tag{4.16}$$

Because we have $0 \leq \tau_p < 1$ for $p \in \mathcal{A}_{\mathrm{pl}}$, we obtain $\omega_p > 0$. Further, we investigate the sign of ζ_p^s defined in (4.8d); the assumption on the choice of c^{pl} and s yields

$$\begin{aligned}\zeta_p^s &= \left(2\mu + a_0^{-2}K - c^{\mathrm{pl}}\right)\left(1 - \tau_p + (\widetilde{\boldsymbol{\eta}}_p^s : \widehat{\boldsymbol{\eta}}_p^{\mathrm{tr}})\right) + c^{\mathrm{pl}} \\ &= (2\mu + a_0^{-2}K) - \underbrace{\left(2\mu + a_0^{-2}K - c^{\mathrm{pl}}\right)}_{\leq 0}\underbrace{\left(\tau_p - (\widetilde{\boldsymbol{\eta}}_p^s : \widehat{\boldsymbol{\eta}}_p^{\mathrm{tr}})\right)}_{\geq 0} \geq (2\mu + a_0^{-2}K) > 0, \end{aligned} \tag{4.17}$$

where the second factor is nonnegative due to the bound $\|\boldsymbol{\eta}_p^s\| \leq \tau_p$ following from (4.13):

$$\tau_p - (\widetilde{\boldsymbol{\eta}}_p^s : \widehat{\boldsymbol{\eta}}_p^{\mathrm{tr}}) = \tau_p - (\boldsymbol{\eta}_p^s : \widehat{\boldsymbol{\eta}}_p^{\mathrm{tr}}) + \chi_{\mathcal{A}_\alpha}\frac{H}{\upsilon a_0^2} \geq \tau_p - \|\boldsymbol{\eta}_p^s\| \geq 0.$$

Using (4.17), we can bound (4.16) from below by

$$\begin{aligned}\mathfrak{I}_p^s &\geq \frac{2\mu\left(a_0^{-2}K\zeta_p^s(1-\tau_p) + c^{\mathrm{pl}}\tau_p\zeta_p^s - 2\mu c^{\mathrm{pl}}\tau_p\right)}{\omega_p\zeta_p^s}\|R_p^{\mathrm{dev}}(\mathbf{v})\|^2 \\ &\geq \frac{2\mu a_0^{-2}K\left(2\mu + a_0^{-2}K + \tau_p\left(c^{\mathrm{pl}} - 2\mu - a_0^{-2}K\right)\right)}{\omega_p\zeta_p^s}\|R_p^{\mathrm{dev}}(\mathbf{v})\|^2 \\ &= \frac{2\mu a_0^{-2}K}{\zeta_p^s}\|R_p^{\mathrm{dev}}(\mathbf{v})\|^2 \end{aligned}$$

which is always nonnegative and strictly positive in case of kinematic hardening, i.e., for $K > 0$. Summation on $p \in \mathcal{A}_{\mathrm{pl}}$ then concludes the proof. □

4.1.2 Numerical results for plasticity

4.1.2.1 Geometry and parameters

For the first set of test, we consider a well-known benchmark problem for plastic computations, namely a quadratic plate of length 2.0 with a circular hole of radius 0.2 in the center. Such a geometry, e.g., arises as the cross-section of a thick plate within a plane strain setting. The numerical simulation is done on the upper right quarter of the plate as depicted in Figure 4.1. On the bottom and left boundaries of the domain, symmetry boundary conditions are posed, whereas on the circular hole and on the right boundary, we have homogeneous Neumann conditions. On the upper boundary, the plate is pulled with the force $f_t \cdot |0.5 - t|$ upwards for $t \in [0, 1]$ and different values of f_t. We use a quasi-static discretization with 17 load steps, a simplicial mesh with around 5200 elements and the material parameters $\varrho = 0$, $H = 200$, $K = 100$, $\sigma_0 \approx 700$, $E = 6.9 \cdot 10^4$, $\nu = 0.33$. The plastic active sets are updated in each Newton iteration, and the active set $\mathcal{A}_{\mathrm{pl}}$ is initialized with the final active set from the previous load step if not stated otherwise. Figure 4.1 illustrates the effective stress $\sigma_{\mathrm{eff}} := \|\operatorname{dev}(\boldsymbol{\sigma})\|$, the active set $\mathcal{A}_{\mathrm{pl}}$ as well as the equivalent plastic strain α at load step t_8.

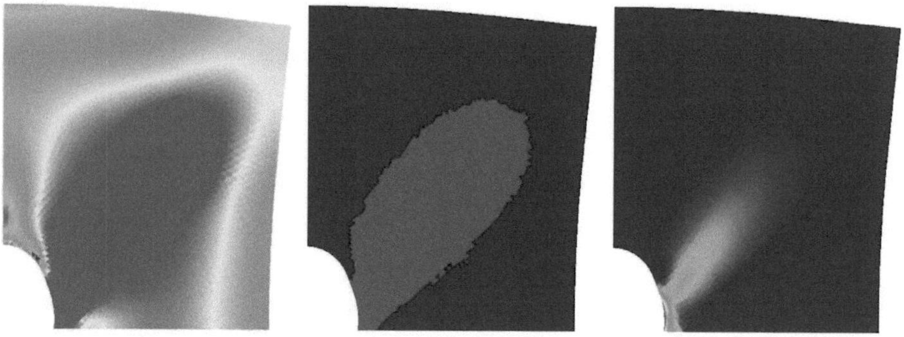

Figure 4.1: Plastic plate at load step t_8 for $f_t = 1100$; left: effective stress σ_{eff}; middle: plastifying elements; right: equivalent plastic strain α. Displacement is amplified by a factor of 10.

4.1.2.2 Influence of s on the convergence

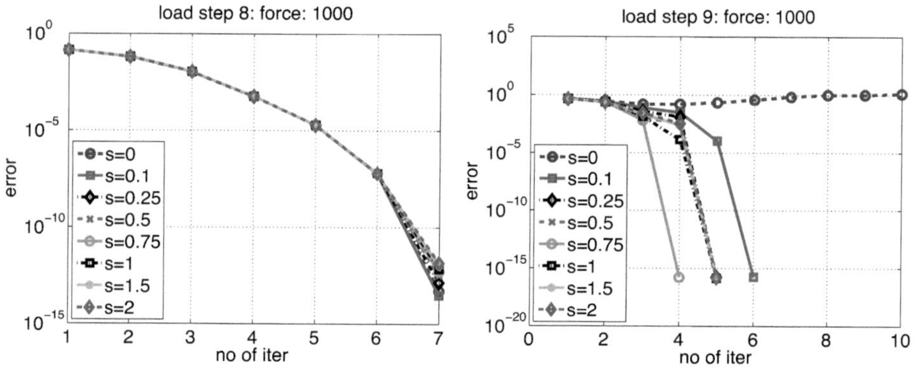

Figure 4.2: Error decay for $f_t = 1000$, $c^{\text{pl}} = 2\mu + a_0^{-2}K$ and different values of s; left: t_8; right: t_9.

First, we investigate the influence of the parameter s on the performance of the semismooth Newton method (4.9). Figure 4.2 displays the decay of the relative error of the Newton scheme for $c^{\text{pl}} = 2\mu + a_0^{-2}K$, different values of s within the interval $[0, 2]$ and the (most critical) load steps t_8 and t_9, respectively. One can see that the scheme does not converge at load step t_9 for the case $s = 0$ corresponding to the radial return method, whereas for $s \geq 0.1$, locally superlinear convergence is obtained. For $s = 0.75$, the relative error goes below the tolerance of 10^{-8} after 4 iterations, and for $s \in \{0.25, 0.5, 1, 1.5, 2\}$ only one more iteration is needed. A too small value of s deteriorates the error decay, especially if we increase the surface load to $f_t = 1100$ where also the schemes with $s \in \{0.1, 0.25\}$ do not converge, as shown on the left of

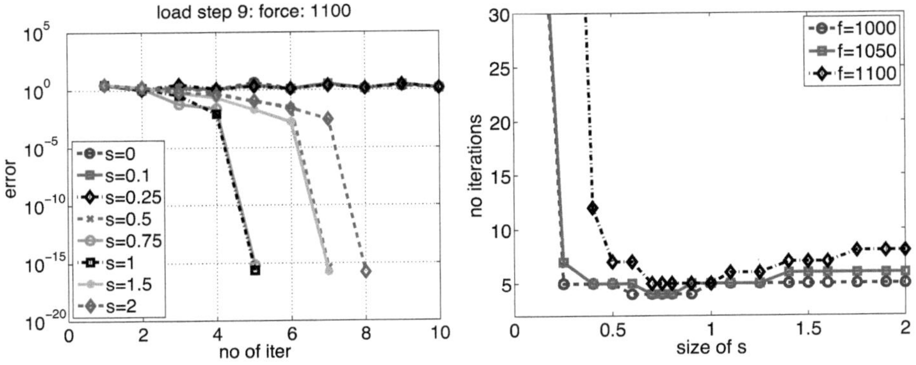

Figure 4.3: Convergence at load step t_9 for $c^{\mathrm{pl}} = 2\mu + a_0^{-2}K$ for different values of s; left: error decay for $f_t = 1100$; right: number of iterations for different forces f_t.

Figure 4.3.

On the right of Figure 4.3, the number of Newton iterations is depicted with respect to s for the load step t_9 and different surface loads. It can be observed that if s is chosen too small, the method does not converge at all. Further, the convergence slightly degrades for $s > 1$; we recall that Lemma 4.3 only covers the case $s \in [0, 1]$. The range of suitable values for s centers around $s = 0.75$ and becomes narrower for a larger load f_t.

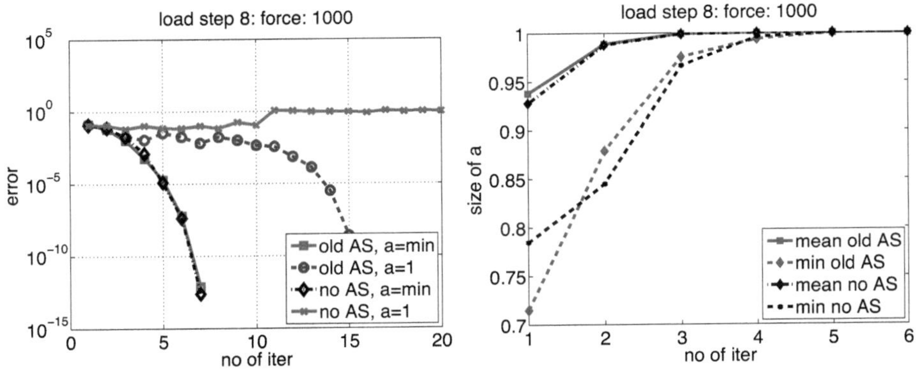

Figure 4.4: Influence of stabilization factor a^{pl} for $s = 1$, $c^{\mathrm{pl}} = 2\mu + a_0^{-2}K$, load $f_t = 1000$, load step t_8; left: relative error decay for different starting solutions; right: mean size and minimum size of a^{pl}.

4.1.2.3 Influence of the stabilization factor a^{pl} on the convergence

In this subsection, we numerically confirm that the stabilization factor (4.13) benefits the convergence behaviour. On the left of Figure 4.4, the error decay at load step t_8 is shown for the load $f_t = 1000$, $s = 1$ and $c^{\text{pl}} = 2\mu + a_0^{-2}K$. For the curves marked "old AS", the final active set $\mathcal{A}_{\text{pl}}$ from the previous load step is taken as the starting solution for the new step, whereas the initialization $\mathcal{A}_{\text{pl}} = \emptyset$ is chosen for those graphs marked with "no AS". "$a = 1$" uses $a^{\text{pl}} = 1$ always, whereas "$a = \min$" employs definition (4.13). One can observe that the Newton method with $a^{\text{pl}} = 1$ needs more iterations and does not converge at all if the starting value is too far away from the solution, whereas the scheme including the factor (4.13) converges faster and is less sensitive to the starting solution.

On the right of Figure 4.4, both the mean value $\frac{1}{|\mathcal{A}_{\text{pl}}|}\sum_{p\in\mathcal{A}_{\text{pl}}} a_p^{\text{pl}}$ and the minimal value $\min_{p\in\mathcal{A}_{\text{pl}}} a_p^{\text{pl}}$ of the stabilization factor a^{pl} are shown. One can see that the use of a^{pl} is very important in the first few iterations and that its value converges rapidly to 1, here after at most 5 iterations.

4.1.2.4 Influence of c^{pl} on the convergence

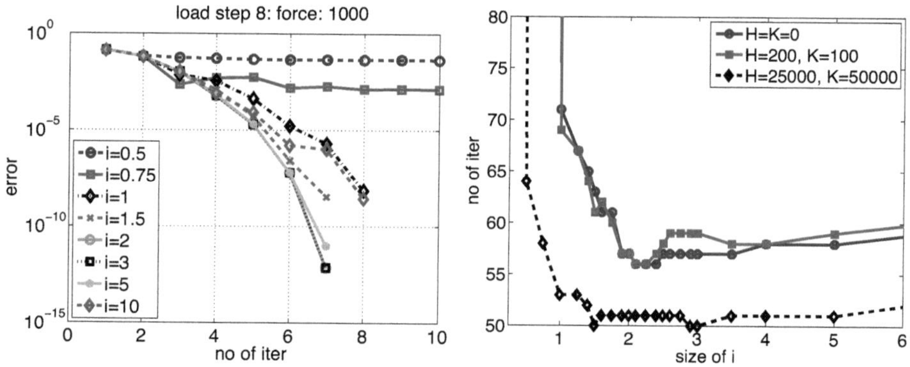

Figure 4.5: Convergence for $f_t = 1000$ and $s = 1$ for different values of $c^{\text{pl}} = i\mu + a_0^{-2}K$; left: relative error decay at time t_8; right: total number of iterations with respect to i for computation with different hardening parameters.

Finally, we employ a^{pl} as defined in (4.13), fix $s = 1$ and vary the size of c^{pl}. The left picture of Figure 4.5 shows the convergence at the load step t_8 for the load $f_t = 1000$ and $c^{\text{pl}} = i\mu + a_0^{-2}K$ with different values of i. The computations for $i \leq 0.75$ do not converge; furthermore, too large values of i tend to deteriorate the convergence rate slightly. The lowest number of iterations is obtained for $i \in [2, 5]$.

The right picture of Figure 4.5 displays the total number of Newton iterations needed for all 17 load steps with respect to i and for different values of the hardening parameters H, K. For the test cases with small or no hardening, one can observe that the scheme does not converge for $i < 1$ and that the number of iterations decreases for increasing $i \in [1, 1.75]$. The fastest convergence is obtained for $i \approx 2$, but for $i \in [2.5, 5]$, the number of total iterations is at most 5% higher. In contrast, the computation with very large hardening parameters yields a better convergence behaviour altogether: the number of iterations is less sensitive to i and the scheme converges for $i \geq 0.5$.

The results of this section can be summarized by the statement that the semismooth quasi-Newton scheme (4.9) including the stabilization factor a^{pl} as defined in (4.13) and the parameters $c^{\mathrm{pl}} = 2\mu + a_0^{-2}K$, $s \in [0.75, 1]$ yields a robust and very efficient algorithm for the numerical simulation of infinitesimal elastoplasticity.

4.2 Application of abstract framework to frictional contact

In this section, we set $\boldsymbol{\varepsilon}^{\mathrm{pl},h} = \mathbf{0}$, $\alpha^h = 0$ and focus on the contact contributions of system (2.21). As before, we use the modified local NCP function $C_p^{\mathrm{co},s}$ which is constructed from (2.18b) by multiplication with the additional factor

$$S_p^{\mathrm{co},s}(\mathbf{u}_p, \boldsymbol{\lambda}_p) := \begin{cases} \mathrm{Id}, & Y^{\mathrm{co}}(\lambda_{p,n}^{\mathrm{tr}}) = 0, \\ \begin{pmatrix} 1 & \mathbf{0} \\ \mathbf{0} & \max\left(Y^{\mathrm{co}}(\lambda_{p,n}^{\mathrm{tr}}), \|\boldsymbol{\lambda}_{p,t}^{\mathrm{tr}}\|\right)^s \mathrm{Id} \end{pmatrix}, & Y^{\mathrm{co}}(\lambda_{p,n}^{\mathrm{tr}}) > 0. \end{cases} \tag{4.18}$$

In Subsection 4.2.1, we present the application of the semismooth Newton scheme to the resulting nonlinear problem, as well as a lemma on the solvability of the tangential system. Subsection 4.2.2 features some numerical tests for a three-dimensional two-body contact problem.

As the displacement $\mathbf{u}^h$ is always taken at time t_{j+1} and the contact stress $\boldsymbol{\lambda}^h$ at time $t_{j+\theta}$, we omit the time index for $\boldsymbol{\lambda}_{j+\theta}^h$ as well as for $\mathbf{u}_{j+1}^h$ in the rest of this section for ease of notation.

4.2.1 Semismooth Newton scheme

The evaluation of $C_p^{\mathrm{co},s}$ implies the following partition of the set of potential contact nodes $\mathcal{N}_{\mathrm{co}}^h$ into the active and inactive sets

$$\mathcal{A}_n := \left\{p \in \mathcal{N}_{\mathrm{co}}^h : \lambda_{p,n}^{\mathrm{tr}} > 0\right\}, \tag{4.19a}$$
$$\mathcal{I}_n := \left\{p \in \mathcal{N}_{\mathrm{co}}^h : \lambda_{p,n}^{\mathrm{tr}} \leq 0\right\}, \tag{4.19b}$$
$$\mathcal{A}_t := \left\{p \in \mathcal{N}_{\mathrm{co}}^h : \|\boldsymbol{\lambda}_{p,t}^{\mathrm{tr}}\| > Y^{\mathrm{co}}(\lambda_{p,n}^{\mathrm{tr}})\right\}, \tag{4.19c}$$
$$\mathcal{I}_t := \left\{p \in \mathcal{N}_{\mathrm{co}}^h : \|\boldsymbol{\lambda}_{p,t}^{\mathrm{tr}}\| \leq Y^{\mathrm{co}}(\lambda_{p,n}^{\mathrm{tr}}),\ Y^{\mathrm{co}}(\lambda_{p,n}^{\mathrm{tr}}) > 0\right\}, \tag{4.19d}$$
$$\mathcal{F}_t := \left\{p \in \mathcal{N}_{\mathrm{co}}^h : \|\boldsymbol{\lambda}_{p,t}^{\mathrm{tr}}\| = Y^{\mathrm{co}}(\lambda_{p,n}^{\mathrm{tr}}) = 0\right\}, \tag{4.19e}$$

where the variables used above are defined in (2.14b), (2.17b). Further, we set

$$\vartheta_p := \frac{Y^{\mathrm{co}}(\lambda_{p,n}^{\mathrm{tr}})}{\|\boldsymbol{\lambda}_{p,t}^{\mathrm{tr}}\|}, \quad \text{for } \|\boldsymbol{\lambda}_{p,t}^{\mathrm{tr}}\| > 0. \tag{4.20}$$

The set $\mathcal{A}_n$ contains the contact nodes with nonpositive normal gap, whereas the nodes in $\mathcal{I}_n$ are free in normal direction. The partition $\mathcal{A}_t$, $\mathcal{I}_t$, $\mathcal{F}_t$ distinguishes between slippy, sticky and free nodes in tangential direction, where the latter case can only occur for pure Coulomb friction with $g_t = 0$. Further, we always have $\mathcal{F}_t \subset \mathcal{I}_n$ due to the definition of Y^{co}.

With the sets (4.19), the system to be solved at each Newton iteration is given in the lemma below. Again, the Newton index $\cdot^{(k)}$ is omitted for ease of notation.

Lemma 4.5. *The application of the semismooth Newton scheme to the contact part of (2.21) with the previous iterate* $(\mathbf{u}^h, \boldsymbol{\lambda}^h) = (\mathbf{u}^h_{j+1}, \boldsymbol{\lambda}^h_{j+\theta})$ *yields the following relations for the primal and dual updates at the contact boundary* $(\delta\mathbf{u}_p, \delta\boldsymbol{\lambda}_p)$, $p \in \mathcal{N}^h_{co}$:

$$\begin{cases} \delta\lambda_{p,n} = -\lambda_{p,n}, & p \in \mathcal{I}_n, \\ \delta u_{p,n} = g_{p,n} - u_{p,n}, & p \in \mathcal{A}_n, \\ \delta\mathbf{u}_{p,t} = -\Delta\mathbf{u}_{p,t}, & p \in \mathcal{I}_t \cap \mathcal{I}_n, \\ \delta\mathbf{u}_{p,t} = -\left(1 + \frac{\mathfrak{F}s}{Y^{co}(\lambda^{tr}_{p,n})}\left(\delta\lambda_{p,n} + c^{co}_n h_p^{-1}\delta u_{p,n}\right)\right)\Delta\mathbf{u}_{p,t}, & p \in \mathcal{I}_t \cap \mathcal{A}_n, \\ \delta\boldsymbol{\lambda}_{p,t} = L^s_p c^{co}_t h_p^{-1}(\delta\mathbf{u}_{p,t}) + \mathbf{q}^s_p, & p \in \mathcal{A}_t \cap \mathcal{I}_n, \\ \delta\boldsymbol{\lambda}_{p,t} = L^s_p c^{co}_t h_p^{-1}(\delta\mathbf{u}_{p,t}) + \mathbf{l}^s_p\left(\delta\lambda_{p,n} + c^{co}_n h_p^{-1}\delta u_{p,n}\right) + \mathbf{q}^s_p, & p \in \mathcal{A}_t \cap \mathcal{A}_n, \\ \delta\boldsymbol{\lambda}_{p,t} = -\boldsymbol{\lambda}_{p,t}, & p \in \mathcal{F}_t. \end{cases} \tag{4.21}$$

Above, we have introduced the abbreviations

$$L^s_p := \frac{1}{1-\vartheta_p}\left(\vartheta_p Id - \frac{1}{\xi^s_p}\boldsymbol{\lambda}^s_{p,t} \otimes \hat{\boldsymbol{\lambda}}^{tr}_{p,t}\right), \tag{4.22a}$$

$$\mathbf{l}^s_p := \frac{\mathfrak{F}}{1-\vartheta_p}\left(\hat{\boldsymbol{\lambda}}^{tr}_{p,t} - \frac{1}{\xi^s_p}\boldsymbol{\lambda}^s_{p,t}\right), \tag{4.22b}$$

$$\mathbf{q}^s_p := \frac{\|\boldsymbol{\lambda}^{tr}_{p,t}\|}{1-\vartheta_p}\left(\vartheta_p\hat{\boldsymbol{\lambda}}^{tr}_{p,t} - \bar{\boldsymbol{\lambda}}_{p,t} + \frac{\left(\bar{\boldsymbol{\lambda}}^s_{p,t} : \hat{\boldsymbol{\lambda}}^{tr}_{p,t}\right) - \vartheta_p}{\xi^s_p}\boldsymbol{\lambda}^s_{p,t}\right), \tag{4.22c}$$

with Y^{co} *from (2.14b),* $\lambda^{tr}_{p,n}$, $\boldsymbol{\lambda}^{tr}_{p,t}$ *from (2.17b),* ϑ_p *from (4.20), and the additional definitions*

$$\hat{\boldsymbol{\lambda}}^{tr}_{p,t} := \frac{\boldsymbol{\lambda}^{tr}_{p,t}}{\|\boldsymbol{\lambda}^{tr}_{p,t}\|}, \qquad \bar{\boldsymbol{\lambda}}_{p,t} := \frac{\boldsymbol{\lambda}_{p,t}}{\|\boldsymbol{\lambda}^{tr}_{p,t}\|}, \tag{4.23a}$$

$$\boldsymbol{\lambda}^s_{p,t} := (1-s)\vartheta_p\hat{\boldsymbol{\lambda}}^{tr}_{p,t} + s a^{co}_p\bar{\boldsymbol{\lambda}}_{p,t}, \qquad \xi^s_p := 1 - \vartheta_p + \left(\boldsymbol{\lambda}^s_{p,t} \cdot \hat{\boldsymbol{\lambda}}^{tr}_{p,t}\right). \tag{4.23b}$$

In (4.23b), we have used the scaling factor $a^{co}_p := 1$.

With $B^h_{co}\delta\boldsymbol{\lambda}^h =: \sum_{p\in\mathcal{N}^h_{co}}\left(B^{co}_{p,n}\delta\lambda_{p,n} + B^{co}_{p,t}\delta\boldsymbol{\lambda}_{p,t}\right)$, *the resulting system to be solved reads*

$$\begin{aligned} K^h\delta\mathbf{u}^h &+ c^{co}_t\sum_{p\in\mathcal{A}_t} h_p^{-1}B^{co}_{p,t}L^s_p\delta\mathbf{u}_{p,t} + c^{co}_n\sum_{p\in\mathcal{A}_t\cap\mathcal{A}_n} h_p^{-1}B^{co}_{p,t}\mathbf{l}^s_p\delta u_{p,n} \\ &+ \sum_{p\in\mathcal{A}_n} B^{co}_{p,n}\delta\lambda_{p,n} + \sum_{p\in\mathcal{I}_t} B^{co}_{p,t}\delta\boldsymbol{\lambda}_{p,t} + \sum_{p\in\mathcal{A}_t\cap\mathcal{A}_n} B^{co}_{p,t}\mathbf{l}^s_p\delta\lambda_{p,n} \\ &= \boldsymbol{\varrho}^h_j - K^h\mathbf{u}^h - \sum_{p\in\mathcal{A}_n} B^{co}_{p,n}\lambda_{p,n} - \sum_{p\in\mathcal{I}_t} B^{co}_{p,t}\boldsymbol{\lambda}_{p,t} - \sum_{p\in\mathcal{A}_t} B^{co}_{p,t}\mathbf{q}^s_p, \end{aligned} \tag{4.24}$$

together with the nodal boundary conditions on $\delta\mathbf{u}^h$ *given by* $(4.21)_{2-4}$.

Remark 4.6. Similar to the plastic case, the variables in (4.22) are only well-defined for $\xi_p^s \neq 0$. This can be ensured by the introduction of appropriate stabilization factors as discussed on page 55; otherwise, the mixed problem (4.24) has to be enlarged by the variables $\delta\boldsymbol{\lambda}_{p,t}$, $p \in \mathcal{A}_t$.

Remark 4.7. By incorporating the conditions $(4.21)_{2-4}$ into the system matrix of (4.24) like explained in [91], a linear system only for the primal variables can be obtained. We refer to [91] for more details.

Proof. First, we deal with the case $p \in \mathcal{F}_t \subset \mathcal{I}_n$, where we enforce the conditions $\delta\lambda_{p,n} = -\lambda_{p,n}$, $\delta\boldsymbol{\lambda}_{p,t} = -\boldsymbol{\lambda}_{p,t}$ following from (1.25).

Next, we consider a node $p \notin \mathcal{F}_t$. If $p \in \mathcal{A}_t$, we have $\|\boldsymbol{\lambda}_{p,t}^{\mathrm{tr}}\| > 0$ such that ϑ_p as given in (4.20) is well-defined. With (4.19) and $Y^{\mathrm{co}}(\lambda_{p,n}^{\mathrm{tr}}) > 0$, the scaled nonlinear complementarity function $C_p^{\mathrm{co},s} = C_p^{\mathrm{co},s}(\mathbf{u}_p, \boldsymbol{\lambda}_p)$ reads

$$C_p^{\mathrm{co},s} = \begin{cases} \begin{pmatrix} \lambda_{p,n} \\ -(Y^{\mathrm{co}}(\lambda_{p,n}^{\mathrm{tr}}))^s c_t^{\mathrm{co}} h_p^{-1} \Delta\mathbf{u}_{p,t} \end{pmatrix}, & p \in \mathcal{I}_n \cap \mathcal{I}_t, \\ \begin{pmatrix} c_n^{\mathrm{co}} h_p^{-1}(g_{p,n} - u_{p,n}) \\ -(Y^{\mathrm{co}}(\lambda_{p,n}^{\mathrm{tr}}))^s c_t^{\mathrm{co}} h_p^{-1} \Delta\mathbf{u}_{p,t} \end{pmatrix}, & p \in \mathcal{A}_n \cap \mathcal{I}_t, \\ \begin{pmatrix} \lambda_{p,n} \\ \|\boldsymbol{\lambda}_{p,t}^{\mathrm{tr}}\|^s \left(\boldsymbol{\lambda}_{p,t} - \vartheta_p \boldsymbol{\lambda}_{p,t}^{\mathrm{tr}}\right) \end{pmatrix}, & p \in \mathcal{I}_n \cap \mathcal{A}_t, \\ \begin{pmatrix} c_n^{\mathrm{co}} h_p^{-1}(g_{p,n} - u_{p,n}) \\ \|\boldsymbol{\lambda}_{p,t}^{\mathrm{tr}}\|^s \left(\boldsymbol{\lambda}_{p,t} - \vartheta_p \boldsymbol{\lambda}_{p,t}^{\mathrm{tr}}\right) \end{pmatrix}, & p \in \mathcal{A}_n \cap \mathcal{A}_t. \end{cases} \tag{4.25}$$

We consider one step of the semismooth Newton method applied to (4.25) scaled with the factor $(S_p^{\mathrm{co},s})^{-1}$:

$$(S_p^{\mathrm{co},s})^{-1}\left(\left(\partial_{\boldsymbol{\lambda}_p} C_p^{\mathrm{co},s}\right)\delta\boldsymbol{\lambda}_p + \left(\partial_{\mathbf{u}_p} C_p^{\mathrm{co},s}\right)\delta\mathbf{u}_p + C_p^{\mathrm{co},s}\right) = \mathbf{0}. \tag{4.26}$$

Using (3.13) and (4.23), we can compute the tangential stiffness matrices in (4.26):

$$(S_p^{\mathrm{co},s})^{-1}\left(\partial_{\mathbf{u}_p} C_p^{\mathrm{co},s}\right) = \begin{cases} -h_p^{-1}\begin{pmatrix} 0 & \mathbf{0} \\ \mathbf{0} & c_t^{\mathrm{co}}\mathrm{Id} \end{pmatrix}, & p \in \mathcal{I}_n \cap \mathcal{I}_t, \\ -h_p^{-1}\begin{pmatrix} c_n^{\mathrm{co}} & \mathbf{0} \\ \frac{\mathfrak{F} c_n^{\mathrm{co}} c_t^{\mathrm{co}} s}{h_p Y^{\mathrm{co}}(\lambda_{p,n}^{\mathrm{tr}})}\Delta\mathbf{u}_{p,t} & c_t^{\mathrm{co}}\mathrm{Id} \end{pmatrix}, & p \in \mathcal{A}_n \cap \mathcal{I}_t, \\ -h_p^{-1}\begin{pmatrix} 0 & \mathbf{0} \\ \mathbf{0} & c_t^{\mathrm{co}}\left(\vartheta_p\mathrm{Id} - \boldsymbol{\lambda}_{p,t}^{s} \otimes \widehat{\boldsymbol{\lambda}}_{p,t}^{\mathrm{tr}}\right) \end{pmatrix}, & p \in \mathcal{I}_n \cap \mathcal{A}_t, \\ -h_p^{-1}\begin{pmatrix} c_n^{\mathrm{co}} & \mathbf{0} \\ \mathfrak{F} c_n^{\mathrm{co}}\widehat{\boldsymbol{\lambda}}_{p,t}^{\mathrm{tr}} & c_t^{\mathrm{co}}\left(\vartheta_p\mathrm{Id} - \boldsymbol{\lambda}_{p,t}^{s} \otimes \widehat{\boldsymbol{\lambda}}_{p,t}^{\mathrm{tr}}\right) \end{pmatrix}, & p \in \mathcal{A}_n \cap \mathcal{A}_t, \end{cases} \tag{4.27a}$$

$$(S_p^{\mathrm{co},s})^{-1}\left(\partial_{\boldsymbol{\lambda}_p} C_p^{\mathrm{co},s}\right) = \begin{cases} \begin{pmatrix} 1 & \mathbf{0} \\ \mathbf{0} & \mathbf{0} \end{pmatrix}, & p \in \mathcal{I}_n \cap \mathcal{I}_t, \\ \begin{pmatrix} 0 & \mathbf{0} \\ -\frac{\mathfrak{F} c_t^{\mathrm{co}} s}{h_p Y^{\mathrm{co}}(\lambda_{p,n}^{\mathrm{tr}})} \Delta \mathbf{u}_{p,t} & \mathbf{0} \end{pmatrix}, & p \in \mathcal{A}_n \cap \mathcal{I}_t, \\ \begin{pmatrix} 1 & \mathbf{0} \\ \mathbf{0} & (1-\vartheta_p)\,\mathrm{Id} + \boldsymbol{\lambda}_{p,t}^s \otimes \widehat{\boldsymbol{\lambda}}_{p,t}^{\mathrm{tr}} \end{pmatrix}, & p \in \mathcal{I}_n \cap \mathcal{A}_t, \\ \begin{pmatrix} 0 & \mathbf{0} \\ -\mathfrak{F} \widehat{\boldsymbol{\lambda}}_{p,t}^{\mathrm{tr}} & (1-\vartheta_p)\,\mathrm{Id} + \boldsymbol{\lambda}_{p,t}^s \otimes \widehat{\boldsymbol{\lambda}}_{p,t}^{\mathrm{tr}} \end{pmatrix}, & p \in \mathcal{A}_n \cap \mathcal{A}_t. \end{cases} \tag{4.27b}$$

One can see that the matrix in (4.27b) is only regular for $p \in \mathcal{I}_n \cap \mathcal{A}_t$; for all other cases, the rank of (4.27b) is less than d. Hence, we cannot always resolve Equation (4.26) for the multiplier update $\delta\boldsymbol{\lambda}_p$. However, due to the local structure of the tangential matrix (4.27a) for the displacement, we can eliminate those components of the primal variable $\delta\mathbf{u}_p$ where the dual matrix (4.27b) is singular. To be more precise, we resolve (4.26) for the dual updates $(\delta\lambda_{p,n})_{p\in\mathcal{I}_n}$, $(\delta\boldsymbol{\lambda}_{p,t})_{p\in\mathcal{A}_t}$ on the one hand and for the primal updates $(\delta u_{p,n})_{p\in\mathcal{A}_n}$, $(\delta\mathbf{u}_{p,t})_{p\in\mathcal{I}_t}$ on the other hand. We sketch the procedure by considering the case $p \in \mathcal{A}_n \cap \mathcal{I}_t$; the substitution of (4.25) and (4.27) into (4.26) leads to

$$\begin{aligned} -c_n^{\mathrm{co}} h_p^{-1} \delta u_{p,n} &= -c_n^{\mathrm{co}} h_p^{-1}(g_{p,n} - u_{p,n}), \\ -\frac{\mathfrak{F} c_t^{\mathrm{co}} s}{h_p Y^{\mathrm{co}}(\lambda_{p,n}^{\mathrm{tr}})} \left(\delta\lambda_{p,n} + c_n^{\mathrm{co}} h_p^{-1} \delta u_{p,n}\right) \Delta\mathbf{u}_{p,t} - c_t^{\mathrm{co}} h_p^{-1} \delta\mathbf{u}_{p,t} &= c_t^{\mathrm{co}} h_p^{-1} \Delta\mathbf{u}_{p,t}. \end{aligned}$$

These equations can be solved for $\delta u_{p,n}$ and $\delta\mathbf{u}_{p,t}$ by

$$\begin{aligned} \delta u_{p,n} &= g_{p,n} - u_{p,n}, \\ \delta\mathbf{u}_{p,t} &= -\left(\frac{\mathfrak{F} s}{Y^{\mathrm{co}}(\lambda_{p,n}^{\mathrm{tr}})} \left(\delta\lambda_{p,n} + c_n^{\mathrm{co}} h_p^{-1} \delta u_{p,n}\right) + 1\right) \Delta\mathbf{u}_{p,t}, \end{aligned}$$

leading to the conditions $(4.21)_{2,4}$.
For the tangential component of the nodes $p \in \mathcal{I}_n \cap \mathcal{A}_t$, the local condition obtained from (4.25), (4.26) and (4.27) reads

$$\left((1-\vartheta_p)\,\mathrm{Id} + \boldsymbol{\lambda}_{p,t}^s \otimes \widehat{\boldsymbol{\lambda}}_{p,t}^{\mathrm{tr}}\right) \delta\boldsymbol{\lambda}_{p,t} - c_t^{\mathrm{co}} h_p^{-1} \left(\vartheta_p \mathrm{Id} - \boldsymbol{\lambda}_{p,t}^s \otimes \widehat{\boldsymbol{\lambda}}_{p,t}^{\mathrm{tr}}\right) \delta\mathbf{u}_{p,t} = \vartheta_p \boldsymbol{\lambda}_{p,t}^{\mathrm{tr}} - \boldsymbol{\lambda}_{p,t}.$$

This equation can be resolved for $\delta\boldsymbol{\lambda}_{p,t}$ if the matrix

$$(1-\vartheta_p)\,\mathrm{Id} + \boldsymbol{\lambda}_{p,t}^s \otimes \widehat{\boldsymbol{\lambda}}_{p,t}^{\mathrm{tr}} \tag{4.28}$$

is nonsingular. One can easily check that the eigenvalues of (4.28) are given by $(1-\vartheta_p)$ and ξ_p^s. For $p \in \mathcal{A}_t$, we always have $(1-\vartheta_p) > 0$, whereas the condition $\xi_p^s \neq 0$ is analysed in more detail after this proof and is assumed to hold here.

Proceeding similarly for each of the other cases, we obtain the expressions given in (4.21). Inserting them into the Newton equation of (2.21a)

$$K^h \delta\mathbf{u}^h + B_{\mathrm{co}}^h \delta\boldsymbol{\lambda}^h = \boldsymbol{\varrho}_j^h - K^h \mathbf{u}^h - B_{\mathrm{co}}^h \boldsymbol{\lambda}^h$$

finally gives the tangential system (4.24). □

Similar considerations as those on page 45 motivate the introduction of a stabilization factor

$$a_p^{\mathrm{co}} := \min\left(1, \frac{Y^{\mathrm{co}}(\lambda_{p,n}^{\mathrm{tr}})}{\|\boldsymbol{\lambda}_{p,t}\|}\right) = \min\left(1, \frac{\vartheta_p}{\|\bar{\boldsymbol{\lambda}}_{p,t}\|}\right) \tag{4.29}$$

in the definition (4.23b) of $\boldsymbol{\lambda}_{p,t}^s$. By this, we can ensure the bound $\|\boldsymbol{\lambda}_{p,t}^s\| \leq \vartheta_p$ for $s \in [0,1]$. Due to the fact that we have $a_p^{\mathrm{co}} \to 1$ as $(\mathbf{u}^h, \boldsymbol{\lambda}^h)$ converges to the solution of (2.21d), the superlinear convergence of the Newton method is expected to hold also for the resulting quasi-Newton scheme.

In the definitions (4.22), we need to assume that ξ_p^s is nonzero for all $p \in \mathcal{A}_t$. Further, as discussed in Lemma 4.8 below, the positive semidefiniteness of the matrix L_p^s defined in (4.22a) is important for the positive definiteness of the tangential stiffness matrix in the first line of (4.24). The eigenvalues of L_p^s are given by

$$\frac{\vartheta_p}{1-\vartheta_p} > 0 \qquad \text{and} \qquad \frac{\vartheta_p \xi_p^s - \boldsymbol{\lambda}_{p,t}^s \cdot \widehat{\boldsymbol{\lambda}}_{p,t}^{\mathrm{tr}}}{\xi_p^s(1-\vartheta_p)} = \frac{\vartheta_p - \boldsymbol{\lambda}_{p,t}^s \cdot \widehat{\boldsymbol{\lambda}}_{p,t}^{\mathrm{tr}}}{1 - \left(\vartheta_p - \boldsymbol{\lambda}_{p,t}^s \cdot \widehat{\boldsymbol{\lambda}}_{p,t}^{\mathrm{tr}}\right)}.$$

The latter eigenvalue is nonnegative if and only if

$$0 \leq \left(\vartheta_p - \boldsymbol{\lambda}_{p,t}^s \cdot \widehat{\boldsymbol{\lambda}}_{p,t}^{\mathrm{tr}}\right) = s\left(\vartheta_p - a_p^{\mathrm{co}} \bar{\boldsymbol{\lambda}}_{p,t} \cdot \widehat{\boldsymbol{\lambda}}_t^{\mathrm{tr}}\right) < 1, \tag{4.30}$$

which in turn also implies $\xi_p^s > 0$. With the definition (4.29) of a_p^{co}, we obtain

$$0 \leq \vartheta_p - a_p^{\mathrm{co}} \bar{\boldsymbol{\lambda}}_{p,t} \cdot \widehat{\boldsymbol{\lambda}}_t^{\mathrm{tr}} \leq 2\vartheta_p < 2, \tag{4.31}$$

such that condition (4.30) is always fulfilled for $s \in [0, \frac{1}{2}]$. But if one wants to use a value of s larger than $\frac{1}{2}$ and guarantee the nonnegativity of the eigenvalues of L_p^s at the same time, the strategy conducted in [91] can be adapted to the present case. In the cited paper, the positive definiteness of L_p^s for $s = 1$ has been ensured by the introduction of an additional factor $\beta_p^{\mathrm{co}} \in (0,1]$. Transferring this approach to the current setting, we obtain

$$\widetilde{L}_p^s := \frac{\beta_p^{\mathrm{co}}}{1 - \beta_p^{\mathrm{co}}\vartheta_p}\left(\vartheta_p \mathrm{Id} - \frac{1}{1 - \beta_p^{\mathrm{co}}\left(\vartheta_p - \boldsymbol{\lambda}_{p,t}^s \cdot \widehat{\boldsymbol{\lambda}}_{p,t}^{\mathrm{tr}}\right)} \boldsymbol{\lambda}_{p,t}^s \otimes \widehat{\boldsymbol{\lambda}}_{p,t}^{\mathrm{tr}}\right) \tag{4.32}$$

with the eigenvalues

$$\frac{\beta_p^{\mathrm{co}}\vartheta_p}{1 - \beta_p^{\mathrm{co}}\vartheta_p} > 0 \qquad \text{and} \qquad \frac{\beta_p^{\mathrm{co}}\left(\vartheta_p - \boldsymbol{\lambda}_{p,t}^s \cdot \widehat{\boldsymbol{\lambda}}_{p,t}^{\mathrm{tr}}\right)}{1 - \beta_p^{\mathrm{co}}\left(\vartheta_p - \boldsymbol{\lambda}_{p,t}^s \cdot \widehat{\boldsymbol{\lambda}}_{p,t}^{\mathrm{tr}}\right)}.$$

Hence, (4.30) becomes

$$0 \leq \beta_p^{\mathrm{co}} s\left(\vartheta_p - a_p^{\mathrm{co}} \bar{\boldsymbol{\lambda}}_{p,t} \cdot \widehat{\boldsymbol{\lambda}}_t^{\mathrm{tr}}\right) < 1, \tag{4.33}$$

which for $s \in [0,1]$ can always be guaranteed if β_p^{co} is defined by

$$\beta_p^{\text{co}} := \begin{cases} \min\left(1, \vartheta_p s^{-1}\left(\vartheta_p - a_p^{\text{co}} \boldsymbol{\lambda}_{p,t} \cdot \hat{\boldsymbol{\lambda}}_t^{\text{tr}}\right)^{-1}\right), & \boldsymbol{\lambda}_{p,t} \cdot \boldsymbol{\lambda}_t^{\text{tr}} < 0, \\ 1, & \text{else.} \end{cases} \tag{4.34}$$

As we have $\boldsymbol{\lambda}_{p,t} \cdot \boldsymbol{\lambda}_{p,t}^{\text{tr}} > 0$ for $p \in \mathcal{A}_t$ at convergence, the consistency condition $\beta_p^{\text{co}} \to 1$ is satisfied. For $0 < s \leq \frac{1}{2}$, the above definition gives $\beta_p^{\text{co}} = 1$ always.

The theoretical importance of β_p^{co} can be seen in the next lemma dealing with the solvability of (4.24) in the case of Tresca friction:

Lemma 4.8. *For $s \in [0,1]$, replace the matrix L_p^s in (4.24) with the modified version $\widetilde{L}_p^s$ from (4.32). For $\mathfrak{F} = 0$, the thus modified system $(4.21)_{2-4}$, (4.24) has a unique solution $(\delta\mathbf{u}^h, \delta\boldsymbol{\lambda}^h)$.*

Proof. For $\mathfrak{F} = 0$, the system (4.24) combined with $(4.21)_{2-4}$ simplifies to the saddle point problem

$$\begin{aligned} \bar{K}\delta\mathbf{u}^h + \bar{B}\delta\boldsymbol{\lambda}^h &= \bar{\mathbf{f}}, \\ \bar{B}^T\delta\mathbf{u}^h &= \bar{\mathbf{g}}, \end{aligned}$$

with some right hand sides $\bar{\mathbf{f}}$, $\bar{\mathbf{g}}$ and the operators

$$\begin{aligned} \langle \bar{K}\delta\mathbf{u}^h, \mathbf{v}^h \rangle &= \langle K^h\delta\mathbf{u}^h, \mathbf{v}^h \rangle + c_t^{\text{co}} \sum_{p\in\mathcal{A}_t} \widetilde{L}_p^s \delta\mathbf{u}_{p,t} \cdot \mathbf{v}_{p,t}, \\ \langle \bar{B}\delta\boldsymbol{\lambda}^h, \mathbf{v}^h \rangle &= \sum_{p\in\mathcal{A}_n} h_p \delta\lambda_{p,n} v_{p,n} + \sum_{p\in\mathcal{I}_t} h_p \delta\boldsymbol{\lambda}_{p,t} \cdot \mathbf{v}_{p,t}. \end{aligned}$$

By construction, $\widetilde{L}_p^s$ is positive semidefinite, yielding the positive definiteness of $\bar{A}$ on $\mathbf{V}_0^h$. The proof now follows by standard arguments like in [21]. □

Remark 4.9. Under some technical assumptions on the regularity of certain submatrices of K^h, the results of Lemma 4.8 can be generalized to the case $\mathfrak{F} > 0$, using techniques from [12, 134]. For the numerical implementation, if (4.24) were not solvable for some intermediate Newton step, one could perform one step of a fixed point iteration where the friction bound is determined by the previous iterate of the contact stress. The new iterate would then be given by the unique solution of the corresponding Tresca friction problem. But for all our numerical examples in Sections 4.2.2 and 4.3.2, the tangential problems have always been solvable.

4.2.2 Numerical results for contact

4.2.2.1 Geometry and parameters

In order to test the numerical performance of the semismooth quasi-Newton scheme (4.24), we simulate the dynamic frictional contact of the two bodies Ω^{ma}, Ω^{sl} depicted in Figure 4.6 and given by

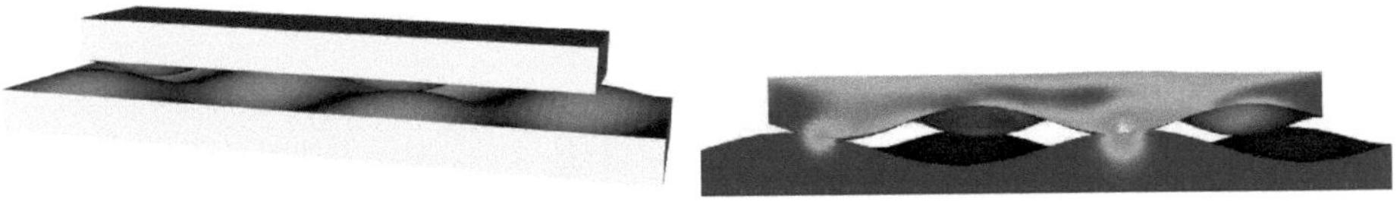

Figure 4.6: Left: geometry of contact example; right: cross-section showing the effective stress at time t_4.

$$\Omega^{\mathrm{ma}} = \left\{ \mathbf{x} \in [0,3] \times [0.5,1.5] \times \mathbb{R} : \quad 0.8 \le x_3 \le 1.0 + 0.08 \cdot \sin\left(\frac{4\pi x_1}{3}\right) \cdot \sin\left(2\pi x_2\right) \right\},$$
$$\Omega^{\mathrm{sl}} = \left\{ \mathbf{x} \in [0.3,2.7] \times [0.7,1.3] \times \mathbb{R} : 1.33 \ge x_3 \ge 1.13 + 0.08 \cdot \sin\left(\frac{4\pi x_1}{2.4}\right) \cdot \sin\left(\frac{2\pi x_2}{0.6}\right) \right\}.$$

The bodies are discretized with over 30000 hexahedral elements, and the dual basis functions $\{\psi_p\}_{p\in\mathcal{N}_{\mathrm{co}}^h}$ at the potential contact nodes are constructed as explained in [59]. The lower (master) body Ω^{ma} is fixed at the bottom ($x_3 = 0.8$) and free otherwise, whereas the upper (slave) domain Ω^{sl} is subject to a surface force of $\left(-5 \cdot 10^4 t,\ 0,\ -500 + 10^5 t\right)^T$ at the top ($x_3 = 1.33$). We compute four time steps with a step size of $\Delta t = 10^{-3}$ and the material parameters $\rho^{\mathrm{ma}} = \rho^{\mathrm{sl}} = 1$, $E^{\mathrm{ma}} = 7 \cdot 10^3$, $E^{\mathrm{sl}} = 7 \cdot 10^4$, $\nu^{\mathrm{ma}} = \nu^{\mathrm{sl}} = 0.3$ as well as $g_t = 0$ and $\mathfrak{F} = 0.5$. In order to avoid spurious oscillations of the contact stress with respect to time, we modify the mass matrix near the contact boundary of Ω^{sl} as described in more detail in Chapter 5.

If not stated differently, the contact constants in (2.17) are chosen as $c_n^{\mathrm{co}} = c_t^{\mathrm{co}} = 4\mu^{\mathrm{sl}}$, and the active sets are initialized with the final active sets from the previous time step, starting from the initialization $\mathcal{A}_n = \mathcal{A}_t = \mathcal{I}_t = \emptyset$ at time t_0. The stabilization factor a_p^{co} as defined in (4.29) is employed in all computations because otherwise we have obtained no convergence. In contrast, the influence of the factor β_p^{co} given in (4.34) is very remote for the dynamic problem considered here; in fact, there were hardly any differences between the results with and without this stabilization. This is most likely due to the fact that each Newton iteration is initialized with the converged solution from the previous time step satisfying $\boldsymbol{\lambda}_{p,t} \cdot \mathbf{u}_{p,t} > 0$ and thus $\boldsymbol{\lambda}_{p,t} \cdot \boldsymbol{\lambda}_{p,t}^{\mathrm{tr}} > 0$ for all slippy nodes $p \in \mathcal{A}_t$. Hence, we do not investigate the effect of including β_p^{co} in detail but refer to [91].

4.2.2.2 Influence of s on the convergence

We start with investigating the influence of the parameter s on the convergence of the quasi-Newton scheme (4.24). On the left side of Figure 4.7, the relative error decay of the method at time t_2 is shown for different values of s within the interval $[0,2]$. The right plot depicts the total number of iterations necessary to reduce the relative error below a tolerance of 10^{-10} for the first four time steps. From these pictures, one can conclude that the best convergence behaviour is obtained for $s \approx 0.75$, whereas the computations for $s \le 0.5$ do not converge. These results are in good agreement with those of Subsection 4.1.2.2 where a similar dependence of the number of iterations on the choice of s has been observed.

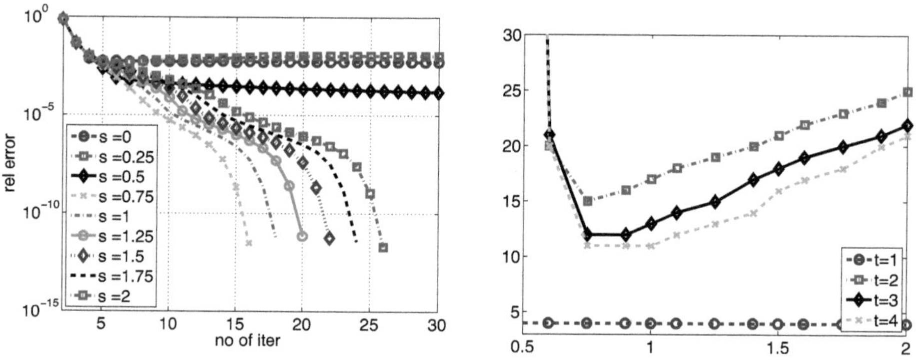

Figure 4.7: Left: error decay of quasi-Newton iteration at time t_2 for different values of s; right: Total number of iterations for t_1 to t_4 with respect to s.

4.2.2.3 Behavior of the iterates at critical nodes

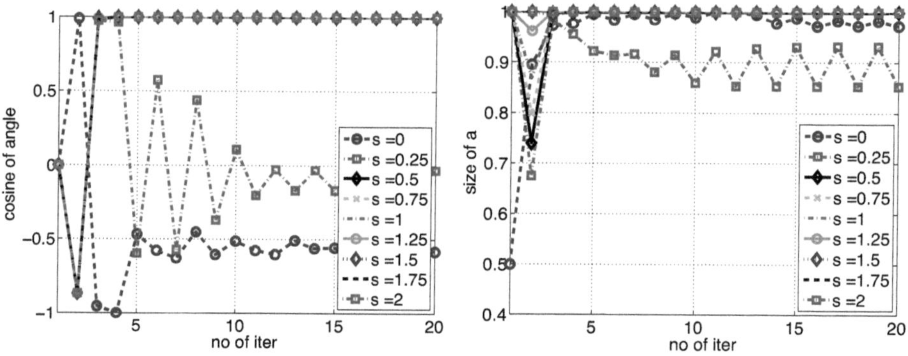

Figure 4.8: Local behaviour at time t_2 for different values of s; left: size of $\widehat{\boldsymbol{\lambda}}_t \cdot \widehat{\boldsymbol{\lambda}}_t^{\mathrm{tr}}$ at node $(2.469, 1.186, 1.060)$; right: size of a^{co} at node $(1.177, 1.186, 1.056)$.

Next, we examine the local behaviour of the iterates at several slippy nodes. The left picture of Figure 4.8 shows the size of the scalar product $\widehat{\boldsymbol{\lambda}}_t \cdot \widehat{\boldsymbol{\lambda}}_t^{\mathrm{tr}} \in [-1, 1]$ at the node $(2.469, 1.186, 1.060)$, corresponding to the cosine of the angle between these two normed vectors. For the exact solution, this value has to be equal to 1. One can see that the product is negative at the second Newton step for any $s \in [0.25, 2]$, but quickly converges to 1 afterwards if $s \geq 0.5$. For $s \in \{0, 0.25\}$, the angle starts oscillating, indicating the fact that these methods do not converge. On the right side of Figure 4.8, the size of the stabilization factor a^{co} defined in (4.29) at the node $(1.177, 1.186, 1.056)$ is displayed. One can observe that the importance of including this factor is restricted to the first few iterates, similar to the results for the plastic case in Subsection 4.1.2.3. For the converging schemes with $s > 0.25$, a^{co} is equal to 1 after at most 4 iterations.

4.2.2.4 Influence of c_n^{co}, c_t^{co} on the convergence

Finally, we fix $s = 1$ and investigate the impact of the size of the constants c_n^{co}, c_t^{co} in (2.17) on the speed of convergence. Table 4.1 states the total number of iterations for $c_n^{\text{co}} = \mu^{\text{sl}} \cdot 10^{i_n}$, $c_t^{\text{co}} = \mu^{\text{sl}} \cdot 10^{i_t}$, $-5 \leq i_n, i_t \leq 3$. These results illustrate that the method indeed converges for a wide range of parameters c_n^{co}, c_t^{co}, but their size has a definite influence on the speed of convergence. In general, it seems to be benefitial to choose rather small values of i_n, i_t satisfying $i_t > i_n$. For larger values of i_n, i_t, a suboptimal but stable convergence behaviour can be observed. We remark that by choosing $i_n = -5$, $i_t = -3$, the number of iterations can be reduced by more than 50% compared to the case $i_n = i_t = 3$.

$i_n \backslash i_t$	−5	−4	−3	−2	−1	0	1	2	3
−5	34	24	22	25	27	27	28	28	28
−4	28	27	23	30	34	35	37	37	37
−3	28	25	27	35	37	40	42	42	43
−2	30	27	30	37	42	44	44	45	45
−1	31	27	30	37	43	45	45	46	46
0	31	27	30	37	43	45	45	46	46
1	31	27	30	37	43	45	45	46	46
2	31	27	30	37	43	45	45	46	46
3	31	27	30	37	43	45	45	46	46

Table 4.1: Total number of iterations for first four time steps for $s = 1$, $c_n^{\text{co}} = \mu^{\text{sl}} \cdot 10^{i_n}$, $c_t^{\text{co}} = \mu^{\text{sl}} \cdot 10^{i_t}$.

4.3 Combination of the schemes

In this section, the semismooth Newton schemes of Sections 4.1 and 4.2 are combined to a general algorithm presented in Subsection 4.3.1. In Subsection 4.3.2, two numerical examples illustrate that this algorithm can efficiently solve dynamic frictional contact problems with three-dimensional elastoplastic bodies, nonmatching meshes at the contact interface, nonlinear isotropic hardening and large differences in the material parameters.

4.3.1 Combined algorithm

Choose a time stepping parameter $\theta \in \left[\frac{1}{2}, 1\right]$. Define the initial solution vector

$$\mathbf{z}_j := (\mathbf{u}_j^h, (\alpha^h, \boldsymbol{\varepsilon}^{\text{pl},h})_j, \boldsymbol{\lambda}_{j-1+\theta}^h)$$

for $j = 0$. Choose appropriate values for s, c^{pl}, c_n^{co}, c_t^{co} and a relative tolerance tol.

Loop on time steps: For $j = 0, \ldots$

1. Initialize $\mathbf{z}_{j+1}^{(0)} = \mathbf{z}_j$. Define the initial active sets $\mathcal{A}_\alpha^{(0)}$, $\mathcal{A}_{\text{pl}}^{(0)}$, $\mathcal{A}_n^{(0)}$ and $\mathcal{A}_t^{(0)}$ according to this vector.

2. Loop on Newton steps: For $k = 1, \ldots$

   a) Assemble the dynamic and the elastoplastic part of the stiffness matrix $\bar{\mathbf{K}}^{(k)}$ and the right hand side $\bar{\mathbf{F}}^{(k)}$ according to (4.9). Add the contact contributions given in (4.24).

   b) Incorporate the nodal boundary conditions $(4.21)_{2-4}$ into the tangential system, leading to $\mathbf{K}^{(k)}$ and $\mathbf{F}^{(k)}$.

   c) Solve the resulting modified system $\mathbf{K}^{(k)}\delta\mathbf{u}^{(k)} = \mathbf{F}^{(k)}$ for $\delta\mathbf{u}^{(k)}$.

   d) Compute the plastic updates $(\delta\alpha^h, \delta\boldsymbol{\varepsilon}^{\text{pl},h})_{j+1}^{(k)}$ from (4.5) and the contact updates $\delta\boldsymbol{\lambda}_{j+\theta}^{h,(k)}$ from $(4.21)_{1,5,6}$ or from the residual of (4.24).

   e) Update the vector $\mathbf{z}_{j+1}^{(k+1)} = \mathbf{z}_{j+1}^{(k)} + \delta\mathbf{z}_{j+1}^{(k)}$.

   f) Update the active sets $\mathcal{A}_{\text{pl}}^{(k+1)}$, $\mathcal{A}_\alpha^{(k+1)}$ according to (4.2) and $\mathcal{A}_n^{(k+1)}$, $\mathcal{A}_t^{(k+1)}$ according to (4.19).

   g) Compute the relative error $e^{(k)} := \frac{\|\delta\mathbf{u}_{j+1}^{(k)}\|}{\|\mathbf{u}_{j+1}^{(k)}\|}$.

   h) If $\mathcal{A}^{(k+1)}_{\cdot} = \mathcal{A}^{(k)}_{\cdot}$ for all active sets and $e^{(k)} < tol$, break.

4.3.2 Numerical results for plastic contact problem

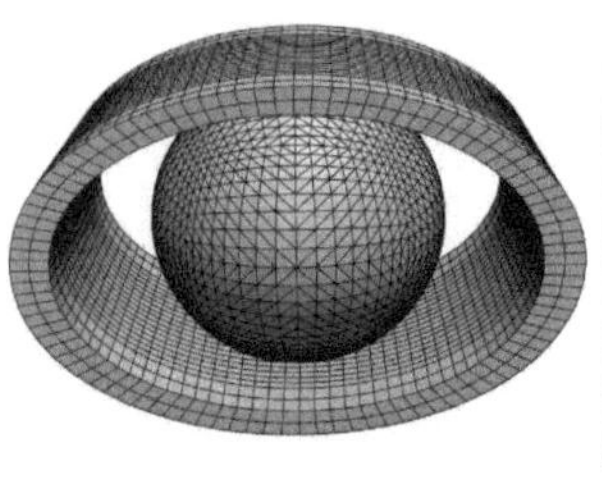

	ball	tube
E	$7 \cdot 10^3$	$7 \cdot 10^5$
ν	0.3	0.3
ϱ	10^{-2}	10^{-2}
H	$5 \cdot 10^3$	$5 \cdot 10^3$
K	$4 \cdot 10^3$	$4 \cdot 10^3$
σ_0	$7 \cdot 10^1$	$7 \cdot 10^4$

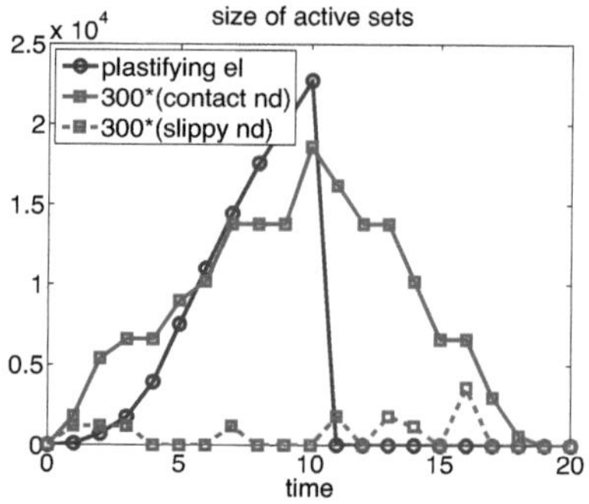

Figure 4.9: Left: geometry of the ball in tube example; middle: material parameters; right: evolution of size of active sets.

As a first example, we consider a ball of radius 0.3 located in a elliptic cylinder with the inner radii 0.3 and 0.45, a thickness of 0.05 and a height of 0.6. The outer boundary of the tube is squeezed from above and below by a compression force given by $\mathbf{g}_N = 10^5 \cdot \min(t, 0.02 - t)\mathbf{n}$, whereas all other boundaries are free. The ball acts as the slave domain and is discretized with around 200000 simplicial elements, whereas the tube consists of 30000 hexahedrals. The volume force as well as the initial velocity are set to zero. We compute 20 time steps of size $\Delta t = 10^{-3}$ with $\theta = 1$ and consider linear isotropic and kinematic hardening as well as Coulomb friction with $\mathfrak{F} = 0.8$. The active sets for the semismooth Newton iteration are initialized with the final active sets of the previous time step and are updated in each Newton iteration. The initial configuration and the material parameters are given in Figure 4.9, as well as the evolution of the size of the active sets $\mathcal{A}_{\mathrm{pl}}$, $\mathcal{A}_{\mathrm{n}}$ and $\mathcal{A}_{\mathrm{t}}$. In Figure 4.10, the deformed geometry is shown at the time steps t_2, t_5, t_8 and t_{10}; the first two rows illustrate the plastification process with a cut through the symmetry axis, whereas the third row shows the active contact nodes on the top of the ball. One can observe that the symmetry of the problem is reproduced by the numerical outcome.

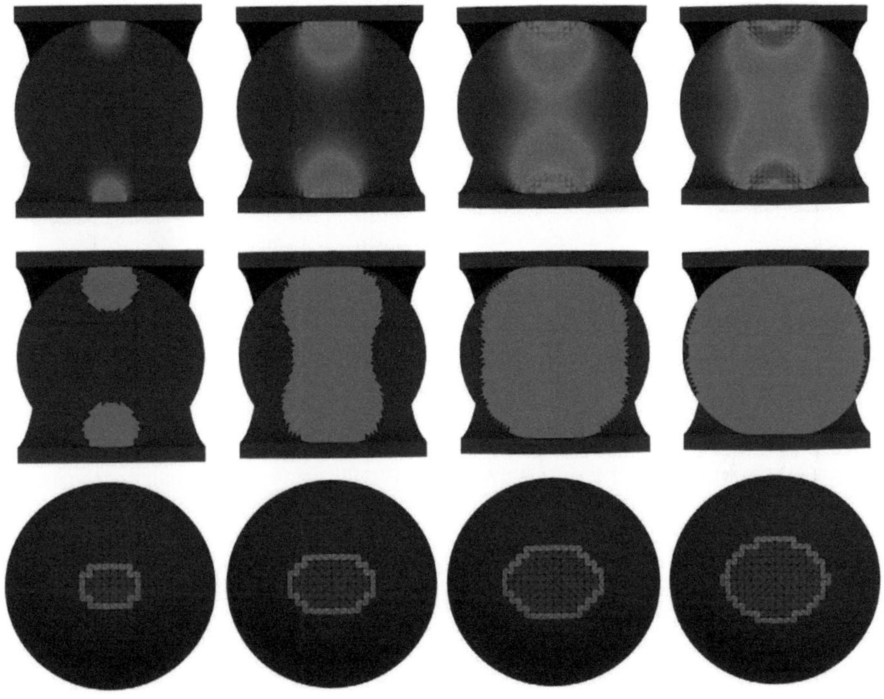

Figure 4.10: Situation at time steps t_2, t_5, t_8, t_{10}; first row: equivalent plastic strain; middle row: plastic active set; lower row: contact active set.

As a second example, we simulate a circular axisymmetric tube of length 4 (oriented in x_1-direction) with inner radius 0.4 and a thickness of 0.1 which is squeezed by two curved tools Ω_1, Ω_2 with

$$\Omega_1 = \left\{ \mathbf{x} \in [-0.8, 0.8] \times [-1.0, -0.2] \times \mathbb{R} : -1.0 \le x_3 \le -0.7 + 1.25 \cdot (x_2 + 1.0)(x_2 + 0.2) \right\},$$
$$\Omega_2 = \left\{ \mathbf{x} \in [-0.8, 0.8] \times [-0.3, 0.5] \times \mathbb{R} : \quad 1.0 \ge x_3 \ge 0.7 - 1.25 \cdot (x_2 + 0.3)(x_2 - 0.5) \right\}$$

from below and above. The tube is fixed at $x_1 = 2$ and free otherwise, whereas the plates are subject to a surface force of $\big(0, 0, -f_z(t)\big)$ on the top of the upper plate and $\big(0, 0, 2 \cdot f_z(t)\big)$ on the bottom of the lower plate, with $f_z(t) = 800 \cdot \min(t, 2.5 \cdot 10^{-3} - t)$. The dynamic computation with around 44000 hexahedral elements nodes is performed in 21 time steps of step size $\Delta t = 2.5 \cdot 10^{-4}$ and $\theta = 1$. The friction constants are $g_t = 0$, $\mathfrak{F} = 0.1$, and the computational parameters are set to $s = 1$, $c^{\mathrm{pl}} = 2\mu + a_0^{-2} K$ and $c_n^{\mathrm{co}} = c_t^{\mathrm{co}} = 4\,\mu_{\mathrm{tube}}$.

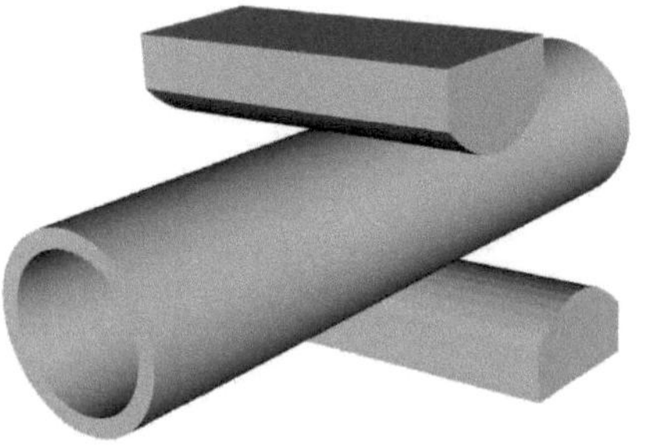

	tube	plates
E	$7 \cdot 10^3$	$7 \cdot 10^6$
ν	0.3	0.3
ϱ	$1 \cdot 10^{-3}$	$1 \cdot 10^{-1}$
H	$2 \cdot 10^3$	$2 \cdot 10^3$
K	$1 \cdot 10^3$	$1 \cdot 10^3$
σ_0	$2 \cdot 10^2$	$5 \cdot 10^4$
σ_∞	$4 \cdot 10^2$	$1 \cdot 10^5$
k_e	17	17

Figure 4.11: Left: geometry of the tube example; middle: material parameters; right: equivalent plastic strain α of upper half of the deformed tube at time t_9.

The geometry and the material parameters are shown in Figure 4.11, as well as computed equivalent plastic strain α after 9 time steps. In order to illustrate the generality of the method, we extend it to the nonlinear hardening function given in (1.21) which affects the definitions (2.14a), (2.18a), (4.1) and (4.3) where Y^{pl} has to be replaced with $Y^{\mathrm{pl}}_{\mathrm{exp}}$. Furthermore, the expressions in (4.5) and (4.6) change but the general procedure sketched in the proof of Lemma 4.1 can be executed for the nonlinear hardening function (1.21) as well.

The locally superlinear convergence of the quasi-Newton scheme can be seen in the left picture of Figure 4.12, where the error decay at time steps t_3, t_6 and t_9 is sketched. The symbols mark the Newton iteration where the correct plastic and contact active sets are found and not changed afterwards.

The evolution of these active sets is illustrated on the right of Figure 4.12 and in Figure 4.13; the former picture displays the total number of active contact nodes/ plastifying elements with respect to time, whereas the latter depicts the position of these nodes/elements for different time steps. The first row of pictures in Figure 4.13 shows the projection of the midpoint of each volume element onto a plane with respect to the x_1-value and the radial angle, the bottom side having the angle 0. Plastifying elements are marked in red, whereas elements without new plastification give a dark blue mark. The second row displays the nodes on the potential

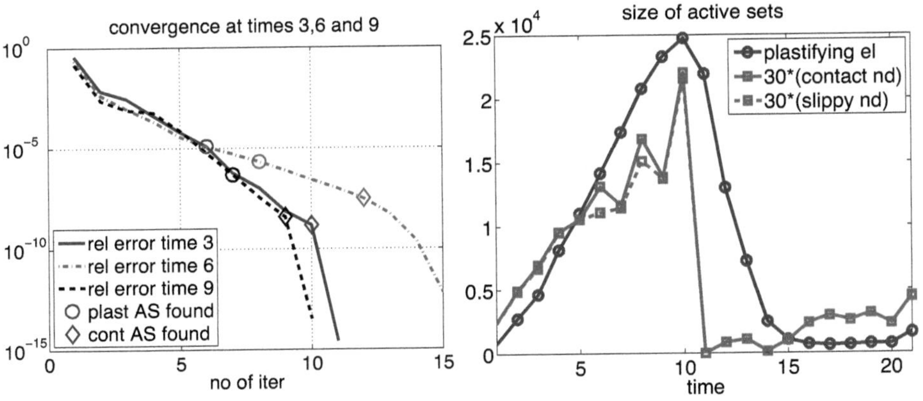

Figure 4.12: Left: error decay at time steps t_3, t_6 and t_9; right: evolution of the plastic and contact active sets during time.

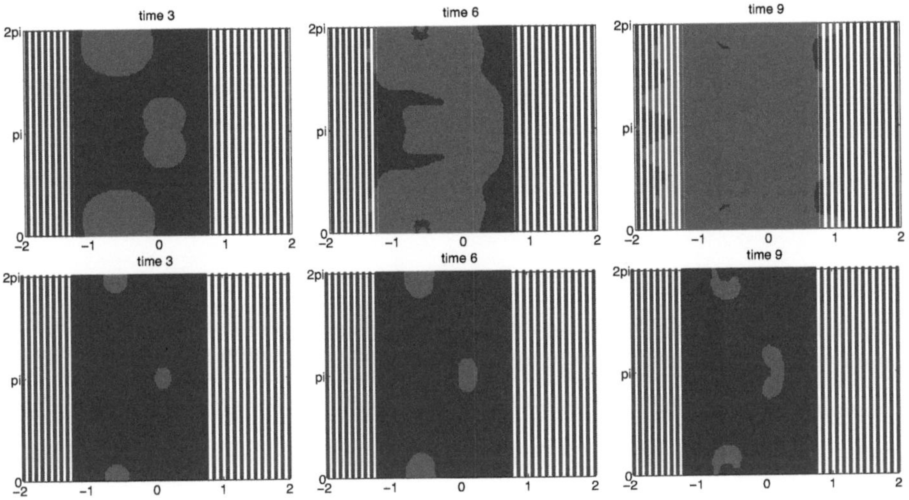

Figure 4.13: Position of the active sets at time steps t_3, t_6, t_9; first row: plasticity; second row: contact.

contact surface, also with respect to the x_1-value and the angle. Inactive nodes are marked in blue and active (slippy) nodes in red. The axial symmetry of the problem formulation can easily be recognized in the numerical results.

Part III

DAE solvers for dynamic normal contact

5 Mass modification techniques

In the previous part, we have described an efficient way of solving dynamic frictional contact problems by means of dual Lagrange multipliers and semismooth Newton methods. However, the values of the multipliers which represent the contact stress often show oscillative behaviour due to the time discretization (see, e.g., [19, 81]). An example is shown in Figure 5.1, where the numerical results for the normal component of the Lagrange multiplier are displayed, computed with a standard lowest order finite element discretization scheme in space and the trapezoidal rule in time.

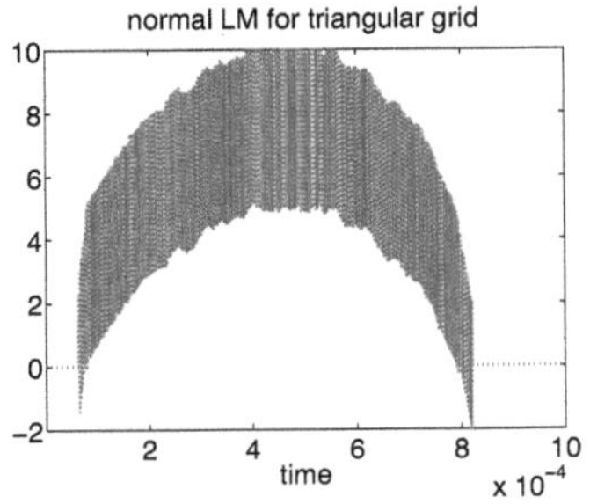

Figure 5.1: Normal Lagrange multiplier for a two-body contact problem given in Section 5.4.2 with standard mass matrix M_h.

In the last years, different approaches have been discussed how to overcome this numerical artifact [50, 102, 103, 106, 110, 148]. In [102, 103], a modified mass matrix is constructed by redistributing the mass from the slave contact nodes to the inner nodes. By this, the original PDE decouples into an algebraic equation in time for the contact nodes and a differential equation in time for the other nodes. In section 5.1, we shortly analyse this technique from a DAE point of view. The algorithm in [50, 106, 110] uses a different approach which is based on a stabilized predictor-corrector scheme where the predictor is additionally projected onto the admissible set. But both techniques give rise to extra computational cost: The projection step in the latter work is equivalent to solving a second contact problem per time step – which is much cheaper than the standard contact problem because of the better condition of the matrix involved, but nevertheless an extra expense – whereas the new mass matrix of the former approach is defined as the solution of a global constrained minimization problem. Furthermore, the mass is only redistributed in the normal direction which results in several recalculations of the mass matrix if the contact normal vector depends on the time – which is the case for most multi-body contact problems.

In this part of the work, we present a different way of defining the modified mass matrix using special quadrature formulas. By this, we get a mass matrix that has zero entries associated with the contact nodes, both in the normal and the tangential direction. We also avoid the expense of solving an additional global minimization problem in order to compute the modified mass matrix, and thus no extra computational cost comes into play. Furthermore, our approach is directly applicable to two-body contact problems and can easily be combined with the methods from [34, 117, 118] to obtain an energy-consistent scheme for frictional contact.

In addition, we analyse the error of the new discretization scheme and prove optimal a priori estimates under suitable regularity assumptions by applying techniques from the theory of variational crimes. We emphasize that in contrast to standard mass lumping techniques, the new method results in a singular mass matrix, such that norm equivalences based on the regularity of the mass matrix do not apply any more. The results have already been published in [70, 73].

As we focus on dynamic contact problems in the next two chapters, all plastic contributions are set to zero and the semi-discrete problem to be solved reads (cf. (2.12), (2.18b))

$$M^h \ddot{\mathbf{u}}^h + A^h \mathbf{u}^h + B^h_{\mathrm{co}} \boldsymbol{\lambda}^h - \mathbf{f}^h = \mathbf{0}, \tag{5.1a}$$

$$\mathbf{C}^{\mathrm{co}}(\mathbf{u}^h, \boldsymbol{\lambda}^h) = \mathbf{0}. \tag{5.1b}$$

Further, we assume for simplicity that the density ϱ is constant on Ω.

The rest of this chapter is organized as follows: First, we present a short analysis of the mass modification technique by means of a simple mass oscillator in Section 5.1. In the following section, we describe the construction of the modified mass matrix via non-standard quadrature formulas and state two general conditions that these formulas have to satisfy. Section 5.3 contains a different interpretation of the new quadrature formulas as a combination of stable interpolation operators and standard quadrature.

In Section 5.4, we give various numerical examples including contact, friction and nonlinear material laws. We demonstrate that the modified mass matrix gives almost the same results as the standard one if no contact is present. For two-body contact problems, we show that the oscillations of the contact stress vanish due to the mass redistribution.

The a priori error of the modified mass matrix approach is analysed in Chapter 6.

5.1 Why mass modification?

5.1.1 Index reduction

As discussed in, e.g., [8], the time-continuous spatially discrete normal contact problem is ill-posed if no further assumption on the behaviour of the system in case of impact is made. In [103], this fact is illustrated by an example of an oscillating system with a single degree of freedom. In addition, as soon as contact occurs and the non-penetration inequality $u_n \leq g_n$ in (1.22) becomes an equality, the second-order ODE (5.1a) has to be solved subject to the corresponding algebraic constraints in (5.1b). The resulting problem is a differential-algebraic equation of index 3 (see [76] for the definition of the index of a DAE). Such systems are challenging to treat numerically, and the multiplier is very sensitive to perturbations in the constraints [154]. Hence, not all algorithms that are suited for solving ODEs give good results for the DAE case. In [116], it is shown that well-known time stepping algorithms like the trapezoidal rule, which is energy-conserving for the linear ODE, can lead to instabilities when applied to the case of a DAE of index 3 (cf. the discussion on page 25).

Hence, we try to improve the regularity and the numerical stability of the solution by reducing the index of the DAE system. For this aim, we follow the idea given in [102, 103] and replace the mass matrix M^h in (5.1a) by a modified mass matrix $\bar{M}^h$ which has no entries associated with the contact nodes $\mathcal{N}^h_{\text{co}}$. By this, the DAE system decouples into purely algebraic equations for the contact nodes and a system of ODEs for the other nodes. To clarify this procedure, we partition the displacement vector into the inner and the contact degrees of freedom by $\mathbf{u}^h = (\mathbf{u}^h_I,\ \mathbf{u}^h_C)^T$. Then, (5.1) reads

$$\begin{pmatrix} M^h_{II} & M^h_{IC} \\ M^h_{CI} & M^h_{CC} \end{pmatrix} \begin{pmatrix} \ddot{\mathbf{u}}^h_I \\ \ddot{\mathbf{u}}^h_C \end{pmatrix} + \begin{pmatrix} A^h_{II} & A^h_{IC} \\ A^h_{CI} & A^h_{CC} \end{pmatrix} \begin{pmatrix} \mathbf{u}^h_I \\ \mathbf{u}^h_C \end{pmatrix} + \begin{pmatrix} \mathbf{0} \\ B^h_C \end{pmatrix} \boldsymbol{\lambda}^h = \begin{pmatrix} \mathbf{f}^h_I \\ \mathbf{f}^h_C \end{pmatrix}, \tag{5.2a}$$

$$\mathbf{C}^{\text{co}}(\mathbf{u}^h_C, \boldsymbol{\lambda}^h) = \mathbf{0}. \tag{5.2b}$$

Replacing the standard mass matrix with the modified one leads to

$$\begin{pmatrix} \bar{M}^h_{II} & \mathbf{0} \\ \mathbf{0} & \mathbf{0} \end{pmatrix} \begin{pmatrix} \ddot{\mathbf{u}}^h_I \\ \ddot{\mathbf{u}}^h_C \end{pmatrix} + \begin{pmatrix} A^h_{II} & A^h_{IC} \\ A^h_{CI} & A^h_{CC} \end{pmatrix} \begin{pmatrix} \mathbf{u}^h_I \\ \mathbf{u}^h_C \end{pmatrix} + \begin{pmatrix} \mathbf{0} \\ B^h_C \end{pmatrix} \boldsymbol{\lambda}^h = \begin{pmatrix} \mathbf{f}^h_I \\ \mathbf{f}^h_C \end{pmatrix}, \tag{5.3a}$$

$$\mathbf{C}^{\text{co}}(\mathbf{u}^h_C, \boldsymbol{\lambda}^h) = \mathbf{0}, \tag{5.3b}$$

thus eliminating the contact accelerations $\ddot{\mathbf{u}}^h_C$ from the system. Furthermore, the contact displacements $\mathbf{u}^h_C$ can be calculated from the inner ones by solving the algebraic system

$$\begin{aligned} A^h_{CC}\mathbf{u}^h_C + B^h_{\text{co}}\boldsymbol{\lambda}^h &= \mathbf{f}_C - A^h_{CI}\mathbf{u}^h_I, \\ \mathbf{C}^{\text{co}}(\mathbf{u}^h_C, \boldsymbol{\lambda}^h) &= \mathbf{0}. \end{aligned}$$

Hence, the resulting system (5.3) is a DAE of index 1, yielding a higher regularity for the contact stress $\boldsymbol{\lambda}^h$ than the original system (5.2), as shown in [102, 103]. Thereby, no spurious oscillations in the numerical computation of $\boldsymbol{\lambda}^h$ occur.

5.1.2 Two-mass oscillating system

To analyse the effect of the mass modification in more detail, we consider the vertical displacement of a spring-mass system consisting of two masses with displacements u^I, u^C and velocities v^I, v^C, each connected with springs with the force constant k. The system is impacting on a fixed obstacle with initial gap g, as depicted on page 70. Using the implicit Newmark scheme (2.13) with parameter θ and time step size Δt, we obtain the following discrete system to be solved at time t_{j+1}: find $\mathbf{z}_{j+1} := (u^I, v^I, u^C, v^C)_{j+1}$, $\lambda_{j+\theta}$ such that

$$B\mathbf{z}_{j+1} + D\lambda_{j+\theta} = G\mathbf{z}_j, \tag{5.4a}$$

$$C^{\text{co}}(u^C_{j+1}, \lambda_{j+\theta}) = 0, \tag{5.4b}$$

with the matrices

$$B = \begin{pmatrix} 2\theta k & \frac{m}{\Delta t} & -\theta k & 0 \\ \frac{1}{\Delta t} & -\frac{1}{2} & 0 & 0 \\ -\theta k & 0 & \theta k & \frac{m}{\Delta t} \\ 0 & 0 & \frac{1}{\Delta t} & -\frac{1}{2} \end{pmatrix}, \quad D = \begin{pmatrix} 0 \\ 0 \\ 1 \\ 0 \end{pmatrix}, \quad G = \begin{pmatrix} 2(\theta-1)k & \frac{m}{\Delta t} & (1-\theta)k & 0 \\ \frac{1}{\Delta t} & \frac{1}{2} & 0 & 0 \\ (1-\theta)k & 0 & (\theta-1)k & \frac{m}{\Delta t} \\ 0 & 0 & \frac{1}{\Delta t} & \frac{1}{2} \end{pmatrix}.$$

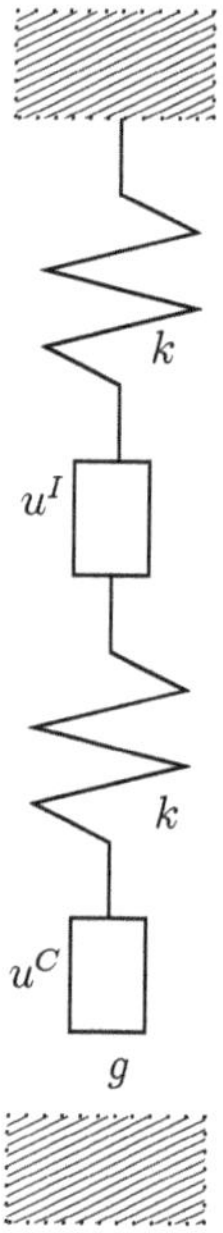

As long as the lower mass is not in contact, the amplification matrix $A = B^{-1}G$ has two complex conjugate pairs of eigenvalues, spanning the eigenmodes depicted in the upper row of Figure 5.2. The blue upper line displays the displacement u^I, and the lower red curve shows the contact displacement u^C. The damping effect of choosing $\theta > \frac{1}{2}$ is clearly visible in the upper right picture.

However, as soon as contact occurs, the augmented amplification matrix constructed by

$$A_C = B_C^{-1} G_C, \quad B_C = \begin{pmatrix} B & D \\ D^T & 0 \end{pmatrix}, \quad G_C = \begin{pmatrix} G & \mathbf{0} \\ \mathbf{0}^T & 0 \end{pmatrix} \tag{5.5}$$

has different eigenmodes depicted in the lower row of Figure 5.2. The additional dashed red line shows the value of the multiplier λ multiplied by a factor of 10^{-4}. One can see that there is a spurious oscillatory mode in the multiplier (and in the velocity v^C as well) that cannot be damped by increasing θ. We remark that the amplitude of these oscillations does not decrease for $\Delta t \to 0$; in contrast, the oscillations become worse.

In order to ameliorate the situation, we modify the mass of the system such that the lower object has zero mass and the upper one a mass of $2m$, leading to the matrices

$$\bar{B} = \begin{pmatrix} 2\theta k & \frac{2m}{\Delta t} & -\theta k & 0 \\ \frac{1}{\Delta t} & -\frac{1}{2} & 0 & 0 \\ -\theta k & 0 & \theta k & 0 \\ 0 & 0 & \frac{1}{\Delta t} & -\frac{1}{2} \end{pmatrix}, \quad \bar{G} = \begin{pmatrix} 2(\theta-1)k & \frac{2m}{\Delta t} & (1-\theta)k & 0 \\ \frac{1}{\Delta t} & \frac{1}{2} & 0 & 0 \\ (1-\theta)k & 0 & (\theta-1)k & 0 \\ 0 & 0 & \frac{1}{\Delta t} & \frac{1}{2} \end{pmatrix}.$$

In Figure 5.3, the same eigenmodes as in Figure 5.2 are shown, this time for the modified amplification matrices $\bar{A}$ and $\bar{A}_C$. The results in the lower row clearly show that the oscillatory part in the multiplier is gone. We remark that the contact velocity v^C is still oscillating but does not affect the other values of the solution of (5.4). However, the middle picture in the first row shows that in the unconstrained case, there exists a generalized eigenvector of $\bar{A}$ having a spurious oscillation in u^C. A more detailed analysis shows that for $\theta \neq \frac{1}{2}$, $\bar{A}$ is diagonalizable with the eigenvalues $\nu_1 = -1$, $\nu_2 = \frac{\theta-1}{\theta}$ and

$$\nu_{3,4} = \frac{8m - 2\Delta t k \pm \Delta t \left((2\theta-1)^2 \Delta t^2 k^2 - 32km\right)^{1/2}}{2\theta \Delta t^2 k + 8m}. \tag{5.6}$$

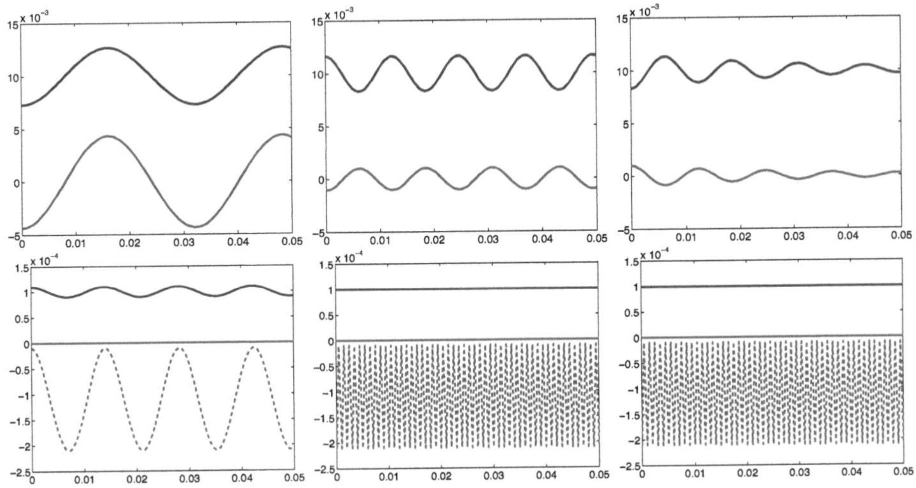

Figure 5.2: Eigenmodes of the unchanged two-mass oscillator for $m = 1$, $k = 10^5$, $\Delta t = 5 \cdot 10^{-4}$; upper row: unconstrained case; lower row: constrained case; left and middle: $\theta = \frac{1}{2}$; right: $\theta = 1$.

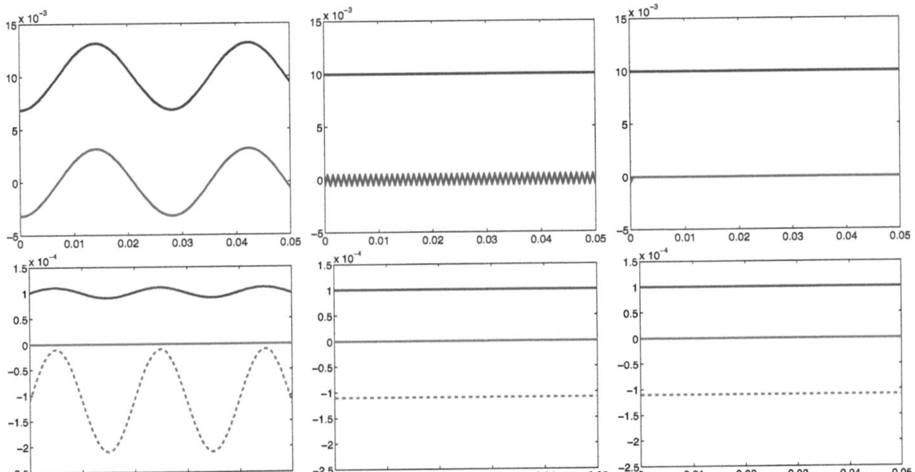

Figure 5.3: Eigenmodes of the modified two-mass oscillator for $m = 1$, $k = 10^5$, $\Delta t = 5 \cdot 10^{-4}$; upper row: unconstrained case; lower row: constrained case; left and middle: $\theta = \frac{1}{2}$; right: $\theta = 1$.

The eigenvectors corresponding to (5.6) span the mode shown on the left of Figure 5.3, whereas those corresponding to ν_1, ν_2 are given by

$$y_1 = \begin{pmatrix} 0 \\ 0 \\ 0 \\ 1 \end{pmatrix}, \quad y_2 = \begin{pmatrix} 0 \\ 0 \\ \left(\frac{1}{2} - \theta\right) \Delta t \\ 1 \end{pmatrix}.$$

The latter vector represents a spurious movement of the displacement u^C. However, for $\theta \in \left(\frac{1}{2}, 1\right]$, we have $|\nu_2| < 1$ and this component is damped or even annihilated for $\theta = 1$, as can be seen from the right upper picture of Figure 5.3.

For the case $\theta = \frac{1}{2}$, the matrix $\bar{A}$ has the double eigenvalue -1 with an one-dimensional eigenspace. A basis of generalized eigenvectors is given by

$$\begin{pmatrix} 0 \\ 0 \\ 0 \\ 1 \end{pmatrix}, \quad \begin{pmatrix} 0 \\ 0 \\ -\frac{1}{4}\Delta t \\ 1 \end{pmatrix}, \quad \begin{pmatrix} -\sqrt{\frac{2m}{k}} i \\ 1 \\ -\sqrt{\frac{2m}{k}} i \\ 1 \end{pmatrix}, \quad \begin{pmatrix} \sqrt{\frac{2m}{k}} i \\ 1 \\ \sqrt{\frac{2m}{k}} i \\ 1 \end{pmatrix}. \tag{5.7}$$

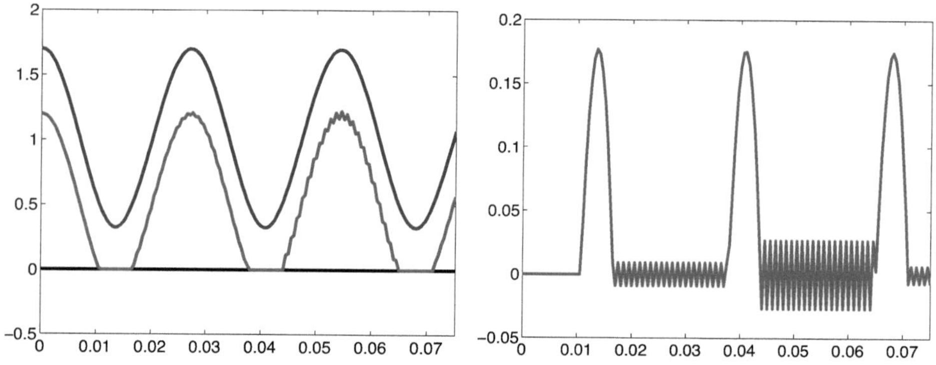

Figure 5.4: Solution of the modified two-mass oscillator for $m = 1$, $k = 10^5$, $\Delta t = 5 \cdot 10^{-4}$; left: evolution of u^I and u^C; right: evolution of difference $(u^I - u^C)$.

The second vector of (5.7) leads to the oscillation in the displacement u^C as illustrated in the middle upper picture of Figure 5.3. This effect can also be seen in numerical computations after contact is released; an example is shown in Figure 5.4 which illustrates the solution of (5.4) with modified mass for $m = 1$, $k = 10^5$, $\Delta t = 5 \cdot 10^{-4}$, $g = 0.5$, $\theta = \frac{1}{2}$ and the initial values $u_0^I = u_0^C = 0.7$, $v_0^I = v_0^C = 0$. The left picture shows the evolution of the displacements, whereas the right picture displays the difference $(u_j^I - u_j^C)$ with respect to time. One can observe that after contact is released, there are spurious oscillations in the displacement u^C. As the second vector in (5.7) is the only one with different entries for u^I and u^C, the amplitude of this oscillation is determined by the value of the difference $(u_j^I - u_j^C)$ at that time t_j where

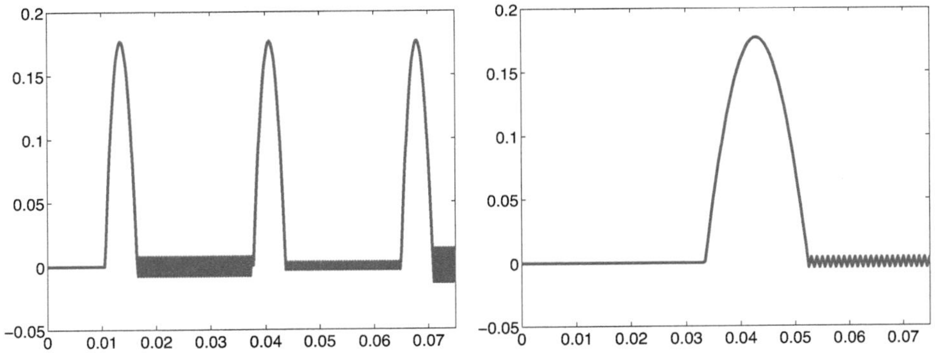

Figure 5.5: Evolution of difference $(u^I - u^C)$ for the modified two-mass oscillator for $m = 1$; left: $k = 10^5$, $\Delta t = 2.5 \cdot 10^{-4}$; right: $k = 10^4$, $\Delta t = 5 \cdot 10^{-4}$.

contact is released, i.e., where this difference becomes negative. As the maximal velocity v^I is proportional to $\sqrt{k/m}$, the size of the difference is bounded by $C\Delta t\sqrt{k/m}$ and thus decreasing for $\Delta t \to 0$. This is illustrated in Figure 5.5 showing the difference $(u_j^I - u_j^C)$ for a smaller time step on the left and for a smaller value of k on the right side. Hence, for simulating stiff bodies with $\theta = \frac{1}{2}$ and a large ratio k/m, the step size Δt needs to be small enough such that this oscillation does not affect the results. However, for two-body contact problems, the effect can be minimized by applying the mass modification to the body with smaller stiffness.

After this example illustrating the effect of the mass modification, we continue in the next section with describing an efficient way of constructing a suitable modified mass matrix.

5.2 Construction of $\bar{M}^h$

As stated in Section 5.1.1, the most important requirement for the modified mass matrix $\bar{M}^h$ is that its entries associated with the nodes in $\mathcal{N}_{\text{co}}^h$ vanish. As this condition does not determine $\bar{M}^h$ uniquely, other sensible demands are that the total mass and/or the first and second order moments of the original system (5.1a) are conserved. These requirements can be written as follows:

M0) total mass $\int_\Omega \varrho \, d\mathbf{x}$: $\mathbf{1}^T \bar{M}^h \mathbf{1} = \mathbf{1}^T M^h \mathbf{1}$,

M1) center of gravity $\int_\Omega \varrho x_i \, d\mathbf{x}$: $\mathbf{1}^T \bar{M}^h \mathbf{x}_i = \mathbf{1}^T M^h \mathbf{x}_i, \quad 1 \le i \le d$,

M2) moments of inertia $\int_\Omega \varrho x_i x_j \, d\mathbf{x}$: $\mathbf{x}_i^T \bar{M}^h \mathbf{x}_j = \mathbf{x}_i^T M^h \mathbf{x}_j, \quad 1 \le i,j \le d$.

Here we use the notation $\mathbf{1} = (1, \dots, 1)^T \in \mathbb{R}^{|\mathcal{N}^h|d}$ and $\mathbf{x}_i = ((\mathbf{x}_p \cdot \mathbf{e}_i)\mathbf{e}_i)_{p \in \mathcal{N}^h} \in \mathbb{R}^{|\mathcal{N}^h|d}$, with $\mathbf{x}_p$ as the vector of the co-ordinates of the vertex p and the i-th unit vector $\mathbf{e}_i \in \mathbb{R}^d$.

However, $\bar{M}^h$ is still not determined uniquely by the above requirements. In [102, 103], $\bar{M}^h$ is chosen as the solution of a global minimization problem, i.e., $\bar{M}^h$ satisfies $\inf_{M'} \|M' - M^h\|$, where M' has to fulfill the constraints M0 to M2. As the solution of this optimization problem is very expensive, we propose a different approach and compute $\bar{M}^h$ as a local modification of M^h using special quadrature formulas. In order to formulate them, we need some preliminary definitions.

As illustrated in Figure 5.6, we denote the union of all elements along Γ_C and its complement by

$$\bar{\Omega}_{C1} := \bigcup_{p \in \mathcal{N}^h_{\text{co}}} (\operatorname{supp} \phi_p), \qquad \Omega_I := \Omega \backslash \bar{\Omega}_{C1}.$$

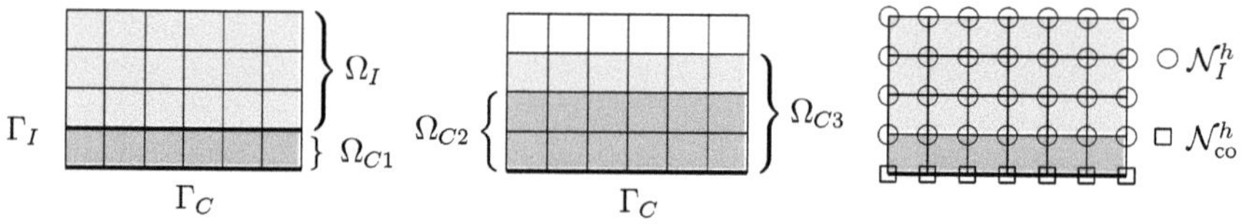

Figure 5.6: Illustration of Ω_{C1}, Ω_{C2}, Ω_{C3} and Ω_I.

The interface $\partial\Omega_{C1} \cap \partial\Omega_I$ is called Γ_I. Further, we introduce layer-like domains

$$\bar{\Omega}_{C2} := \bigcup_{\substack{p \subset \bar{\Omega}_{C1},\\ p \in \mathcal{N}^h}} (\operatorname{supp} \phi_p), \quad \bar{\Omega}_{C3} := \bigcup_{\substack{p \subset \bar{\Omega}_{C2},\\ p \in \mathcal{N}^h}} (\operatorname{supp} \phi_p).$$

Instead of (2.5a), we now define a modified bilinear form for discrete functions $\boldsymbol{\chi}^h, \boldsymbol{\eta}^h \in \mathbf{V}^h$,

$$\bar{m}^h_i(\boldsymbol{\chi}^h, \boldsymbol{\eta}^h) := Q^i(\varrho \boldsymbol{\chi}^h \cdot \boldsymbol{\eta}^h) \tag{5.8}$$

with a quadrature rule Q^i that is composed of local quadrature formulas Q^i_K, $K \in \mathcal{T}^h$. Intuitively, one can see that the mass matrix has zero entries associated with the nodes in $\mathcal{N}^h_{\text{co}}$ if no quadrature point is located in Ω_{C1}. Further, the number of conserved moments M0 to M2 depends on the degree of exactness of the quadrature rule.

In the following two subsections, we present two representative examples for suitable quadrature rules Q^i, $i \in \{0, 1\}$. The first one can be applied to unstructured simplicial meshes, whereas the second one needs some assumptions on the structure of the mesh near Γ_C.

5.2.1 Quadrature rule Q^0

On each element $K \in \mathcal{T}^h$, $K \subset \Omega_I$, we take any standard quadrature formula of sufficient accuracy. The only modification is on elements $K \subset \Omega_{C1}$ which satisfy by definition $\bar{K} \cap \Gamma_C \neq \emptyset$. Depending on the shape of the element and its position relative to Γ_C, different situations can occur. In Figures 5.7 and 5.8, several cases are depicted; for each situation, the corresponding quadrature nodes and weights with respect to the reference element $\widehat{K}$ ($[0,1]^d$ or the unit simplex, respectively) are given. The intersection of $\partial\widehat{K}$ with the boundary Γ_C is on the bottom side of each element and marked in gray, whereas the quadrature nodes are indicated by filled black bullets. It can easily be verified that all quadrature rules can integrate constant functions on $\widehat{K}$ exactly; further, with $|\widehat{K}|$ replaced by $|K|$, they are also exact of degree zero on each actual element $K \in \mathcal{T}^h$.

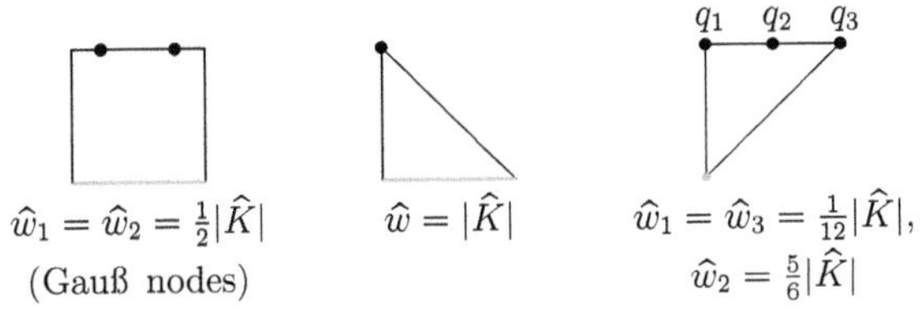

Figure 5.7: Nodes and weights of the quadrature rule $Q^0_{\hat{K}}$ in 2D for $K \subset \Omega_{C1}$.

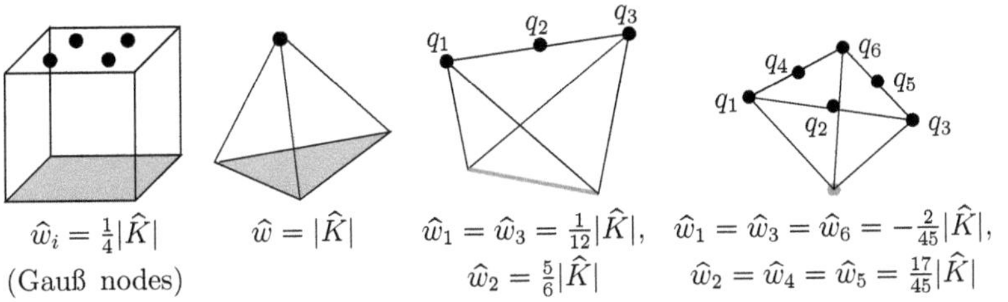

Figure 5.8: Nodes and weights of the quadrature rule $Q^0_{\hat{K}}$ in 3D for $K \subset \Omega_{C1}$.

5.2.2 Quadrature rule Q^1

On each element $K \in \mathcal{T}^h$, $K \subset \Omega \backslash \Omega_{C2}$, we again take any standard quadrature formula of sufficient accuracy. Modifications occur on elements $K \subset \Omega_{C2}$. In order to describe the construction of the quadrature formula, we need the notion of so-called macro-elements:

We assume that there exists a second triangulation $\mathcal{T}_1^h$ possibly with hanging nodes such that each element of $\mathcal{T}_1^h$ can be written as the union of elements in $\mathcal{T}^h$. Moreover if $K \in \mathcal{T}_1^h$ with $K \subset \Omega_I$ then $K \in \mathcal{T}^h$. Each $K \in \mathcal{T}_1^h \backslash \mathcal{T}^h$ contains at least one element of $\mathcal{T}^h$ being in Ω_{C1} (see the shaded parts in Figures 5.9 and 5.10) and exactly one element $K_1 \in \mathcal{T}^h$ with $K_1 \subset \Omega_I$. We note that such triangulation $\mathcal{T}_1^h$ exists for any $\mathcal{T}^h$ as no further conditions on the shape of the macro-elements are imposed, but it is not uniquely defined. If $\mathcal{T}^h$ is obtained from $\mathcal{T}^{2h}$ by uniform refinement near the contact boundary, $\mathcal{T}_1^h$ can easily be constructed. Thus, the following formulas are applicable in many cases because we usually need a fine mesh near the contact zone.

Next, we construct a quadrature formula on a given macro-element $K \in \mathcal{T}_1^h$ such that no quadrature node is located in Ω_{C1} and all functions in $\boldsymbol{\mathcal{P}}_2(K)$ can be integrated exactly. For an arbitrary mesh, we do not know a priori how the macro-element K looks like. Hence, we have to define a quadrature formula on the unique element $K_1 \in \mathcal{T}^h$ with $T_1 \subset \Omega_I \cap K$ and compute the weights locally with respect to the shape of K. But if we have a locally regular mesh like in Figure 5.6, we can compute the weights on a reference macro-element $\hat{K}$ and transform them onto K, similar to Subsection 5.2.1.

In Figure 5.9, a representative quadrilateral reference macro-element $\hat{K}$ in 2D is depicted, with two examples of suitable quadrature formulas. The rules exhibit a tensor product structure with Gauß nodes in the direction tangential to the contact part of $\partial\hat{K}$ and equidistant nodes in the remaining direction; thus, the formulas can easily be generalized to the 3D case. One can directly verify that the formula on the left side of 5.9 is exact for $\boldsymbol{\mathcal{Q}}_3(\hat{K})$. By this, we

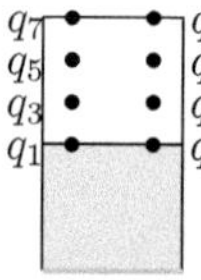

$w_1 = w_2 = 2|\widehat{K}|,$
$w_3 = w_4 = -\frac{15}{4}|\widehat{K}|,$
$w_5 = w_6 = 3|\widehat{K}|,$
$w_7 = w_8 = -\frac{3}{4}|\widehat{K}|.$

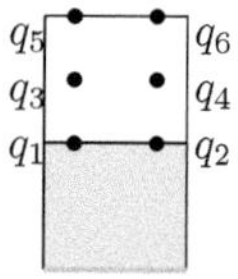

$w_1 = w_2 = \frac{1}{3}|\widehat{K}|,$
$w_3 = w_4 = -\frac{2}{3}|\widehat{K}|,$
$w_5 = w_6 = \frac{5}{6}|\widehat{K}|.$

Figure 5.9: Nodes and weights for the quadrature rule $Q^1_{\widehat{K}}$ on a quadrilateral macro-element $\widehat{K}$ in 2D.

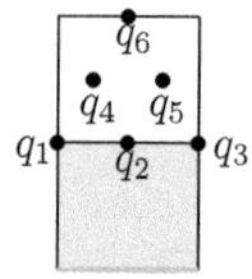

$w_1 = \frac{1}{3}|\widehat{K}|,\quad w_2 = |\widehat{K}|,$
$w_3 = \frac{1}{3}|\widehat{K}|,\quad w_4 = -\frac{2}{3}|\widehat{K}|,$
$w_5 = -\frac{2}{3}|\widehat{K}|,\quad w_6 = \frac{2}{3}|\widehat{K}|.$

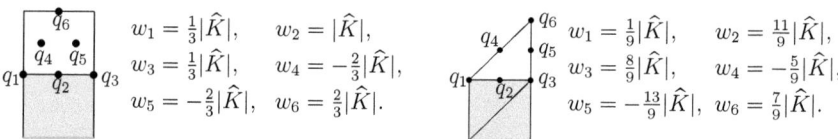

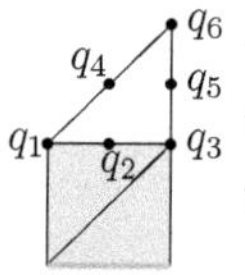

$w_1 = \frac{1}{9}|\widehat{K}|,\quad w_2 = \frac{11}{9}|\widehat{K}|,$
$w_3 = \frac{8}{9}|\widehat{K}|,\quad w_4 = -\frac{5}{9}|\widehat{K}|,$
$w_5 = -\frac{13}{9}|\widehat{K}|,\quad w_6 = \frac{7}{9}|\widehat{K}|.$

Figure 5.10: Nodes and weights for a quadrature rule which is exact for $\mathcal{P}_2(\widehat{K})$ on suitable macro-elements $\widehat{K}$ in 2D.

can guarantee that any function $\chi \in \mathcal{Q}_2(K)$ is integrated exactly as long as K is the image of $\widehat{K}$ under a transformation F_K with $\det(F_K^{-1}) \in \mathcal{Q}_1(\widehat{K})$, which is the case for isoparametric $\mathcal{Q}_1$-elements. If the transformation is linear affine and $\det(F_K^{-1})$ is constant, we can even reduce the number of quadrature nodes and use the formula on the right of Figure 5.9, which is exact for $\mathcal{Q}_2(\widehat{K})$.

For the conservation of the zeroth, first and second order moments, the quadrature formula only needs a local exactness of $\mathcal{P}_2(K)$ instead of $\mathcal{Q}_2(K)$ (cf. Lemma 5.1 below). Hence, on the left side of Figure 5.10, we have given another formula for the quadrilateral case which is exact for $\mathcal{P}_2(\widehat{K})$ but not for $\mathcal{Q}_2(\widehat{K})$. If F_K is linear affine, the transformed formula is still exact for $\mathcal{P}_2(K)$; however, we will see later on page 78 and in Section 5.4.3 that this degree of exactness is not sufficient for stability if quadrilaterals/hexahedrals are considered. For the simplicial case, in contrast, stability can be guaranteed if the quadrature rule is exact for $\mathcal{P}_2(K)$; a possible example is depicted on the right of Figure 5.9.

5.2.3 Properties of the quadrature rules

Using $\bar{m}_i^h(\cdot,\cdot)$, $i = 0, 1$, from (5.8) for the definition of the mass matrix (now denoted by $\bar{M}_i^h$), we obtain the modified version of (5.1):

$$\bar{M}_i^h \ddot{\mathbf{u}}^h + A^h \mathbf{u}^h + B_{\mathrm{co}}^h \boldsymbol{\lambda}^h = \mathbf{f}^h, \tag{5.9a}$$

$$\mathbf{C}^{\mathrm{co}}(\mathbf{u}^h, \boldsymbol{\lambda}^h) = \mathbf{0}. \tag{5.9b}$$

The solutions of (5.1) and (5.9) are not the same; but from now on $(\mathbf{u}^h, \boldsymbol{\lambda}^h)$ only refers to the solution of (5.9), also neglecting the index i in order to keep the notation simple.

The examples presented in Subsections 5.2.1 and 5.2.2 can be considered as prototypes for two different ways of constructing the modified mass matrix $\bar{M}_i^h$; the former is associated to the triangulation $\mathcal{T}_0^h := \mathcal{T}^h$, whereas the latter is based on the macro-triangulation $\mathcal{T}_1^h$. Written concisely, we can define the formula Q^i, $i \in \{0,1\}$, locally with respect to the triangulation $\mathcal{T}_i^h$ as follows:

$$Q^i(\chi^h \cdot \eta^h) := \sum_{K \in \mathcal{T}_i^h} Q_K^i(\chi^h \cdot \eta^h), \quad \chi^h, \eta^h \in \mathbf{V}^h. \tag{5.10}$$

Below, we summarize the main features of the examples in Subsections 5.2.1 and 5.2.2 as requirements which a suitable quadrature formula Q^i has to satisfy:

$Q1)$ No quadrature point of Q^i is placed in $\bar{\Omega}_{C1} \setminus \bar{\Gamma}_I$.

$Q2)$ For $K \in \mathcal{T}_i^h$, the local formula Q_K^i is exact for all functions in $\mathcal{P}_{2i}(K)$ for the simplicial or for all functions in $\mathcal{Q}_{2i}(K)$ for the quadrilateral/hexahedral case.

The following lemma is a direct consequence of condition $Q2$:

Lemma 5.1. *If the quadrature formula Q^i, $i \in \{0,1\}$, is chosen according to condition $Q2$, the new mass matrix $\bar{M}_i^h$ in (5.9) conserves the total mass, i.e., $\mathbf{1}^T \bar{M}_i^h \mathbf{1} = \mathbf{1}^T M^h \mathbf{1}$ holds with the vector $\mathbf{1} = (1, \dots, 1)^T \in \mathbb{R}^{d|\mathcal{N}^h|}$. Moreover, the mass matrix $\bar{M}_1^h$ conserves the zeroth, first and second order moments of the original system (5.1a) (i.e. the total mass, the center of gravity and the moments of inertia).*

Proof. The functions $\varrho(\mathbf{1} \cdot \mathbf{1})$, $\varrho(\mathbf{1} \cdot \mathbf{x}_i)$ and $\varrho(\mathbf{x}_i \cdot \mathbf{x}_j)$, $1 \le i, j \le d$, all lie within the space $\mathcal{P}_2(\Omega)$. Hence, Q2 implies for, e.g., the second order moments and the quadrature formula Q^1:

$$\bar{m}_1^h(\mathbf{x}_i, \mathbf{x}_j) = \sum_{K \in \mathcal{T}_1^h} Q_K^1(\varrho(\mathbf{x}_i \cdot \mathbf{x}_j)) = \sum_{K \in \mathcal{T}_1^h} \int_K \varrho(\mathbf{x}_i \cdot \mathbf{x}_j) ds = m(\mathbf{x}_i, \mathbf{x}_j).$$

M0 and M1 follow analogously, with M0 needing only an exact integration of constant functions which is satisfied for both Q^0 and Q^1. □

Remark 5.2. The modification of the mass matrix M^h can be seen as a generalization of the widely used technique of mass lumping. There, M^h is replaced by a diagonal matrix M_L^h whose entries are given by $\sum_q (M^h)_{pq}$. M_L^h can for example be obtained when the integral $\int_T \phi_p \, \phi_q \, d\mathbf{x}$ on an element $T \in \mathcal{T}^h$ is replaced by the quadrature formula $\sum_{\mathbf{z}} \phi_p(\mathbf{z}) \, \phi_q(\mathbf{z}) \, |T| / \mathcal{N}_T$, where $\mathcal{N}_T$ is the number of vertices of T and $\sum_{\mathbf{z}}$ is the sum over all vertices of T. But as replacing M^h by M_L^h preserves the total mass and the center of gravity, the second order moments are in general not conserved. Thus, the modification $\bar{M}_1^h$ introduces a smaller altering of the physical behaviour of the underlying body than the standard mass lumping technique. The reason why we speak of a generalization is that $\bar{M}_i^h$, $i \in \{0,1\}$, does not keep the positive definiteness of M^h like M_L^h does. Nevertheless, as shown in Chapter 6, the a priori error estimates have the same asymptotic rate as those of the lumped mass technique.

5.3 Different interpretation of $\bar{m}_i^h(\cdot,\cdot)$

In this section, we look at the modified bilinear form $\bar{m}_i^h(\cdot,\cdot)$ defined in (5.8) from a different mathematical point of view. Let us assume that there exists an interpolation operator I_i^h, $i \in \{0,1\}$, on $\mathbf{V}^h$ satisfying the following three conditions

*P*1) I_i^h is $\mathbf{L}^2(\Omega_{C2})$-stable, i.e., $\|I_i^h\boldsymbol{\chi}^h\|_{0,\Omega_{C2}} \leq c\|\boldsymbol{\chi}^h\|_{0,\Omega_{C2}}, \quad \boldsymbol{\chi}^h \in \mathbf{V}^h$.

*P*2) $I_i^h\boldsymbol{\chi}^h|_{\bar{\Omega}_I} = \boldsymbol{\chi}^h|_{\bar{\Omega}_I}, \quad \boldsymbol{\chi}^h \in \mathbf{V}^h$.

*P*3) simplicial case: $I_i^h\boldsymbol{\chi}^h|_K = \boldsymbol{\chi}^h|_K, \quad \boldsymbol{\chi}^h|_K \in \boldsymbol{\mathcal{P}}_i(K), \ K \in \mathcal{T}_i^h$,
quadrilateral/hexahedral case: $I_i^h\boldsymbol{\chi}^h|_K = \boldsymbol{\chi}^h|_K, \quad \boldsymbol{\chi}^h|_K \in \boldsymbol{\mathcal{Q}}_i(K), \ K \in \mathcal{T}_i^h$,

and

$$\bar{m}_i^h(\boldsymbol{\chi}^h, \boldsymbol{\eta}^h) = m(I_i^h\boldsymbol{\chi}^h, I_i^h\boldsymbol{\eta}^h), \quad \boldsymbol{\chi}^h, \boldsymbol{\eta}^h \in \mathbf{V}^h. \tag{5.11}$$

Remark 5.3. Condition *P*1 is a common assumption for a well-defined interpolation operator, whereas *P*2 is motivated by the observation that $\bar{m}_i^h(\boldsymbol{\chi}^h, \boldsymbol{\eta}^h) = m(\boldsymbol{\chi}^h, \boldsymbol{\eta}^h)$ for $\boldsymbol{\chi}^h, \boldsymbol{\eta}^h \in \mathbf{V}^h$ with $\operatorname{supp}(\boldsymbol{\chi}^h) \subset \bar{\Omega}_I$ or $\operatorname{supp}(\boldsymbol{\eta}^h) \subset \bar{\Omega}_I$. Requirement *P*3 reflects the exactness condition *Q*2.

Remark 5.4. As shown in [70], conditions *Q*1, *P*1, *P*2 and the relation (5.11) imply the inequalities

$$c\|\boldsymbol{\chi}^h\|_{0,\Omega_I}^2 \leq \bar{m}_i^h(\boldsymbol{\chi}^h, \boldsymbol{\chi}^h) \leq C\|\boldsymbol{\chi}^h\|_{0,\Omega_I}^2, \quad \boldsymbol{\chi}^h \in \mathbf{V}^h. \tag{5.12}$$

Hence we can state that $|\cdot|_{h,i} := \sqrt{\bar{m}_i^h(\cdot,\cdot)}$ is equivalent to the $\mathbf{L}^2(\Omega_I)$-norm and thus a seminorm on $\mathbf{V}^h$. Condition *P*1 and (5.11) implicate that $\bar{m}_i^h(\cdot,\cdot)$ is continuous on $\mathbf{V}^h$.

By condition *Q*1, all quadrature points of Q^i are placed in $\bar{\Omega}_I$, and thus we get by (5.8), (5.10) and *P*2:

$$\bar{m}_i^h(\boldsymbol{\chi}^h, \boldsymbol{\eta}^h) = \sum_{K\in\mathcal{T}_i^h} Q_K^i(\varrho\boldsymbol{\chi}^h \cdot \boldsymbol{\eta}^h) = \sum_{K\in\mathcal{T}_i^h} Q_K^i(\varrho I_i^h\boldsymbol{\chi}^h \cdot I_i^h\boldsymbol{\eta}^h) = \bar{m}_i^h(I_i^h\boldsymbol{\chi}^h, I_i^h\boldsymbol{\eta}^h).$$

From (5.11), we find

$$\sum_{K\in\mathcal{T}_i^h} Q_K^i(\varrho I_i^h\boldsymbol{\chi}^h \cdot I_i^h\boldsymbol{\eta}^h) = \sum_{K\in\mathcal{T}_i^h} \int_K \left(\varrho I_i^h\boldsymbol{\chi}^h \cdot I_i^h\boldsymbol{\eta}^h\right) d\mathbf{x}. \tag{5.13}$$

This formula holds if the quadrature formula Q_K^i is exact on each element $K \in \mathcal{T}_i^h$ for the product of the functions $(I_i^h\boldsymbol{\chi}^h \cdot I_i^h\boldsymbol{\eta}^h)$, $\boldsymbol{\chi}^h, \boldsymbol{\eta}^h \in \mathbf{V}^h$. This requirement is satisfied due to condition *Q*2. Hence, Remark 5.4 implies the positive semidefiniteness of the mass matrix for all quadrature formulas presented in Subsections 5.2.1, 5.2.2 except for the one on the left of Figure 5.10. Indeed, the local mass matrix of the latter formula is indefinite with the smallest eigenvalue $\lambda_{\min} = -0.0468$. In Section 5.4.3, we will numerically confirm that this quadrature rule leads to numerical instability.

For a given quadrature formula Q^i, we can define a corresponding interpolation operator I_i^h via the relation (5.13). In the following, we return to the prototypes presented in Subsections 5.2.1 and 5.2.2 and construct appropriate operators I_0^h and I_1^h.

5.3.1 Interpolation operator I_0^h

The action of the linear operator I_0^h is characterized by the images of the basis functions ϕ_p, $p \in \mathcal{N}^h$. Hence, we compute the right hand side of (5.13) for functions

$$\boldsymbol{\chi}^h, \boldsymbol{\eta}^h \in \{\mathbf{e}_j\phi_p, 1 \le j \le d, p \subset K \in \mathcal{T}_0^h\}.$$

Equation (5.13) and condition $Q1$ imply that $I_0^h(\mathbf{e}_j\phi_p) = 0$ has to hold for any $p \in \mathcal{N}_{\text{co}}^h$. For nodes $p \in \mathcal{N}_I^h$, we know from $Q1$ that for any modified basis function ϕ_p^0 constructed according to the specification

$$\phi_p^0 := \begin{cases} \phi_p & \text{on } K \subset \Omega_I, \\ \phi_p + \sum_{q \in \mathcal{N}_{\text{co}}^h, q \subset K} \beta_{pq}\phi_q & \text{on } K \subset \Omega_{C1}, \end{cases} \tag{5.14}$$

with $\beta_{pq} \in \mathbb{R}$, we have

$$Q_K^0(\mathbf{e}_j\phi_p \cdot \mathbf{e}_k\phi_q) = \delta_{jk}Q_K^0(\phi_p\phi_q) = \delta_{jk}Q_K^0(\phi_p^0\phi_q^0).$$

Hence we are looking for scalar values β_{pq} in (5.14) such that

$$Q_K^0(\phi_p^0\phi_q^0) = \begin{cases} 0, & p \in \mathcal{N}_{\text{co}}^h \text{ or } q \in \mathcal{N}_{\text{co}}^h, \\ \int_K \phi_p^0\phi_q^0 \, d\mathbf{x}, & \text{else}, \end{cases} \tag{5.15}$$

and define

$$I_0^h\mathbf{e}_j\phi_p := \mathbf{e}_j\phi_p^0, \quad 1 \le j \le d, \, p \in \mathcal{N}^h. \tag{5.16}$$

The first condition of (5.15) is valid for any $\beta_{pq} \in \mathbb{R}$, whereas the second relation depends on the quadrature formula Q_K^0 and is only fulfilled for certain choices of β_{pq}. Figures 5.11 and 5.12 display several representatives of the functions ϕ_p^0, $p \in \Gamma_I$, corresponding to the quadrature rules in Figures 5.7 and 5.8. They span a modified finite element space $\bar{\mathbf{V}}_0^h$:

$$\bar{\mathbf{V}}_0^h := \left\{ \sum_{p \in \mathcal{N}_I^h} \boldsymbol{\alpha}_p\phi_p^0 : \boldsymbol{\alpha}_p \in \mathbb{R}^d \right\}. \tag{5.17}$$

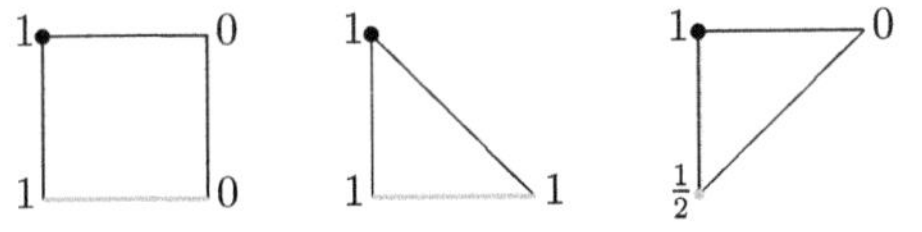

Figure 5.11: Nodal values of the modified basis functions ϕ_p^0 on $\hat{K}$ for $K \subset \Omega_{C1}$ for the 2D case.

One can easily verify that the operator I_0^h given by (5.16) satisfies conditions $P1$ to $P3$ and by construction (5.11).

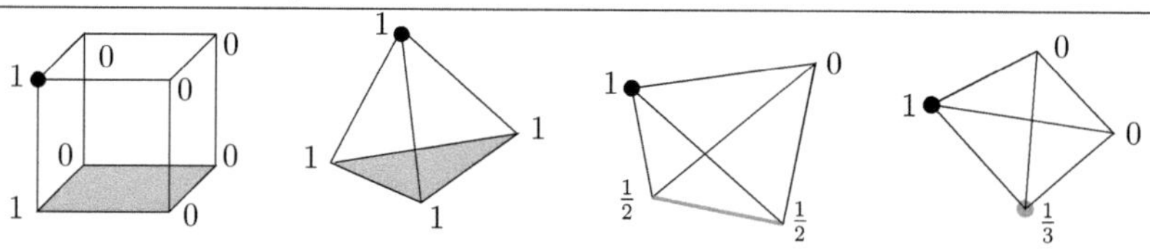

Figure 5.12: Nodal values of the modified basis functions ϕ_p^0 on $\widehat{K}$ for $K \subset \Omega_{C1}$ for the 3D case.

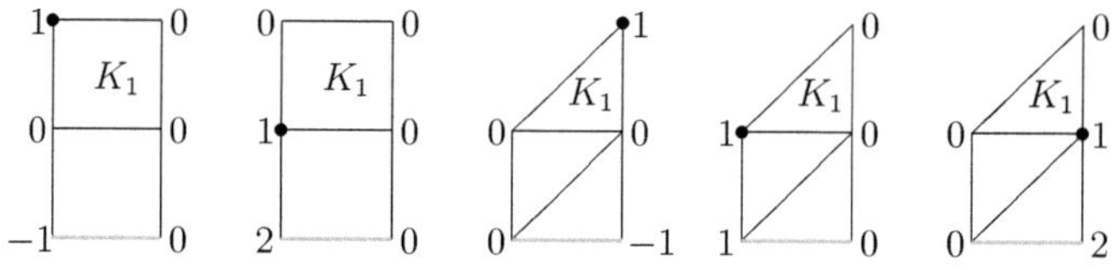

Figure 5.13: Nodal values of the modified basis functions ϕ_p^1 on $\widehat{K}$ for $K \subset \Omega_{C2}$ for the 2D case.

5.3.2 Interpolation operator I_1^h

We proceed analogeously to the previous subsection and look for modified local basis functions ϕ_p^1, $p \in \mathcal{N}_I^h$, defined according to (5.14) such that

$$Q_K^1(\mathbf{e}_j\phi_p \cdot \mathbf{e}_k\phi_q) = \delta_{jk}Q_K^1\left(\phi_p^1\phi_q^1\right) = \begin{cases} 0, & p \in \mathcal{N}_{\text{co}}^h \text{ or } q \in \mathcal{N}_{\text{co}}^h, \\ \delta_{jk}\int_K \phi_p^1\phi_q^1\,d\mathbf{x}, & \text{else.} \end{cases}$$

A possible construction of the functions ϕ_p^1, $p \in \mathcal{N}_I^h$, is done as follows: For each $K \in \mathcal{T}_1^h$, let $K_1 \in \mathcal{T}^h$ be the unique element satisfying $K_1 \subset K \cap \Omega_I$. For each vertex $p \subset \bar{K}_1$, we define ϕ_p^1 as the polynomial extension of $\phi_p|_{K_1}$ onto K. This implies $\phi_p^1|_{K'} = \phi_p|_{K'}$ for $K' \in \mathcal{T}^h \cap \mathcal{T}_1^h$. The resulting functions for the 2D case are shown in Figure 5.13.

Now, we can define the interpolation operator I_1^h and the modified finite element space $\bar{\mathbf{V}}_1^h$ as in (5.16) and (5.17), respectively, with the upper index 0 replaced by 1. The conditions $P1$ to $P3$ are easy to verify.

Remark 5.5. We note that the modified basis functions ϕ_p^i, $i \in \{0,1\}$, for $p \subset \Gamma_I$ are in general not globally continuous because of their elementwise definition. This leads to $\bar{\mathbf{V}}_i^h \not\subset \mathbf{V}^h$. If one wants to avoid such nonconforming situation or if the mesh is strongly anisotropic, the nodal values of the modified basis functions can be adapted to be continuous or to respect the anisotropy. However, this results in $\operatorname{supp}\phi_p^i \neq \operatorname{supp}\phi_p$ in general, such that an elementwise definition of the interpolation operator and the corresponding quadrature formula is not possible.

The characterization of the bilinear form $\bar{m}_i^h(\cdot,\cdot)$, $i \in \{0,1\}$, in terms of the interpolation operator I_i^h will be used in Chapter 6, where the discretization error of the modified discrete system (5.9) is analysed.

5.4 Numerical results

In this section, we perform several numerical tests which show the improvement of the dynamic contact computation due to the modified mass matrix. If not stated otherwise, all problems are discretized in time using the trapezoidal rule corresponding to $\theta = \frac{1}{2}$ in (2.21a). Furthermore, we use the energy-consistent normal contact condition (2.16) and incorporate the friction conditions as described in Section 4.2 with $s = 1$, leading to:

$$\left(\frac{2}{\Delta t^2}\bar{M}^h + \frac{1}{2}A^h\right)\mathbf{u}^h_{j+1} + B^h_{\mathrm{co}}\boldsymbol{\lambda}^h_{j+1/2} = \mathbf{f}^h_{j+1/2} + \frac{2}{\Delta t^2}\bar{M}^h\left(\mathbf{u}^h_j + \Delta t\mathbf{v}^h_j\right) - \frac{1}{2}A^h\mathbf{u}^h_j, \tag{5.18a}$$

$$\mathbf{v}^h_{j+1/2} - \Delta\mathbf{u}^h_{j+1} = \mathbf{0}, \tag{5.18b}$$

$$\mathbf{C}^{\mathrm{en}}_n(\mathbf{u}_{j+1}, \boldsymbol{\lambda}_{j+1/2}) = \mathbf{0}, \tag{5.18c}$$

$$\mathbf{C}^{\mathrm{co},1}_t(\mathbf{u}_{j+1}, \boldsymbol{\lambda}_{j+1/2}) = \mathbf{0}. \tag{5.18d}$$

5.4.1 Nonlinear beam in 2D

First, we test if the modified mass matrix method yields the same results as the standard computation as long as we are not in contact. To this end, we consider the cross section of an elastic beam of length 10 and height 1. We assume that there act no body forces on the beam and use the geometrically nonlinear Saint Venant–Kirchhoff material (1.7) which is discretized in time as described in [156].

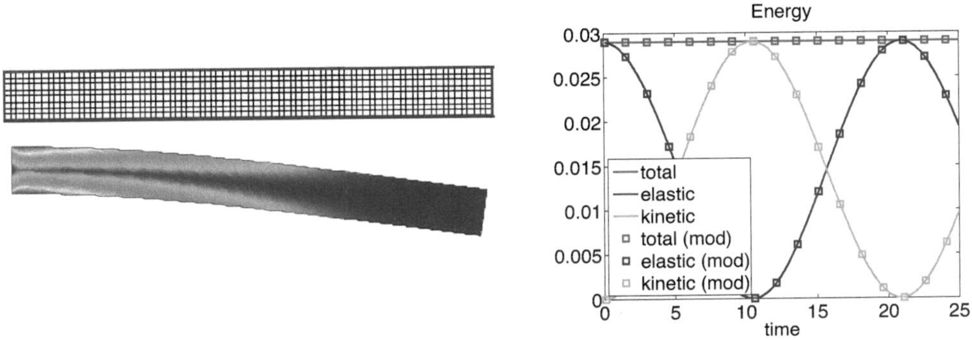

Figure 5.14: Left: initial grid and effective stress at time t_1; right: energy conservation.

As material parameters, we use the elasticity module $E = 200$, the Poisson ratio $\nu = 0.3$ and the density $\varrho = 1.0$. We assume that the left side of the beam is clamped ($\mathbf{u} = \mathbf{0}$), whereas all other parts of the boundary are free ($\sigma(\mathbf{u})\mathbf{n} = \mathbf{0}$). The initial displacement $\mathbf{u}_0$ is computed such that it satisfies the same boundary conditions as above apart from $\sigma(\mathbf{u}_0)\mathbf{n} = (0, -0.002x_1)^T$ on the bottom boundary Γ_C, where we locally modify the mass matrix using the quadrature rule Q^1 given on the left of Figure 5.9. The initial velocity $\mathbf{v}_0$ is set to zero, and we compute 250 time steps with a step size of $\Delta t = 0.1$. For the spatial discretization, we use a regular quadrilateral grid with $h = 0.125$ shown on the left side in Figure 5.14, together with a plot of the deformed geometry after the first time step. On the right, the evolution of the discrete kinetic, elastic and total energy defined in (2.15) are displayed. One can see that the energy conservation property of the time stepping scheme with $\theta = \frac{1}{2}$ is not affected by the modification of the mass matrix.

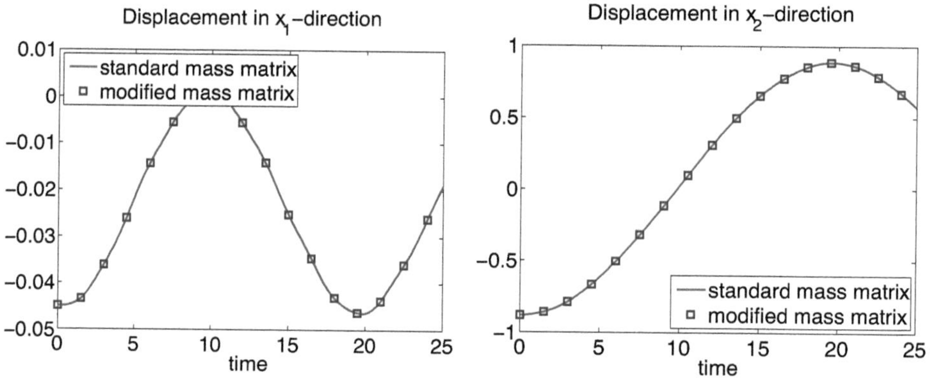

Figure 5.15: Displacement of the node (10|0.5) on the tip of the beam.

Figure 5.15 presents the displacement in x_1- and x_2-direction of the node (10|0.5) which is located at the tip of the beam. We get the same results for both kinds of mass matrix, leading to the conclusion that the quality of the numerical results for the modified matrix is as good as for the usual one if no contact occurs. In Chapter 6, we analyse this situation from a theoretical point of view and give a rigorous proof of the former statement.

Next, we compare the results of the standard mass matrix with the modified one in the case of contact. To this end, we consider various examples of two-body contact problems.

5.4.2 Frictionless two-body contact in 2D

The first setting we consider is the contact of two linear elastic discs, each with the same radius $R = 8$ and the data $E = 100$, $\nu = 0.3$, $\varrho = 2.0 \cdot 10^{-8}$. The initial distance is 0.1, and both circles move at a speed of 800 towards each other. We only consider the case $\mathfrak{F} = 0$ as the frictional work is very small for this example. The grid is refined near the potential contact boundary Γ_C as shown in Figure 5.16, and the size of the time step is $\Delta t = 10^{-6}$. The effective stress σ_{eff} after the impact is visualized on the right of Figure 5.16.

We assemble the mass matrix with two distinct grids: The standard matrix M^h and the mass matrix $\bar{M}_0^h$ are computed for a triangular as well as a quadrilateral mesh, whereas the matrix $\bar{M}_1^h$ is constructed for the quadrilateral grid only. We use the quadrature formulas depicted in Figures 5.7 and 5.9 on the appropriate reference elements and map them onto the actual elements. The computed energies for $\bar{M}_0^h$ (const) and M^h (stand) are displayed in Figure 5.17.

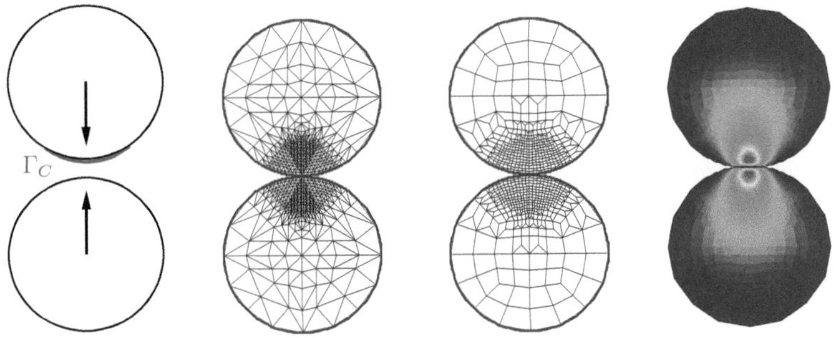

Figure 5.16: Two circle frictionless contact problem: initial grids and effective stress.

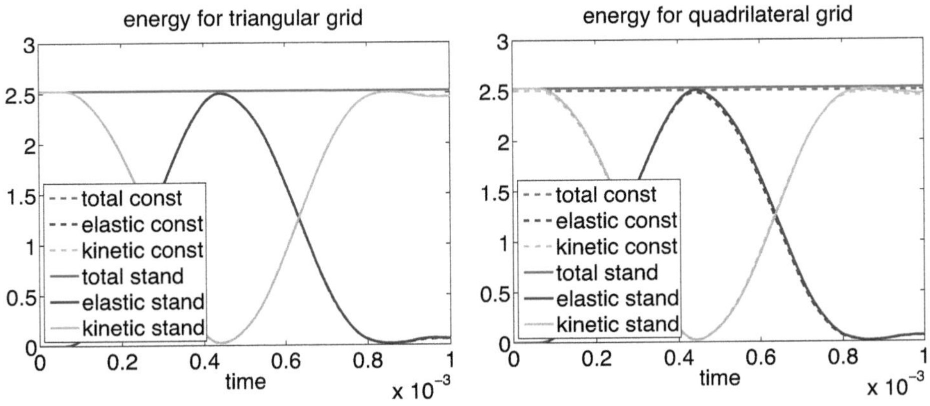

Figure 5.17: Two circle problem: energy results on triangular and quadrilateral grid.

We obtain the same total energy for all computations except for $\bar{M}_0^h$ on the quadrilateral grid. This is due to the linear determinant of the local transformation that is not necessarily integrated exactly (cf. the discussion on page 76). Hence, we rather use the quadrature formula Q^1 if we are dealing with a quadrilateral mesh and curvilinear boundaries. The computed energies for $\bar{M}_1^h$ are exactly the same as those for the standard mass matrix and are thus omitted in Figure 5.17.

Figure 5.18 shows the results for the normal contact stress $\lambda_{j+1/2,n}$ at the bottom slave node. One can see that the standard method exhibits unphysical oscillations which can even lead to negative values of the Lagrange multiplier, thus violating the KKT conditions (2.16). Because of this, the primal-dual active set strategy described in Section 4.2 does not converge any more, such that we have to enforce the artificially averaged condition

$$\lambda_{j,n} := \frac{1}{2}(\lambda_{j-1/2,n} + \lambda_{j+1/2,n}) \geq 0 \tag{5.19}$$

instead of $\lambda_{j+1/2,n} \geq 0$ in (2.16) for the results using the standard mass matrix.

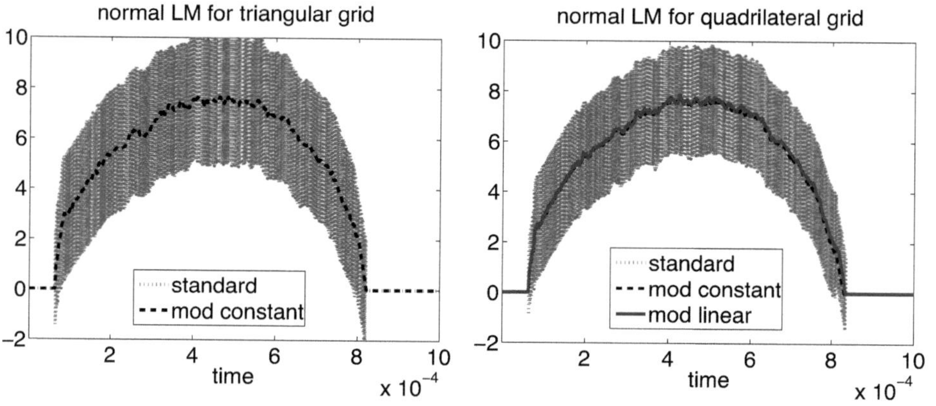

Figure 5.18: Two circle problem: normal Lagrange multiplier for simplicial and quadrilateral grid at bottom slave node.

In contrast, the modification of the mass matrix leads to smoother results, both for the simplicial and the quadrilateral grid. We observe that the beginning and the end of the contact period are the same for all calculations.

5.4.3 Frictional two-body contact in 2D

As a second example, we compute the contact of two rectangular blocks sketched in Figure 5.19. The upper block, taking the role of the slave side, is given by $\Omega^{\mathrm{sl}} = [-0.5,\, 0.5] \times [0.025,\, 1.025]$ and the lower one by $\Omega^{\mathrm{ma}} = [-1,\, 1] \times [-0.5,\, 0]$. Both blocks consist of the same linear elastic material given by $E = 300$, $\nu = 0.3$ and $\varrho = 2 \cdot 10^{-9}$, and we apply Coulomb friction with $g_t = 0$ and the friction coefficient $\mathfrak{F} = 0.1$.

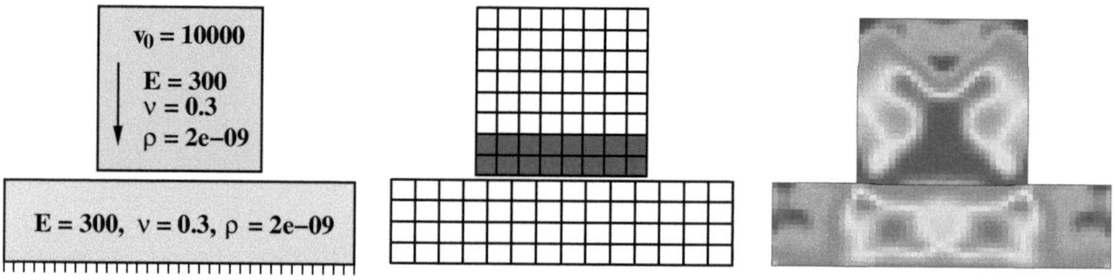

Figure 5.19: Two block frictional contact problem: initial configuration, coarse mesh and effective stress.

The lower block is subject to homogeneous Dirichlet conditions at the bottom and is at rest at $t = 0$, whereas the upper block moves downwards at a constant velocity of $\mathbf{v}_0 = (0,\, -10^4)^T$. For the time discretization, we use the time step size $\Delta t = 5 \cdot 10^{-8}$ and compute 300 time steps.

The initial configuration, the quadrilateral mesh and the distorted bodies with the effective stress during contact are shown in Figure 5.19. For the elements in Ω_{C2} (indicated by the shadowed region), we use the quadrature formula given on the left of Figure 5.9 to compute $\bar{M}^h$. The computed energy is the same for the standard and the modified mass matrix and is shown in Figure 5.20. The left picture indicates a slight loss of energy which can be seen more clearly in the right image. It shows that the total energy is conserved apart from the contact work

$$\mathbb{E}_j^{\text{co}} := \sum_{k=0}^{j-1} \sum_{p \in \mathcal{N}_{\text{co}}^h} \Delta \mathbf{u}_{p,k}^h \cdot \boldsymbol{\lambda}_{p,k+1/2}^h \tag{5.20}$$

that dissipates due to the friction. Thus, the modified method reflects the correct physical behaviour of the energy.

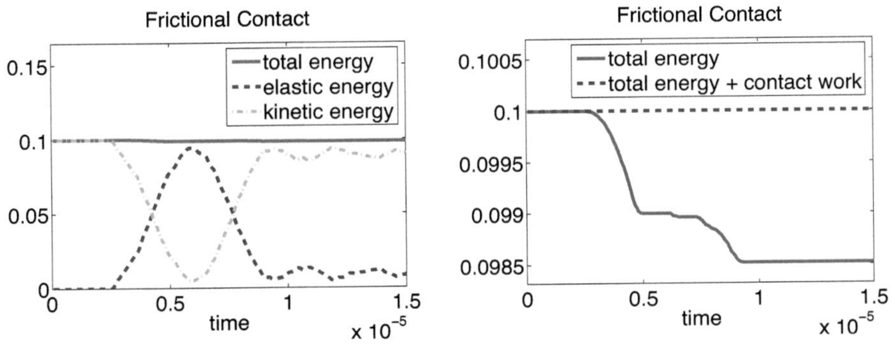

Figure 5.20: Two block problem: energy and contact work.

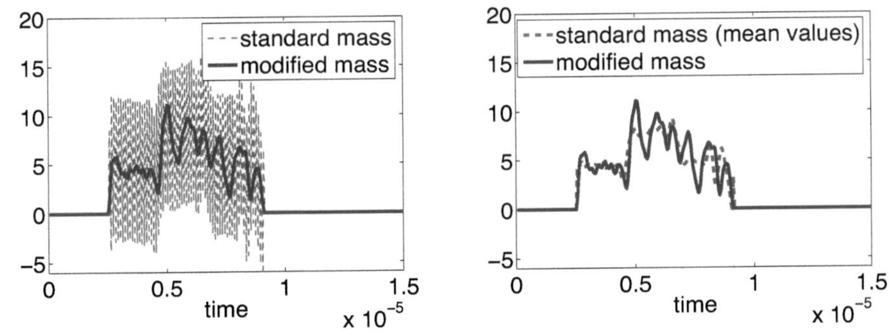

Figure 5.21: Two block problem: normal Lagrange multiplier at midpoint.

In Figure 5.21, the results for the normal Lagrange multiplier at the midpoint (0, 0.025) using the modified mass matrix are compared with those for the standard method. As in Subsection

5.4.2, one can observe that the latter approach leads to algorithmic oscillations in the Lagrange multiplier $\lambda_{j+1/2,n}$ with respect to time. In the right picture of Figure 5.21, the mean values $\lambda_{j,n}$ as computed in (5.19) for the standard mass matrix are compared with the values $\lambda_{j+1/2,n}$ for the modified mass matrix. Using this interpretation, both approaches lead to a similar behaviour; however, (5.19) introduces an artificial history dependence in the Lagrange multiplier which is not present in the original system.

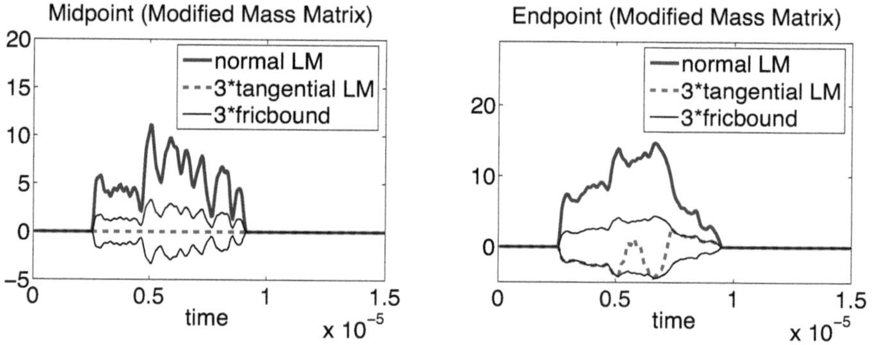

Figure 5.22: Two block problem with modified mass matrix: Lagrange multipliers.

In Figure 5.22, we plot the evolution of the normal and tangential components of the Lagrange multiplier at the midpoint $(0, 0.025)$ and the end point $(0.5, 0.025)$. As there are no non-physical oscillations in these values, the computed multiplier seems to represent the correct value of the contact stress.

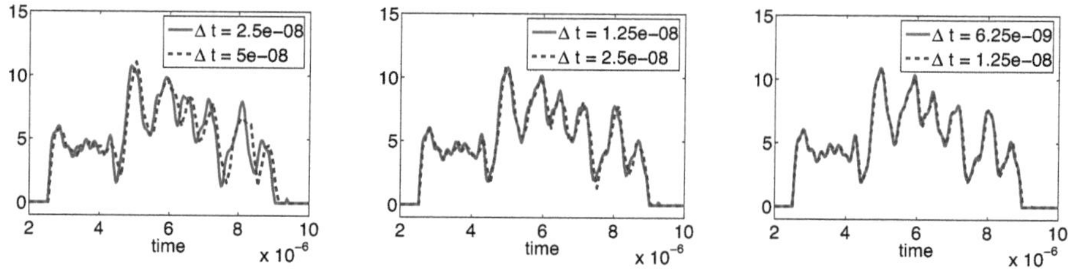

Figure 5.23: Two block problem: Convergence of multipliers for $\Delta t \to 0$.

Further, we illustrate the behaviour of the normal Lagrange multiplier $\lambda_{j+1/2,n}$ for the modified mass matrix for $h \to 0$ and $\Delta t \to 0$. The results for the multiplier located at the midpoint of Γ_C are shown in Figures 5.23 and 5.24. For $\Delta t \to 0$, the values converge to a fixed curve, in contrast to the computation with the standard mass matrix (not shown here) where the amplitude of the oscillations increases for decreasing Δt. From Figure 5.24, we can observe that for $h \to 0$, the frequency of the physical oscillations increases whereas their amplitude decreases.

Finally, we investigate the influence of the quadrature formula on the results. We use the same problem setting as before, except that we set $E = 30$, $\Delta t = 5 \cdot 10^{-7}$ and consider frictionless

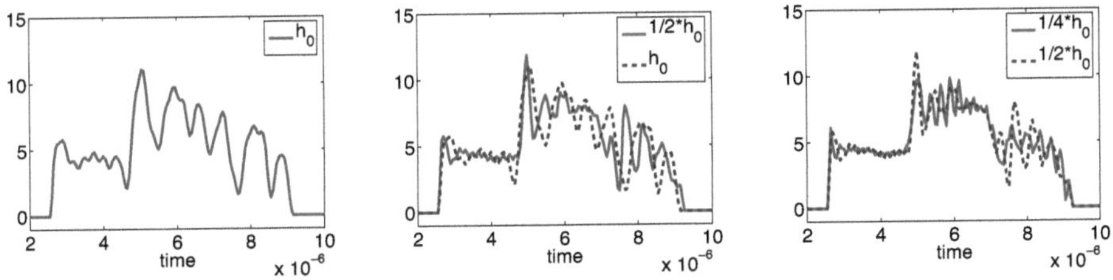

Figure 5.24: Two block problem: Behaviour of multipliers for $h \to 0$.

contact. The two tensor product formulas depicted in Figure 5.9 both lead to the same correct results as the local transformations are all linear affine mappings. But if we use the quadrature formula with six nodes depicted on the left of Figure 5.10 which does not satisfy condition $P2$, the displacements at the nodes at the top of Ω_{C2} start to oscillate. This comes from the fact that the basis functions associated with these vertices are not integrated correctly. In the end, we get extremely large velocities at these nodes which give a negative contribution to the kinetic energy due to the negative eigenvalues of the local mass matrices (see Section 5.2). After $t = 10^{-5}$, rounding errors lead to a numerical growth of the energy and finally to the divergence of the algorithm. This shows the importance of condition $P2$ for getting a stable integration scheme.

5.4.4 Comparison with stabilized predictor-corrector scheme

As already mentioned in the beginning of this chapter, there is a different approach to avoid the spurious oscillations in the contact stress. In [50, 110], a stabilized predictor-corrector scheme is proposed in which the predicted displacement is projected onto the admissible set for the normal contact conditions. In the following, we shortly quote the main idea of this algorithm and refer to [50, 106, 110] for details.

Using the standard mass matrix M^h and a fully implicit treatment of the non-penetration constraints (2.13c) [50, 99], the system (5.18) can be written in a predictor–corrector notation as follows:

$$\mathbf{u}^{\text{pred}}_{j+1} = \mathbf{u}^h_j + \Delta t\, \mathbf{v}^h_j, \tag{5.21a}$$

$$\left(\frac{2}{\Delta t^2} M^h + \frac{1}{2} A^h\right) \mathbf{u}^h_{j+1} + B^h_{\text{co}} \boldsymbol{\lambda}^h_{j+1/2} = \mathbf{f}^h_{j+1/2} + \frac{2}{\Delta t^2} M^h \mathbf{u}^{\text{pred}}_{j+1} - \frac{1}{2} A^h \mathbf{u}^h_j, \tag{5.21b}$$

$$\mathbf{C}^{\text{co}}_n(\mathbf{u}^h_{j+1}, \boldsymbol{\lambda}^h_{j+1/2}) = \mathbf{0}, \tag{5.21c}$$

$$\mathbf{v}^h_{j+1} = -\mathbf{v}^h_j + \frac{2}{\Delta t}(\mathbf{u}^h_{j+1} - \mathbf{u}^{\text{pred}}_j). \tag{5.21d}$$

As we expect $u_{p,j+1,n} = u_{p,j,n}$ for any active node $p \in \mathcal{A}^h_n$ in contact, (5.21d) yields $v_{p,j+1,n} = -v_{p,j,n}$ which leads to oscillations in the velocity and thus in the Lagrange multiplier. This can be avoided by enforcing the normal contact conditions not only on $\mathbf{u}^h_{j+1}$ but also on $\mathbf{u}^{\text{pred}}_{j+1}$.

Hence, (5.21a) is replaced by

$$M^h \mathbf{u}_{j+1}^{\text{pred}} + B_{\text{co}}^h \boldsymbol{\lambda}_{j+1/2}^{\text{pred}} = M^h \left(\mathbf{u}_j^h + \Delta t \mathbf{v}_j^h\right), \tag{5.22a}$$

$$\mathbf{C}_n^{\text{co}}(\mathbf{u}_{j+1}^{\text{pred}}, \boldsymbol{\lambda}_{j+1/2}^{\text{pred}}) = \mathbf{0}. \tag{5.22b}$$

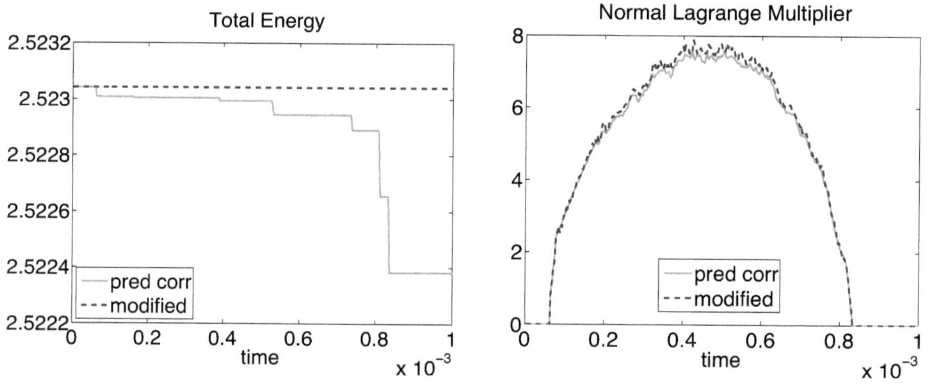

Figure 5.25: Energy and Lagrange multipliers for the stabilized scheme.

The value of $\boldsymbol{\lambda}^{\text{pred}}$ in (5.22) does not have to be computed explicitly as it does not enter the succeeding computation. In [110], it is stated that the scheme (5.22) leads to a stable, dissipative algorithm. We now compare its results with those of (5.18) using the modified mass matrix $\bar{M}^h$. As the former algorithm is not formulated for the frictional case, we choose the frictionless setting considered in Subsection 5.4.2 for comparison. In Figure 5.25, we see that the stabilized predictor-corrector scheme loses energy when the active set changes, i.e., when nodes come into contact or are released, whereas the modified mass matrix method conserves the total energy. The computed Lagrange multipliers in the normal direction are also shown in Figure 5.25; we remark that both methods yield reasonable results.

The comparison shows that although the stabilized predictor-corrector scheme looks promising, the method described in this chapter has some advantages: The total energy is conserved, and we do not need to solve the additional problem (5.22).

5.4.5 Frictional two-body contact in 3D

Finally, we show two numerical simulations of dynamic two-body contact problems with Coulomb friction in the three-dimensional case using the modified mass matrix.

First, we consider a bowl having an inital velocity $\mathbf{v}_0 = (0,\, 0,\, -5)^T$ downwards and impinging on a brick being at rest. The kinetic energy of the bowl is transferred to the brick such that the brick moves downwards. At time $t_0 = 0$, the midpoint of the bowl is $(0,\, 0,\, 0)^T$ and its radius is 0.6, whereas the brick is modeled by the domain $(-0.7, 0.7) \times (-0.7, 0.7) \times (-1.3, -0.7)$. The material is assumed to be linear elastic, the mesh consists of tetrahedra, and the friction

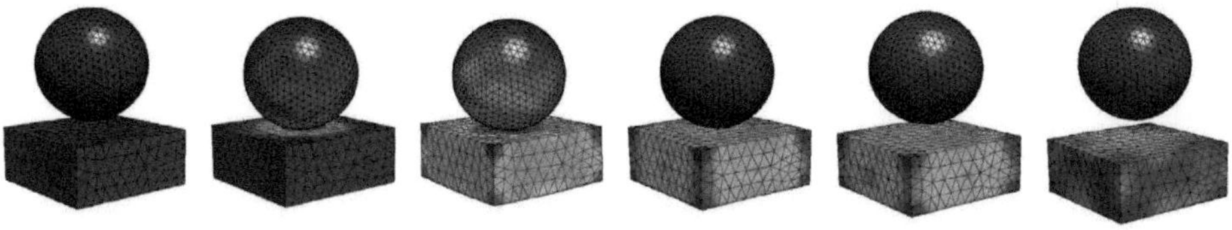

Figure 5.26: Bowl brick problem: effective stress at t_0, t_{50}, t_{75}, t_{100}, t_{150} and t_{200}.

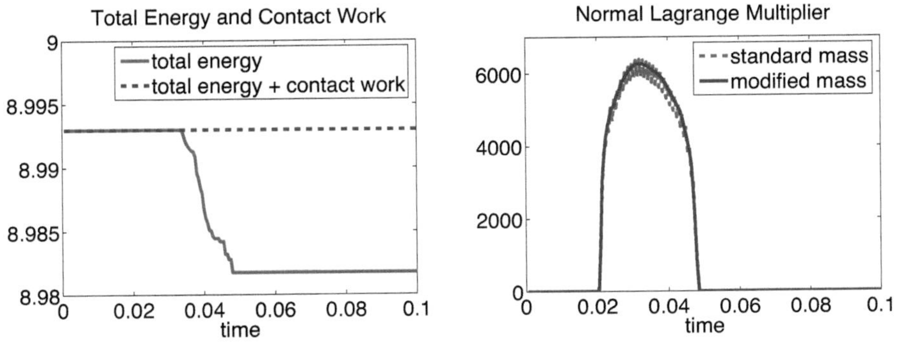

Figure 5.27: Bowl brick problem: total energy and normal Lagrange multiplier.

coefficient $\mathfrak{F}$ is set to 0.5. We assume the bowl to be the slave side with $E^{\text{sl}} = 6 \cdot 10^4$, $\nu^{\text{sl}} = 0.3$ and $\varrho^{\text{sl}} = 0.8$. For the brick, we set $E^{\text{ma}} = 3 \cdot 10^4$, $\nu^{\text{ma}} = 0.3$ and $\varrho^{\text{ma}} = 1.0$. We use the time step $\Delta t = 5 \cdot 10^{-4}$ and compute the discrete dynamical solution for $t \in (0, 0.1]$. Figure 5.26 shows the deformed configuration and the effective stress σ_{eff} at the time steps t_0, t_{50}, t_{75}, t_{100}, t_{150} and $t_{200} = 0.1$. The evolution of the energies (2.15) and the contact work (5.20) is presented in the left picture of Figure 5.27. Again, we observe no numerical dissipation. The right plot in Figure 5.27 shows the value of the Lagrange multipliers in normal direction at the lowest point of the bowl with respect to time. Due to the use of the modified mass matrix, no algorithmic oscillations occur.

As a final example, we compute the impact of an elastic torus with inner radius 3.5 and outer radius 6.5 onto a block with a wavelike surface, both discretized with hexahedral elements as shown in Figure 5.28. The block, which is assumed to be the slave side, is fixed at the bottom, whereas all other boundaries are free. We impose a gravitational volume force of $\mathbf{l} = (0, 0, -0.5)^T$. The initial velocity of the torus has a rotational and a translational part given by

$$\mathbf{v}_0(\mathbf{x}) = 30 \begin{pmatrix} 0 \\ 1 \\ 0 \end{pmatrix} \times \mathbf{x} + \begin{pmatrix} 15 \\ 0 \\ -30 \end{pmatrix}, \quad \mathbf{x} \in \Omega^{\text{ma}}.$$

Due to the large rotations, we employ a nonlinear Neo–Hooke material which is given by (1.9) with $c_m = 0$, $\nu = 0.3$, $E^{\text{ma}} = 5 \cdot 10^3$ for the torus and $E^{\text{sl}} = 10^6$ for the wave. The mass

Figure 5.28: Torus wave problem: effective stress at times t_{50}, t_{100}, t_{150} and t_{200}; upper row: $\mathfrak{F} = 0.02$; lower row: $\mathfrak{F} = 0.5$.

densities are $\varrho^{\mathrm{ma}} = 10^{-4}$, $\varrho^{\mathrm{sl}} = 10^{-2}$, and the mass modification is done using the quadrature rule Q^1 given in Figure 5.9. We compute 200 time steps with a step size of $\Delta t = 5 \cdot 10^{-4}$, use $\theta = 1$ and assemble the contact conditions in an updated Lagrangian manner.

Figure 5.28 shows the deformed geometry and the effective stress at several intermediate time steps for two different values of the friction coefficient $\mathfrak{F}$; the upper row is computed with a very low coefficient of $\mathfrak{F} = 0.02$, whereas the lower row uses $\mathfrak{F} = 0.5$. The difference in the results is clearly visible. In the former setting, only slippy nodes occur in tangential direction, such that the rotation hardly slows down and has only a weak influence on the translational movement. In contrast, the higher friction coefficient leads to sticky nodes in tangential direction, thus turning the rotational movement into a lateral one.

6 A priori error estimates

In the previous chapter, we have presented the construction of the modified mass matrix and have illustrated its benefit for the numerical treatment of dynamic contact problems. In this chapter, we investigate the a priori error of the resulting modified space discretization. Because we need sufficient regularity of the corresponding continuous problem later on, we assume the contact stress on Γ_C to be known and include them within the right hand side function f. In addition, we assume homogeneous Dirichlet boundary conditions on Γ_D for simplicity, i.e., $\mathbf{V}_D = \mathbf{V}_0$ with the definitions from (2.1).

For the characterization of the necessary regularity of $\mathbf{u}$, we use the following definitions from [55]: let $p \in [1, \infty]$ and a Banach space $\mathbf{X}$ be given. The time-dependent vector space $L^p(0, T; \mathbf{X})$ consists of those functions $\mathbf{u} : [0, T] \to \mathbf{X}$ such that the respective norm

$$\|\mathbf{u}\|_{L^p(0,T;\mathbf{X})} := \left(\int_0^T \|\mathbf{u}(t)\|_{\mathbf{X}}^p \, dt \right)^{1/p}, \quad p \in [1, \infty), \tag{6.1a}$$

$$\|\mathbf{u}\|_{L^\infty(0,T;\mathbf{X})} := \text{ess sup}_{t \in [0,T]} \|\mathbf{u}(t)\|_{\mathbf{X}}, \tag{6.1b}$$

is finite. Further, we define $W^{l,p}(0, T; \mathbf{X})$, $l \in \mathbb{N}_0$, as the space of all functions $\mathbf{u} \in L^p(0, T; \mathbf{X})$ with weak time derivatives $\frac{\partial^m \mathbf{u}}{\partial t^m} \in L^p(0, T; \mathbf{X})$, $m \in \{0, \dots, l\}$; the corresponding norm is given by

$$\|\mathbf{u}\|_{W^{l,p}(0,T;\mathbf{X})} := \left(\int_0^T \sum_{m=0}^{l} \left\| \frac{\partial^m \mathbf{u}}{\partial t^m}(t) \right\|_{\mathbf{X}}^p dt \right)^{1/p}, \quad p \in [1, \infty), \tag{6.2}$$

and analogeously to (6.1b) for $p = \infty$. For the case $\mathbf{X} = \mathbf{H}^k(\omega)$, we use the abbreviations $\| \cdot \|_{k,\omega} := \| \cdot \|_{\mathbf{H}^k(\omega)}$ for the static and $\| \cdot \|_{l;k,\omega} := \| \cdot \|_{H^l(0,T;\mathbf{H}^k(\omega))}$, $\| \cdot \|_{\infty;k,\omega} := \| \cdot \|_{L^\infty(0,T;\mathbf{H}^k(\omega))}$ for the time-dependent norms from now on.

With these definitions, we can formulate the continuous problem we consider in this chapter: find $\mathbf{u} \in L^2((0,T); \mathbf{V}_0)$ such that $\dot{\mathbf{u}} \in L^2((0,T); \mathbf{L}^2(\Omega))$, $\ddot{\mathbf{u}} \in L^2((0,T); \mathbf{V}_0')$ and

$$\begin{aligned} m(\ddot{\mathbf{u}}, \chi) + a(\mathbf{u}, \chi) &= f(\chi), && \chi \in \mathbf{V}_0, \quad t \in (0, T], \\ (\mathbf{u}|_{t=0}, \chi) &= (\mathbf{u}_0, \chi), && \chi \in \mathbf{V}_0, \\ (\dot{\mathbf{u}}|_{t=0}, \chi) &= (\mathbf{v}_0, \chi), && \chi \in \mathbf{V}_0. \end{aligned} \tag{6.3}$$

The rest of this chapter is organized as follows: After stating some preliminary lemmas, we prove a priori estimates for the semi-discrete system in Section 6.1 and for the fully discrete system in Section 6.2. Section 6.3 contains a numerical example confirming the error reduction properties proved in the previous sections.

In this chapter, c and C denote generic constants independent of the discretization parameters h, Δt, but possibly dependent on Ω, T or the material parameters.

6.1 Semi-discrete system

Using the modified mass bilinear form characterized by (5.11) and the finite element spaces introduced in Section 2.2, (6.3) is approximated by the semi-discrete problem: find $\mathbf{u} \in H^2((0,T), \mathbf{V}_0^h)$ such that

$$\begin{aligned} \bar{m}_i^h(\ddot{\mathbf{u}}^h, \boldsymbol{\chi}^h) + a(\mathbf{u}^h, \boldsymbol{\chi}^h) &= f(\boldsymbol{\chi}^h), \qquad \chi^h \in \mathbf{V}_0^h, \quad t \in (0,T], \\ \left(\mathbf{u}^h|_{t=0}, \boldsymbol{\chi}^h\right) &= \left(\mathbf{u}_0^h, \boldsymbol{\chi}^h\right), \quad \chi^h \in \mathbf{V}_0^h, \\ \left(\dot{\mathbf{u}}^h|_{t=0}, \boldsymbol{\chi}^h\right) &= \left(\mathbf{v}_0^h, \boldsymbol{\chi}^h\right), \quad \chi^h \in \mathbf{V}_0^h, \end{aligned} \tag{6.4}$$

with suitable discrete initial values $\mathbf{u}_0^h$, $\mathbf{v}_0^h$ specified below.

In order to obtain an a priori error estimate for the solution of (6.4), we proceed along the lines of [7] where the effect of quadrature errors on the numerical solution has been studied. A similar analysis for the lumped mass matrix method can be found in [160].

For the proof, we need a projection operator onto the discrete space $\mathbf{V}^h$ which preserves the homogeneous Dirichlet boundary conditions on Γ_D and allows for local approximation and stability estimates. An example is the Scott–Zhang operator Z^h proposed in [153] that satisfies for each element $K \in \mathcal{T}^h$:

$$0 \le l \le k \le 2,\ k > \frac{1}{2}: \quad \|Z^h\boldsymbol{\chi} - \boldsymbol{\chi}\|_{l,K} \le Ch^{k-l}|\boldsymbol{\chi}|_{k,\omega_K}, \quad \boldsymbol{\chi} \in \mathbf{H}^k(\Omega), \tag{6.5a}$$

$$|Z^h\boldsymbol{\chi}|_{1,K} \le C|\boldsymbol{\chi}|_{1,\omega_K}, \quad \boldsymbol{\chi} \in \mathbf{H}^1(\Omega), \tag{6.5b}$$

$$\|Z^h\boldsymbol{\chi}\|_{0,K} \le C\|\boldsymbol{\chi}\|_{0,\omega_K}, \quad \boldsymbol{\chi} \in \mathbf{L}^2(\Omega). \tag{6.5c}$$

where $\bar{\omega}_K := \bigcup\{\bar{T} : T \in \mathcal{T}^h,\ \bar{T} \cap \bar{K} \neq \emptyset\}$ is the patch of elements around K.

Using the layer-like domains Ω_{Cj}, $j = 1,2,3$, given in Figure 5.6, we obtain the following lemma on the approximation properties of the interpolation operator I_i^h, $i \in \{0,1\}$, introduced in (5.11):

Lemma 6.1. *Let I_i^h satisfy conditions P1 to P3 given in Section 5.3. Then the following inequalities hold:*

$$\|I_i^h\boldsymbol{\chi}^h - \boldsymbol{\chi}^h\|_{0,\Omega} \le Ch\|\boldsymbol{\chi}^h\|_{1,\Omega_{C2}}, \quad \boldsymbol{\chi}^h \in \mathbf{V}^h, \tag{6.6a}$$

$$\|I_i^h Z^h\boldsymbol{\chi} - \boldsymbol{\chi}\|_{0,\Omega_{C1}} \le Ch\|\boldsymbol{\chi}\|_{1,\Omega_{C3}}, \quad \boldsymbol{\chi} \in \mathbf{V}. \tag{6.6b}$$

Proof. The estimate (6.6a) follows directly from $P2$ and $P3$, whereas (6.6b) results from the triangle inequality, (6.5a), (6.5b) and (6.6a). □

Another tool we are going to use in the following is the Sobolev embedding theorem [40] which states that for an interval $[a,b] \subset \mathbb{R}$, $L^\infty([a,b])$ is continuously embedded in $H^1([a,b])$. A corollary of this theorem is the following lemma given in [46]:

Lemma 6.2. *Let $\boldsymbol{\eta} \in L^2(0,T;\mathbf{X})$ and $\dot{\boldsymbol{\eta}} \in L^2(0,T;\mathbf{X}')$ for some Banach space $\mathbf{X}$. Then we have $\boldsymbol{\eta} \in C^0([0,T];\mathbf{X})$ and*

$$\sup_{t\in[0,T]} \|\boldsymbol{\eta}(t)\|_{\mathbf{X}} \le C\Big(\|\boldsymbol{\eta}\|_{L^2(0,T;\mathbf{X})} + \|\dot{\boldsymbol{\eta}}\|_{L^2(0,T;\mathbf{X}')}\Big).$$

For functions $\boldsymbol{\chi}^h, \boldsymbol{\eta}^h \in \mathbf{V}^h$, we define the quadrature error of the modified bilinear form using (5.11)

$$\varepsilon_i^h(\boldsymbol{\chi}^h, \boldsymbol{\eta}^h) := \bar{m}_i^h(\boldsymbol{\chi}^h, \boldsymbol{\eta}^h) - m(\boldsymbol{\chi}^h, \boldsymbol{\eta}^h) = m(I_i^h\boldsymbol{\chi}^h, I_i^h\boldsymbol{\eta}^h) - m(\boldsymbol{\chi}^h, \boldsymbol{\eta}^h).$$

The next lemma gives a bound on this bilinear form:

Lemma 6.3. *Let $\boldsymbol{\eta} \in \mathbf{V}$, $\boldsymbol{\chi}^h \in \mathbf{V}^h$ and assume that I_i^h meets conditions P1 to P3. Then the following estimate holds:*

$$|\varepsilon_i^h(Z^h\boldsymbol{\eta}, \boldsymbol{\chi}^h)| \le Ch\left(\|\boldsymbol{\eta}\|_{1,\Omega_{C3}}\|\boldsymbol{\chi}^h\|_{0,\Omega_{C2}} + \|\boldsymbol{\eta}\|_{0,\Omega_{C2}}\|\boldsymbol{\chi}^h\|_{1,\Omega_{C2}}\right). \tag{6.7}$$

For $\boldsymbol{\eta}, \boldsymbol{\chi} \in \mathbf{H}^2(\Omega)$, we obtain

$$|\varepsilon_i^h(Z^h\boldsymbol{\eta}, Z^h\boldsymbol{\chi})| \le Ch^2\left(\|\boldsymbol{\eta}\|_{2,\Omega}\|\boldsymbol{\chi}\|_{1,\Omega} + \|\boldsymbol{\eta}\|_{1,\Omega}\|\boldsymbol{\chi}\|_{2,\Omega}\right). \tag{6.8}$$

Proof. Recalling condition $P2$ which implies that $(I_i^h - \mathrm{Id})\boldsymbol{\chi}^h$ vanishes on $\bar{\Omega}_I$ for $\boldsymbol{\chi}^h \in \mathbf{V}^h$, we consider (6.7):

$$\begin{aligned}|\varepsilon_i^h(Z^h\boldsymbol{\eta}, \boldsymbol{\chi}^h)| &= |m((I_i^h - \mathrm{Id})Z^h\boldsymbol{\eta}, I_i^h\boldsymbol{\chi}^h) + m(Z^h\boldsymbol{\eta}, (I_i^h - \mathrm{Id})\boldsymbol{\chi}^h)| \\ &\le C(\|(I_i^h - \mathrm{Id})Z^h\boldsymbol{\eta}\|_{0,\Omega_{C1}}\|I_i^h\boldsymbol{\chi}^h\|_{0,\Omega_{C1}} + \|Z^h\boldsymbol{\eta}\|_{0,\Omega_{C1}}\|(I_i^h - \mathrm{Id})\boldsymbol{\chi}^h\|_{0,\Omega_{C1}}) \\ &\le Ch\left(\|Z^h\boldsymbol{\eta}\|_{1,\Omega_{C2}}\|\boldsymbol{\chi}^h\|_{0,\Omega_{C2}} + \|\boldsymbol{\eta}\|_{0,\Omega_{C2}}\|\boldsymbol{\chi}^h\|_{1,\Omega_{C2}}\right) \\ &\le Ch\left(\|\boldsymbol{\eta}\|_{1,\Omega_{C3}}\|\boldsymbol{\chi}^h\|_{0,\Omega_{C2}} + \|\boldsymbol{\eta}\|_{0,\Omega_{C2}}\|\boldsymbol{\chi}^h\|_{1,\Omega_{C2}}\right),\end{aligned}$$

where we made use of $P1$, (6.5b), (6.5c) and (6.6a).

Now we turn to (6.8). Due to the Sobolev embedding theorem and standard arguments, we get for $\boldsymbol{\chi} \in \mathbf{V}$

$$\|\boldsymbol{\chi}\|_{0,\Omega_{Cj}}^2 \le Ch\|\boldsymbol{\chi}\|_{1,\Omega}^2, \quad j \in \{1,2,3\}.$$

Here, we have employed the fact that the width of Ω_{Cj} is bounded in terms of a constant multiple of h. Similarly, we obtain $\|\boldsymbol{\chi}\|_{1,\Omega_{C1}}^2 \le Ch\|\boldsymbol{\chi}\|_{2,\Omega}^2$ for $\boldsymbol{\chi} \in \mathbf{H}^2(\Omega)$ which leads to

$$\begin{aligned}|\varepsilon_i^h(Z^h\boldsymbol{\eta}, Z^h\boldsymbol{\chi})| &\le \|(I_i^h - \mathrm{Id})Z^h\boldsymbol{\eta}\|_{0,\Omega_{C1}}\|I_i^hZ^h\boldsymbol{\chi}\|_{0,\Omega_{C1}} \\ &\quad + \|Z^h\boldsymbol{\eta}\|_{0,\Omega_{C1}}\|(I_i^h - \mathrm{Id})Z^h\boldsymbol{\chi}\|_{0,\Omega_{C1}} \\ &\le Ch\left(\|Z^h\boldsymbol{\eta}\|_{1,\Omega_{C2}}\|Z^h\boldsymbol{\chi}\|_{0,\Omega_{C2}} + \|Z^h\boldsymbol{\eta}\|_{0,\Omega_{C1}}\|Z^h\boldsymbol{\chi}\|_{1,\Omega_{C2}}\right) \\ &\le Ch^2\left(\|\boldsymbol{\eta}\|_{2,\Omega}\|\boldsymbol{\chi}\|_{1,\Omega} + \|\boldsymbol{\eta}\|_{1,\Omega}\|\boldsymbol{\chi}\|_{2,\Omega}\right).\end{aligned}$$

□

Further, we state another auxiliary result for the time derivative of the discrete operator Z^h; the proof directly follows from the definition of Z^h given in [153] or Section 7.2.3.

Lemma 6.4. *For $\boldsymbol{\eta} \in \mathbf{V}$ we have $Z^h\dot{\boldsymbol{\eta}} = (\dot{Z^h\boldsymbol{\eta}})$.*

Now we are able to derive a priori bounds on the difference $\mathbf{u}^h - \mathbf{u}$. We start with a $\mathbf{H}^1(\Omega)$-estimate which can be shown for I_i^h satisfying conditions $P1$ to $P3$ without additional assumptions on the regularity of (6.3). To this end, we introduce $\mathbf{w}^h \in L^2(0,T;\mathbf{V}_0^h)$ as a suitable elliptic projection of $\mathbf{u}$ onto $\mathbf{V}_0^h$:

$$a(\mathbf{w}^h, \boldsymbol{\chi}^h) = f(\boldsymbol{\chi}^h) - \bar{m}_i^h(Z^h\ddot{\mathbf{u}}, \boldsymbol{\chi}^h), \quad \boldsymbol{\chi}^h \in \mathbf{V}_0^h, \quad 0 \le t \le T. \tag{6.9}$$

For $\mathbf{V}_0^h \subset \mathbf{V}_0$ and a $\mathbf{V}_0$-coercive bilinear form $a(\cdot,\cdot)$, this equation uniquely defines $\mathbf{w}^h$. Now we bound the difference between $\mathbf{u}$ and $\mathbf{w}^h$:

Lemma 6.5. *Choose I_i^h such that conditions P1 to P3 hold. If the solution* $\mathbf{u}$ *of (6.3) meets the requirements* $\frac{\partial^k \mathbf{u}}{\partial t^k} \in L^2(0,T;\mathbf{H}^2(\Omega))$, $\frac{\partial^{k+2}\mathbf{u}}{\partial t^{k+2}} \in L^2(0,T;\mathbf{V})$ *for some* $k \in \{0,1,2\}$, *then the* k*-th time derivative of the solution* $\mathbf{w}^h$ *of (6.9) satisfies*

$$\left\| \frac{\partial^k \mathbf{w}^h}{\partial t^k} - \frac{\partial^k \mathbf{u}}{\partial t^k} \right\|_{0;1,\Omega} \leq Ch \left(\left\| \frac{\partial^k \mathbf{u}}{\partial t^k} \right\|_{0;2,\Omega} + \left\| \frac{\partial^{k+2} \mathbf{u}}{\partial t^{k+2}} \right\|_{0;1,\Omega} \right). \tag{6.10}$$

Proof. We only give the proof for $k=0$, as the other estimates follow with Lemma 6.4 and differentiation with respect to time.

We start with (6.3) and (6.9) for a test function $\boldsymbol{\chi}^h \in L^2(0,T;\mathbf{V}_0^h)$:

$$\begin{aligned} a(\mathbf{w}^h - Z^h\mathbf{u}, \boldsymbol{\chi}^h) &= a(\mathbf{u} - Z^h\mathbf{u}, \boldsymbol{\chi}^h) + m(\ddot{\mathbf{u}} - Z^h\ddot{\mathbf{u}}, \boldsymbol{\chi}^h) \\ &\quad + m(Z^h\ddot{\mathbf{u}}, \boldsymbol{\chi}^h) - \bar{m}_i^h(Z^h\ddot{\mathbf{u}}, \boldsymbol{\chi}^h). \end{aligned} \tag{6.11}$$

We choose $\boldsymbol{\chi}^h = \mathbf{w}^h - Z^h\mathbf{u}$; then we get with the $\mathbf{V}_0$-coercivity of $a(\cdot,\cdot)$, the continuity of $m(\cdot,\cdot)$, integration from 0 to T and the division of both sides by $\|\mathbf{w}^h - Z^h\mathbf{u}\|_{0;1,\Omega}$:

$$\begin{aligned} \|\mathbf{w}^h - Z^h\mathbf{u}\|_{0;1,\Omega} &\leq C\left(\|\ddot{\mathbf{u}} - Z^h\ddot{\mathbf{u}}\|_{0;0,\Omega} + \|\mathbf{u} - Z^h\mathbf{u}\|_{0;1,\Omega} \right) \\ &\quad + \sup_{\boldsymbol{\chi}^h \in L^2(0,T;\mathbf{V}_0^h)\setminus\{\mathbf{0}\}} \frac{C}{\|\boldsymbol{\chi}^h\|_{0;1,\Omega}} \int_0^T \left| \varepsilon^h(Z^h\ddot{\mathbf{u}}, \boldsymbol{\chi}^h) \right| dt. \end{aligned}$$

With Lemma 6.3 and the Cauchy–Schwarz inequality, we obtain

$$\int_0^T \left| \varepsilon^h(Z^h\ddot{\mathbf{u}}, \boldsymbol{\chi}^h) \right| dt \leq Ch \|\ddot{\mathbf{u}}\|_{0;1,\Omega} \|\boldsymbol{\chi}^h\|_{0;1,\Omega}.$$

In terms of the approximation properties (6.5a) and the above estimates, we find

$$\|\mathbf{w}^h - \mathbf{u}\|_{0;1,\Omega} \leq \|\mathbf{w}^h - Z^h\mathbf{u}\|_{0;1,\Omega} + \|Z^h\mathbf{u} - \mathbf{u}\|_{0;1,\Omega} \leq Ch\left(\|\mathbf{u}\|_{0;2,\Omega} + \|\ddot{\mathbf{u}}\|_{0;1,\Omega} \right).$$

□

With this lemma, we are able to prove the following result for the $\mathbf{H}^1(\Omega)$-norm of the error:

Theorem 6.6. *Assume that the solution* $\mathbf{u}$ *of (6.3) meets the smoothness requirements* $\mathbf{u}, \dot{\mathbf{u}}, \ddot{\mathbf{u}} \in L^2(0,T;\mathbf{H}^2(\Omega))$ *and* $\frac{\partial^3 \mathbf{u}}{\partial t^3}, \frac{\partial^4 \mathbf{u}}{\partial t^4} \in L^2(0,T;\mathbf{V})$. *Let* I_i^h *satisfy P1 to P3. Take the initial conditions* $\mathbf{u}^h(0) = Z^h\mathbf{u}_0$ *and* $\dot{\mathbf{u}}^h(0) = Z^h\mathbf{v}_0$. *Then the following estimate holds:*

$$\left\|\dot{\mathbf{u}}^h - \dot{\mathbf{u}}\right\|_{\infty;0,\Omega_I} + \|\mathbf{u}^h - \mathbf{u}\|_{\infty;1,\Omega} \leq Ch\left(\sum_{s=0}^{2} \left\| \frac{\partial^s \mathbf{u}}{\partial t^s} \right\|_{0;2,\Omega} + \sum_{s=3}^{4} \left\| \frac{\partial^s \mathbf{u}}{\partial t^s} \right\|_{0;1,\Omega} \right). \tag{6.12}$$

Proof. Let $\mathbf{u}^h, \mathbf{w}^h$ be defined by (6.4) and (6.9), respectively. Setting $\boldsymbol{\theta}^h := \mathbf{u}^h - \mathbf{w}^h$ gives for any test function $\boldsymbol{\chi}^h \in L^2(0,T;\mathbf{V}_0^h)$:

$$\bar{m}_i^h(\ddot{\boldsymbol{\theta}}^h, \boldsymbol{\chi}^h) + a(\boldsymbol{\theta}^h, \boldsymbol{\chi}^h) = \bar{m}_i^h(Z^h\ddot{\mathbf{u}}, \boldsymbol{\chi}^h) - \bar{m}_i^h(\ddot{\mathbf{w}}^h, \boldsymbol{\chi}^h). \tag{6.13}$$

We choose $\boldsymbol{\chi}^h = \dot{\boldsymbol{\theta}}^h$ in (6.13) and thus obtain

$$\frac{1}{2}\frac{d}{dt}\left(|\dot{\boldsymbol{\theta}}^h|_{h,i}^2 + a(\boldsymbol{\theta}^h, \boldsymbol{\theta}^h) \right) = \bar{m}_i^h(Z^h\ddot{\mathbf{u}} - \ddot{\mathbf{w}}^h, \dot{\boldsymbol{\theta}}^h),$$

with $|\cdot|_{h,i} = \sqrt{\bar{m}_i^h(\cdot,\cdot)}$ as defined in Remark 5.4. Using the Cauchy–Schwarz and Young's inequality and adding the nonnegative term $a(\boldsymbol{\theta}^h, \boldsymbol{\theta}^h)$ on the right hand side, we arrive at

$$\frac{d}{dt}\left(|\dot{\boldsymbol{\theta}}^h|_{h,i}^2 + a(\boldsymbol{\theta}^h, \boldsymbol{\theta}^h)\right) \leq C|Z^h\ddot{\mathbf{u}} - \ddot{\mathbf{w}}^h|_{h,i}^2 + \left(|\dot{\boldsymbol{\theta}}^h|_{h,i}^2 + a(\boldsymbol{\theta}^h, \boldsymbol{\theta}^h)\right), \tag{6.14}$$

From (6.14), we conclude with Gronwall's lemma, the seminorm equivalence (5.12) and the $\mathbf{V}_0$-coercivity of $a(\cdot,\cdot)$

$$\begin{aligned}\|\dot{\boldsymbol{\theta}}^h\|_{0,\Omega_I}^2 + \|\boldsymbol{\theta}^h\|_{1,\Omega}^2 &\leq C\left(|\dot{\boldsymbol{\theta}}^h|_{h,i}^2 + a(\boldsymbol{\theta}^h, \boldsymbol{\theta}^h)\right) \\ &\leq C\left(\|\dot{\boldsymbol{\theta}}^h(0)\|_{0,\Omega_I}^2 + \|\boldsymbol{\theta}^h(0)\|_{1,\Omega}^2 + \|Z^h\ddot{\mathbf{u}} - \ddot{\mathbf{w}}^h\|_{0;0,\Omega_I}^2\right).\end{aligned} \tag{6.15}$$

This inequality holds for all $t \in (0, T]$ and hence also for the supremum, such that it only remains to bound the terms on the right hand side. Firstly, we obtain

$$\begin{aligned}\|Z^h\ddot{\mathbf{u}} - \ddot{\mathbf{w}}^h\|_{0;0,\Omega_I} &\leq \|Z^h\ddot{\mathbf{u}} - \ddot{\mathbf{u}}\|_{0;0,\Omega_I} + \|\ddot{\mathbf{u}} - \ddot{\mathbf{w}}^h\|_{0;0,\Omega_I} \\ &\leq Ch\left(\|\ddot{\mathbf{u}}\|_{0;2,\Omega} + \left\|\frac{\partial^4\mathbf{u}}{\partial t^4}\right\|_{0;1,\Omega}\right),\end{aligned}$$

where we made use of (6.5a) and Lemma 6.5 for $k = 2$. Secondly, we get

$$\begin{aligned}\|\dot{\boldsymbol{\theta}}^h(0)\|_{0,\Omega_I} &\leq \|Z^h\mathbf{v}_0 - \mathbf{v}_0\|_{0,\Omega_I} + \|\mathbf{v}_0 - \dot{\mathbf{w}}^h(0)\|_{0,\Omega_I} \\ &\leq C\left(h\|\mathbf{v}_0\|_{1,\Omega} + \|\dot{\mathbf{u}} - \dot{\mathbf{w}}^h\|_{0;0,\Omega_I} + \|\ddot{\mathbf{u}} - \ddot{\mathbf{w}}^h\|_{0;0,\Omega_I}\right) \\ &\leq Ch\left(\sum_{s=1}^{2}\left\|\frac{\partial^s\mathbf{u}}{\partial t^s}\right\|_{0;2,\Omega} + \sum_{s=3}^{4}\left\|\frac{\partial^s\mathbf{u}}{\partial t^s}\right\|_{0;1,\Omega}\right),\end{aligned} \tag{6.16}$$

where we applied Lemmas 6.2 and 6.5. Finally, we estimate similarly:

$$\|\boldsymbol{\theta}^h(0)\|_{1,\Omega} \leq Ch\left(\sum_{s=0}^{1}\left\|\frac{\partial^s\mathbf{u}}{\partial t^s}\right\|_{0;2,\Omega} + \sum_{s=2}^{3}\left\|\frac{\partial^s\mathbf{u}}{\partial t^s}\right\|_{0;1,\Omega}\right). \tag{6.17}$$

Inserting all these results into (6.15), we conclude

$$\|\dot{\boldsymbol{\theta}}^h\|_{\infty;0,\Omega_I} + \|\boldsymbol{\theta}^h\|_{\infty;1,\Omega} \leq Ch\left(\sum_{s=0}^{2}\left\|\frac{\partial^s\mathbf{u}}{\partial t^s}\right\|_{0;2,\Omega} + \sum_{s=3}^{4}\left\|\frac{\partial^s\mathbf{u}}{\partial t^s}\right\|_{0;1,\Omega}\right). \tag{6.18}$$

Now we recall $\mathbf{u}^h - \mathbf{u} = \boldsymbol{\theta}^h + \mathbf{w}^h - \mathbf{u}$ and get

$$\begin{aligned}&\|\dot{\mathbf{u}}^h - \dot{\mathbf{u}}\|_{\infty;0,\Omega_I} + \|\mathbf{u}^h - \mathbf{u}\|_{\infty;1,\Omega} \\ &\leq \|\dot{\boldsymbol{\theta}}^h\|_{\infty;0,\Omega_I} + \|\boldsymbol{\theta}^h\|_{\infty;1,\Omega} + \|\dot{\mathbf{u}} - \dot{\mathbf{w}}^h\|_{\infty;0,\Omega_I} + \|\mathbf{u} - \mathbf{w}^h\|_{\infty;1,\Omega}.\end{aligned}$$

Due to Lemmas 6.2 and 6.5, we can estimate $\|\mathbf{u} - \mathbf{w}^h\|_{\infty;1,\Omega}$ and $\|\dot{\mathbf{u}} - \dot{\mathbf{w}}^h\|_{\infty;0,\Omega_I}$ such that we arrive at (6.12). □

Next, we turn to the estimate in the $\mathbf{L}^2(\Omega)$-norm, where we need a further assumption, namely the $\mathbf{H}^2(\Omega)$-regularity of the boundary value problem (6.3) [21].

Lemma 6.7. *Let (6.3) be* $\mathbf{H}^2(\Omega)$*-regular and assume* $\frac{\partial^k \mathbf{u}}{\partial t^k}, \frac{\partial^{k+2}\mathbf{u}}{\partial t^{k+2}} \in L^2(0,T;\mathbf{H}^2(\Omega))$ *for some* $k \in \{0,1\}$*. Let the interpolation operator* I_i^h *satisfy conditions P1 to P3. Then the following estimate holds:*

$$\left\| \frac{\partial^k \mathbf{w}^h}{\partial t^k} - \frac{\partial^k \mathbf{u}}{\partial t^k} \right\|_{0;0,\Omega} \le Ch^2 \left(\left\| \frac{\partial^k \mathbf{u}}{\partial t^k} \right\|_{0;2,\Omega} + \left\| \frac{\partial^{k+2}\mathbf{u}}{\partial t^{k+2}} \right\|_{0;2,\Omega} \right). \tag{6.19}$$

Proof. Again, it suffices to consider the case $k = 0$. We start with the observation that for $\boldsymbol{\psi} \in L^2(0,T;\mathbf{L}^2(\Omega))$, we get

$$\|\boldsymbol{\psi}\|_{0;0,\Omega} = \sup_{\mathbf{g}\in L^2(0,T;\mathbf{L}^2(\Omega))\setminus\{0\}} \frac{\left|\int_0^T (\boldsymbol{\psi},\mathbf{g})dt\right|}{\|\mathbf{g}\|_{0;0,\Omega}}, \tag{6.20}$$

as $L^2(0,T;\mathbf{L}^2(\Omega))$ is its own dual space. Now we choose $\boldsymbol{\varphi} \in L^2(0,T;\mathbf{V}_0)$ as the solution of the following auxiliary problem for any fixed $\mathbf{g} \in L^2(0,T;\mathbf{L}^2(\Omega))$:

$$a(\boldsymbol{\varphi},\mathbf{v}) = (\mathbf{g},\mathbf{v}), \quad \mathbf{v} \in \mathbf{V}_0, \quad 0 \le t \le T. \tag{6.21}$$

By our assumptions on the regularity of the boundary value problem (6.21), we obtain $\boldsymbol{\varphi} \in L^2(0,T;\mathbf{V}_0 \cap \mathbf{H}^2(\Omega))$ (see [21]) and

$$\|\boldsymbol{\varphi}\|_{0;2,\Omega} \le C\|\mathbf{g}\|_{0;0,\Omega}. \tag{6.22}$$

If we choose $\mathbf{v} = \mathbf{w}^h - \mathbf{u}$ in (6.21) and use (6.11), we get for $\boldsymbol{\chi}^h \in L^2(0,T;\mathbf{V}_0^h)$:

$$(\mathbf{w}^h - \mathbf{u},\mathbf{g}) = a(\mathbf{w}^h - \mathbf{u}, \boldsymbol{\varphi} - \boldsymbol{\chi}^h) + m(\ddot{\mathbf{u}} - Z^h\ddot{\mathbf{u}}, \boldsymbol{\chi}^h) + \varepsilon_i^h(Z^h\ddot{\mathbf{u}}, \boldsymbol{\chi}^h).$$

Now we take $\boldsymbol{\chi}^h = Z^h\boldsymbol{\varphi}$ and integrate the above equality from 0 to T:

$$\left|\int_0^T (\mathbf{w}^h - \mathbf{u},\mathbf{g})dt\right| \le \int_0^T \left|a(\mathbf{w}^h - \mathbf{u}, \boldsymbol{\varphi} - Z^h\boldsymbol{\varphi})\right| dt + \int_0^T \left|m(\ddot{\mathbf{u}} - Z^h\ddot{\mathbf{u}}, Z^h\boldsymbol{\varphi})\right| dt$$
$$+ \int_0^T \left|\varepsilon_i^h(Z^h\ddot{\mathbf{u}}, Z^h\boldsymbol{\varphi})\right| dt. \tag{6.23}$$

For the first term on the right hand side, we get with the continuity of $a(\cdot,\cdot)$, the approximation property of Z^h (6.5a) and Lemma 6.5:

$$\int_0^T \left|a(\mathbf{w}^h - \mathbf{u}, \boldsymbol{\varphi} - Z^h\boldsymbol{\varphi})\right| dt \le Ch^2 \left(\|\mathbf{u}\|_{0;2,\Omega} + \|\ddot{\mathbf{u}}\|_{0;1,\Omega}\right) \|\boldsymbol{\varphi}\|_{0;2,\Omega}.$$

The second and third term are estimated by means of (6.5a), (6.5c) and Lemma 6.3:

$$\int_0^T \left|m(\ddot{\mathbf{u}} - Z^h\ddot{\mathbf{u}}, Z^h\boldsymbol{\varphi})\right| dt + \int_0^T \left|\varepsilon_i^h(Z^h\ddot{\mathbf{u}}, Z^h\boldsymbol{\varphi})\right| dt \le Ch^2\|\ddot{\mathbf{u}}\|_{0;2,\Omega}\|\boldsymbol{\varphi}\|_{0;2,\Omega}.$$

Using (6.20), (6.22) and these estimates, we can conclude

$$\begin{aligned}\|\mathbf{w}^h - \mathbf{u}\|_{0;0,\Omega} &= \sup_{\mathbf{g}\in L^2(0,T;\mathbf{L}^2(\Omega))\setminus\{0\}} \frac{\left|\int_0^T (\mathbf{w}^h - \mathbf{u},\mathbf{g})dt\right|}{\|\mathbf{g}\|_{0;0,\Omega}} \\ &\le Ch^2 \left(\|\mathbf{u}\|_{0;2,\Omega} + \|\ddot{\mathbf{u}}\|_{0;2,\Omega}\right).\end{aligned}$$

□

This lemma implies the following estimate:

Theorem 6.8. *Let the problem (6.3) be $\mathbf{H}^2(\Omega)$-regular and assume $\frac{\partial^s \mathbf{u}}{\partial t^s} \in L^2(0,T;\mathbf{H}^2(\Omega))$ for all $s \in \{0,\dots,3\}$, $\frac{\partial^4 \mathbf{u}}{\partial t^4} \in L^2(0,T;\mathbf{V})$. Choose I_i^h according to P1 to P3 and take the initial conditions $\mathbf{u}^h(0) = Z^h\mathbf{u}_0$ and $\dot{\mathbf{u}}^h(0) = Z^h\mathbf{v}_0$. Then the following inequality holds:*

$$\|\mathbf{u}^h - \mathbf{u}\|_{\infty;0,\Omega} \leq Ch^2 \left(\sum_{s=0}^{3} \left\| \frac{\partial^s \mathbf{u}}{\partial t^s} \right\|_{0;2,\Omega} + \left\| \frac{\partial^4 \mathbf{u}}{\partial t^4} \right\|_{0;1,\Omega} \right). \tag{6.24}$$

Proof. Following the proof of [7, Theorem 4.1], using the seminorm equivalence (5.12) and Lemma 6.7, we find

$$\|\mathbf{u}^h - \mathbf{u}\|_{\infty;0,\Omega_I} \leq Ch^2 \sum_{s=0}^{3} \left\| \frac{\partial^s \mathbf{u}}{\partial t^s} \right\|_{0;2,\Omega}. \tag{6.25}$$

A Poincaré-type estimate yields for $\mathbf{v} \in \mathbf{V}$

$$\|\mathbf{v}\|_{0,\Omega_{C1}} \leq C(\|\mathbf{v}\|_{0,\Omega_I} + h|\mathbf{v}|_{1,\Omega}). \tag{6.26}$$

Equation (6.25) and Theorem 6.6 then conclude the proof.

□

Remark 6.9. The regularity assumptions posed on the exact solution $\mathbf{u}$ in Theorem 6.8 are a little more strict than those from [7, Theorem 4.1]. For Theorem 6.6, we need less regularity for the higher time derivatives of $\mathbf{u}$ than [7, Theorem 4.2], but we can only bound the $\mathbf{L}^2(\Omega_I)$-norm of the error $(\dot{\mathbf{u}} - \dot{\mathbf{u}}^h)$ instead of the norm on the full domain Ω. Still, Theorem 6.8 provides an error estimate of $\mathcal{O}(h^2)$ with respect to the $\mathbf{L}_2(\Omega)$-norm.

6.2 Fully discrete system

In this section, we look at the time-discretized version of (6.4) using the trapezoidal rule as in (5.18). Let $M\Delta t := T$ for some integer M. Set $\mathbf{w}_j := \mathbf{w}(t_j)$ for $j \in \mathbb{N}_0$, $\mathbf{w} \in C(\bar{\Omega} \times [0,T])$, $t_j = j\Delta t$, and denote, similar as in Section 2.3,

$$\partial_{\Delta t}\mathbf{w}_{j+1} := \frac{1}{\Delta t}(\mathbf{w}_{j+1} - \mathbf{w}_j), \qquad \mathbf{w}_{j+1/2} := \frac{1}{2}(\mathbf{w}_{j+1} + \mathbf{w}_j). \tag{6.27}$$

The fully discrete problem we consider reads: find sequences $(\mathbf{u}_j^h)_{j=0}^M$, $(\mathbf{v}_j^h)_{j=0}^M$ in $\mathbf{V}_0^h$ such that $\mathbf{u}_0^h = Z^h\mathbf{u}_0$, $\mathbf{v}_0^h = Z^h\mathbf{v}_0$ and

$$\begin{aligned} m_i^h(\partial_{\Delta t}\mathbf{v}_{j+1}^h, \boldsymbol{\chi}^h) + a(\mathbf{u}_{j+1/2}^h, \boldsymbol{\chi}^h) &= (\mathbf{f}_{j+1/2}, \boldsymbol{\chi}^h), \quad \boldsymbol{\chi}^h \in \mathbf{V}_0^h, \quad 0 \leq j \leq M-1, \\ \partial_{\Delta t}\mathbf{u}_{j+1}^h &= \mathbf{v}_{j+1/2}^h. \end{aligned} \tag{6.28}$$

The main results are the following two theorems which are proved afterwards in the remaining part of this section.

Theorem 6.10. *Under the assumptions of Theorem 6.6, we have*

$$\max_{0\le j\le M-1}\left(\|\partial_{\Delta t}(\mathbf{u}_{j+1}^h-\mathbf{u}_{j+1})\|_{0,\Omega_I}+\|\mathbf{u}_{j+1/2}^h-\mathbf{u}_{j+1/2}\|_{1,\Omega}\right) \tag{6.29}$$
$$\le C\left(h\left(\sum_{s=0}^{2}\left\|\frac{\partial^s\mathbf{u}}{\partial t^s}\right\|_{0;2,\Omega}+\sum_{s=3}^{4}\left\|\frac{\partial^s\mathbf{u}}{\partial t^s}\right\|_{0;1,\Omega}\right)+\Delta t^2\sum_{s=3}^{4}\left\|\frac{\partial^s\mathbf{u}}{\partial t^s}\right\|_{0;0,\Omega}\right).$$

Theorem 6.11. *Let the assumptions of Theorem 6.8 be satisfied and let $\frac{\partial^4\mathbf{u}}{\partial t^4}\in L^2(0,T;\mathbf{H}^2(\Omega))$. Then the following estimate holds:*

$$\max_{0\le j\le M}\|\mathbf{u}_j^h-\mathbf{u}_j\|_{0,\Omega_I}\le C\left(h^2\sum_{s=0}^{3}\left\|\frac{\partial^s\mathbf{u}}{\partial t^s}\right\|_{0;2,\Omega}+h^2\Delta t^2\left\|\frac{\partial^4\mathbf{u}}{\partial t^4}\right\|_{0;2,\Omega}+\Delta t^2\sum_{s=3}^{4}\left\|\frac{\partial^s\mathbf{u}}{\partial t^s}\right\|_{0;0,\Omega}\right). \tag{6.30}$$

The $\mathbf{L}_2(\Omega)$-error can be bound as follows:

Corollary 6.12. *Under the assumptions of Theorems 6.10, 6.11, we obtain*

$$\max_{0\le j\le M-1}\|\mathbf{u}_{j+1/2}^h-\mathbf{u}_{j+1/2}\|_{0,\Omega} \tag{6.31}$$
$$\le C\left(h^2\sum_{s=0}^{3}\left\|\frac{\partial^s\mathbf{u}}{\partial t^s}\right\|_{0;2,\Omega}+h^2\left\|\frac{\partial^4\mathbf{u}}{\partial t^4}\right\|_{0;1,\Omega}+h^2\Delta t^2\left\|\frac{\partial^4\mathbf{u}}{\partial t^4}\right\|_{0;2,\Omega}+\Delta t^2\sum_{s=3}^{4}\left\|\frac{\partial^s\mathbf{u}}{\partial t^s}\right\|_{0;0,\Omega}\right).$$

Proof. The result follows from the Poincaré-type estimate (6.26), the triangle inequality and Theorems 6.10, 6.11. □

In order to prove Theorem 6.10, we need some technical lemmas and further notation. We set $\boldsymbol{\theta}^h := Z^h\mathbf{u}-\mathbf{w}^h$ and write

$$\boldsymbol{\xi}_j^h := \mathbf{u}_j^h-\mathbf{w}_j^h,\quad 0\le j\le M. \tag{6.32}$$

Further, we define sequences $(\boldsymbol{\rho}_j^h)_{j=0}^{M-1}$, $(\boldsymbol{\pi}_j^h)_{j=0}^{M-1}$, $(\boldsymbol{\varepsilon}_j^h)_{j=0}^{M-1}$ and $(\boldsymbol{\nu}_j^h)_{j=0}^{M-2}$ in $\mathbf{V}^h$ by

$$\boldsymbol{\rho}_j^h := Z^h\ddot{\mathbf{u}}_{j+1/2}-\partial_{\Delta t}\left(Z^h\dot{\mathbf{u}}_{j+1}\right), \tag{6.33a}$$
$$\boldsymbol{\pi}_j^h := \partial_{\Delta t}(Z^h\mathbf{u}_{j+1})-Z^h\dot{\mathbf{u}}_{j+1/2}, \tag{6.33b}$$
$$\boldsymbol{\varepsilon}_0^h := \partial_{\Delta t}\boldsymbol{\theta}_1^h-\boldsymbol{\pi}_0^h+\frac{\Delta t}{2}\boldsymbol{\rho}_0^h, \tag{6.33c}$$
$$\boldsymbol{\varepsilon}_j^h := \partial_{\Delta t}\boldsymbol{\theta}_{j+1}^h-\boldsymbol{\pi}_j^h+\frac{\Delta t}{2}\left(\sum_{k=0}^{j}\boldsymbol{\rho}_k^h+\sum_{k=0}^{j-1}\boldsymbol{\rho}_k^h\right),\quad 1\le j\le M-1, \tag{6.33d}$$
$$\boldsymbol{\nu}_j^h := \partial_{\Delta t}(\partial_{\Delta t}\boldsymbol{\theta}_{j+2}^h)-\partial_{\Delta t}\boldsymbol{\pi}_{j+1}^h+\boldsymbol{\rho}_{j+1/2}^h. \tag{6.33e}$$

The following two Lemmas are used for the proof of Theorem 6.10:

Lemma 6.13. *The following inequality holds:*

$$\max_{0\le j\le (M-1)}\left(\|\partial_{\Delta t}\boldsymbol{\xi}_{j+1}^h\|_{0,\Omega_I}^2+\|\boldsymbol{\xi}_{j+1/2}^h\|_{1,\Omega}^2\right)$$
$$\le C\left(\Delta t\sum_{j=0}^{M-2}\|\boldsymbol{\nu}_j^h\|_{0,\Omega_I}^2+\|\partial_{\Delta t}\boldsymbol{\xi}_1^h\|_{0,\Omega_I}^2+\|\boldsymbol{\xi}_{1/2}^h\|_{1,\Omega}^2\right). \tag{6.34}$$

Proof. Using the definitions, we obtain (for details see [7, proof of Lemma 5.1]):

$$m_i^h\left(\partial_{\Delta t}(\partial_{\Delta t}\boldsymbol{\xi}_{j+2}^h), \boldsymbol{\chi}^h\right) + \frac{1}{2}a\left(\boldsymbol{\xi}_{j+1/2}^h + \boldsymbol{\xi}_{j+3/2}^h, \boldsymbol{\chi}^h\right) = m_i^h\left(\boldsymbol{\nu}_j^h, \boldsymbol{\chi}^h\right). \tag{6.35}$$

We choose the test function $\boldsymbol{\chi}^h = \partial_{\Delta t}\boldsymbol{\xi}_{j+3/2}^h \in \mathbf{V}^h$ in (6.35) and arrive at

$$\begin{aligned}&|\partial_{\Delta t}\boldsymbol{\xi}_{j+2}^h|_{h,i}^2 - |\partial_{\Delta t}\boldsymbol{\xi}_{j+1}^h|_{h,i}^2 + a\left(\boldsymbol{\xi}_{j+3/2}^h, \boldsymbol{\xi}_{j+3/2}^h\right) - a\left(\boldsymbol{\xi}_{j+1/2}^h, \boldsymbol{\xi}_{j+1/2}^h\right)\\ &\le 2\Delta t\, m_i^h\left(\boldsymbol{\nu}_j^h, \partial_{\Delta t}\boldsymbol{\xi}_{j+3/2}^h\right).\end{aligned}$$

Summing from 0 to $(l-1)$ for $1 \le l \le (M-1)$ and the use of Young's inequality gives for any $\alpha > 0$:

$$\begin{aligned}&|\partial_{\Delta t}\boldsymbol{\xi}_{l+1}^h|_{h,i}^2 - |\partial_{\Delta t}\boldsymbol{\xi}_1^h|_{h,i}^2 + a\left(\boldsymbol{\xi}_{l+1/2}^h, \boldsymbol{\xi}_{l+1/2}^h\right) - a\left(\boldsymbol{\xi}_{1/2}^h, \boldsymbol{\xi}_{1/2}^h\right)\\ &\le 2\Delta t\left(\alpha\sum_{j=0}^{l-1}|\boldsymbol{\nu}_j^h|_{h,i}^2 + \frac{1}{\alpha}\sum_{j=0}^{l-1}\left|\partial_{\Delta t}\boldsymbol{\xi}_{j+3/2}^h\right|_{h,i}^2\right).\end{aligned} \tag{6.36}$$

We set $A := \max_{0\le j\le(M-1)}\left|\partial_{\Delta t}\boldsymbol{\xi}_{j+1}^h\right|_{h,i}$ which leads to

$$\sum_{j=0}^{l-1}\left|\partial_{\Delta t}\boldsymbol{\xi}_{j+3/2}^h\right|_{h,i}^2 \le A^2 M. \tag{6.37}$$

Choosing $\alpha = 4T$ in (6.36) and using (6.37), we get

$$|\partial_{\Delta t}\boldsymbol{\xi}_{l+1}^h|_{h,i}^2 + a\left(\boldsymbol{\xi}_{l+1/2}^h, \boldsymbol{\xi}_{l+1/2}^h\right) \le 8\Delta t T\sum_{j=0}^{M-2}|\boldsymbol{\nu}_j^h|_{h,i}^2 + \frac{1}{2}A^2 + |\partial_{\Delta t}\boldsymbol{\xi}_1^h|_{h,i}^2 + a\left(\boldsymbol{\xi}_{1/2}^h, \boldsymbol{\xi}_{1/2}^h\right).$$

This holds for each l with $1 \le l \le (M-1)$, hence we can take the maximum of the left hand side. Subtracting $\frac{1}{2}A^2$ on both sides as well as using the $\mathbf{V}_0$-coercivity, the continuity of $a(\cdot,\cdot)$ and the seminorm equivalence (5.12), we finally arrive at (6.34). □

The estimation of the first term on the right hand side of (6.34) is carried out in the next lemma:

Lemma 6.14. *Let the solution* $\mathbf{u}$ *of (6.3) satisfy* $\ddot{\mathbf{u}} \in L^2(0,T;\mathbf{H}^2(\Omega))$, $\frac{\partial^4\mathbf{u}}{\partial t^4} \in L^2(0,T;\mathbf{V})$ *and let* I_i^h *fulfill conditions P1 to P3. Then the following inequality holds:*

$$\Delta t\sum_{j=0}^{M-2}\|\boldsymbol{\nu}_j^h\|_{0,\Omega_I}^2 \le C\left(h^2\left(\|\ddot{\mathbf{u}}\|_{0;2,\Omega}^2 + \left\|\frac{\partial^4\mathbf{u}}{\partial t^4}\right\|_{0;1,\Omega}^2\right) + \Delta t^4\left\|\frac{\partial^4\mathbf{u}}{\partial t^4}\right\|_{0;0,\Omega}^2\right). \tag{6.38}$$

Proof. By the definition of $\boldsymbol{\nu}_j^h$ (6.33e), we get

$$\|\boldsymbol{\nu}_j^h\|_{0,\Omega_I}^2 \le \|\partial_{\Delta t}(\partial_{\Delta t}\boldsymbol{\theta}_{j+2}^h)\|_{0,\Omega_I}^2 + \|\boldsymbol{\rho}_{j+1/2}^h - \partial_{\Delta t}\boldsymbol{\pi}_{j+1}^h\|_{0,\Omega_I}^2. \tag{6.39}$$

Using Taylor expansion, we obtain for the first term on the right hand side (see [7, proof of Lemma 5.2] for details):

$$\begin{aligned}\partial_{\Delta t}(\partial_{\Delta t}\boldsymbol{\theta}_{j+2}^h) &= \frac{1}{\Delta t^2}\left(\boldsymbol{\theta}_{j+2}^h - 2\boldsymbol{\theta}_{j+1}^h + \boldsymbol{\theta}_j^h\right)\\ &= \frac{1}{\Delta t^2}\left(\int_{(j+1)\Delta t}^{(j+2)\Delta t}((j+2)\Delta t - t)\ddot{\boldsymbol{\theta}}^h(t)dt + \int_{j\Delta t}^{(j+1)\Delta t}(t - j\Delta t)\ddot{\boldsymbol{\theta}}^h(t)dt\right).\end{aligned}$$

Hence, we arrive at

$$\|\partial_{\Delta t}(\partial_{\Delta t}\boldsymbol{\theta}^h_{j+2})\|^2_{0,\Omega_I} \le C\frac{1}{\Delta t}\int_{j\Delta t}^{(j+2)\Delta t}\left\|\ddot{\boldsymbol{\theta}}^h(t)\right\|^2_{0,\Omega_I} dt.$$

Recalling the definition of $\boldsymbol{\theta}^h$, we conclude with Lemma 6.5 and (6.5a):

$$\Delta t\sum_{j=0}^{M-2}\|\partial_{\Delta t}(\partial_{\Delta t}\boldsymbol{\theta}^h_{j+2})\|^2_{0,\Omega_I} \le Ch^2\left(\|\ddot{\mathbf{u}}\|^2_{0;2,\Omega} + \left\|\frac{\partial^4\mathbf{u}}{\partial t^4}\right\|^2_{0;1,\Omega}\right). \tag{6.40}$$

The second term on the right hand side of (6.39) gives with (6.33a) and (6.33b):

$$\boldsymbol{\rho}^h_{j+1/2} - \partial_{\Delta t}\boldsymbol{\pi}^h_{j+1} = Z^h\left(\frac{1}{4}\ddot{\mathbf{u}}_{j+2} + \frac{1}{2}\ddot{\mathbf{u}}_{j+1} + \frac{1}{4}\ddot{\mathbf{u}}_j - \frac{1}{\Delta t^2}\mathbf{u}_{j+2} + \frac{2}{\Delta t^2}\mathbf{u}_{j+1} - \frac{1}{\Delta t^2}\mathbf{u}_j\right).$$

Taylor expansion at $t = (j+1)\Delta t$, summation over j and (6.5c) lead to

$$\Delta t\sum_{j=0}^{M-2}\|\boldsymbol{\rho}^h_{j+1/2} - \partial_{\Delta t}\boldsymbol{\pi}^h_j\|^2_{0,\Omega_I} \le C\Delta t^4\left\|\frac{\partial^4\mathbf{u}}{\partial t^4}\right\|^2_{0;0,\Omega}. \tag{6.41}$$

(6.40) and (6.41) conclude the proof. □

Now we are able to prove Theorem 6.10:

Proof. With $\mathbf{u}^h_j - \mathbf{u}_j = \boldsymbol{\xi}^h_j + (\mathbf{w}^h_j - \mathbf{u}_j)$ we can estimate

$$\begin{aligned}
&\max_{0\le j\le M-1}\left(\|\partial_{\Delta t}(\mathbf{u}^h_{j+1} - \mathbf{u}_{j+1})\|_{0,\Omega_I} + \|\mathbf{u}^h_{j+1/2} - \mathbf{u}_{j+1/2}\|_{1,\Omega}\right) \\
&\le \max_{0\le j\le M-1}\left(\|\partial_{\Delta t}(\mathbf{w}^h_{j+1} - \mathbf{u}_{j+1})\|_{0,\Omega_I} + \|\mathbf{w}^h_{j+1/2} - \mathbf{u}_{j+1/2}\|_{1,\Omega}\right. \\
&\qquad\left. + \|\partial_{\Delta t}\boldsymbol{\xi}^h_{j+1}\|_{0,\Omega_I} + \|\boldsymbol{\xi}^h_{j+1/2}\|_{1,\Omega}\right).
\end{aligned} \tag{6.42}$$

Using Taylor expansion as well as Lemmas 6.2 and 6.5, the first term gives

$$\begin{aligned}
\|\partial_{\Delta t}(\mathbf{w}^h_{j+1} - \mathbf{u}_{j+1})\|^2_{0,\Omega_I} &\le \|\partial_{\Delta t}(\mathbf{w}^h_j - \mathbf{u}_j) - (\dot{\mathbf{w}}^h - \dot{\mathbf{u}})_{j+1/2}\|^2_{0,\Omega_I} + \|(\dot{\mathbf{w}}^h - \dot{\mathbf{u}})_{j+1/2}\|^2_{0,\Omega_I} \\
&\le \Delta t\int_0^{\Delta t}\|\ddot{\mathbf{w}}^h(t) - \ddot{\mathbf{u}}(t)\|^2_{0,\Omega_I}dt + \|(\dot{\mathbf{w}}^h - \dot{\mathbf{u}})_{j+1/2}\|^2_{0,\Omega_I} \\
&\le C\left(\|\ddot{\mathbf{w}}^h - \ddot{\mathbf{u}}\|^2_{0;0,\Omega_I} + \|\dot{\mathbf{w}}^h - \dot{\mathbf{u}}\|^2_{0;0,\Omega_I}\right) \\
&\le Ch^2\left(\sum_{s=0}^{1}\left\|\frac{\partial^s\mathbf{u}}{\partial t^s}\right\|^2_{0;2,\Omega} + \sum_{s=2}^{3}\left\|\frac{\partial^s\mathbf{u}}{\partial t^s}\right\|^2_{0;1,\Omega}\right).
\end{aligned}$$

The second term yields the same upper bound, due to Lemmas 6.2 and 6.5.

For the last two terms of (6.42), we employ Lemmas 6.13 and 6.14 to obtain

$$\begin{aligned}
&\max_{0\le j\le(M-1)}\left(\|\partial_{\Delta t}\boldsymbol{\xi}^h_{j+1}\|^2_{0,\Omega_I} + \|\boldsymbol{\xi}^h_{j+1/2}\|^2_{1,\Omega}\right) \le C\left(\|\partial_{\Delta t}\boldsymbol{\xi}^h_1\|^2_{0,\Omega_I} + \|\boldsymbol{\xi}^h_{1/2}\|^2_{1,\Omega}\right) \\
&\quad + C\left(h^2\left(\|\ddot{\mathbf{u}}\|^2_{0;2,\Omega} + \left\|\frac{\partial^4\mathbf{u}}{\partial t^4}\right\|^2_{0;1,\Omega}\right) + \Delta t^4\left\|\frac{\partial^4\mathbf{u}}{\partial t^4}\right\|^2_{0;0,\Omega}\right).
\end{aligned} \tag{6.43}$$

Only the initial terms remain to be bounded. By the definitions (6.32) and (6.33c), we get

$$m_i^h(\partial_{\Delta t}\boldsymbol{\xi}_1^h, \boldsymbol{\chi}^h) + \frac{\Delta t}{2} a(\boldsymbol{\xi}_{1/2}^h, \boldsymbol{\chi}^h) = m_i^h(\boldsymbol{\varepsilon}_0^h, \boldsymbol{\chi}^h).$$

Hence, by choosing the test function $\boldsymbol{\chi}^h = \partial_{\Delta t}\boldsymbol{\xi}_1^h$, we obtain with Young's inequality

$$|\partial_{\Delta t}\boldsymbol{\xi}_1^h|_{h,i}^2 + \frac{1}{4}\left(a(\boldsymbol{\xi}_1^h, \boldsymbol{\xi}_1^h) - a(\boldsymbol{\xi}_0^h, \boldsymbol{\xi}_0^h)\right) = m_i^h(\boldsymbol{\varepsilon}_0^h, \partial_{\Delta t}\boldsymbol{\xi}_1^h) \leq C|\boldsymbol{\varepsilon}_0^h|_{h,i}^2 + \frac{1}{2}|\partial_{\Delta t}\boldsymbol{\xi}_1^h|_{h,i}^2.$$

Using the seminorm equivalence (5.12), this leads to

$$\|\partial_{\Delta t}\boldsymbol{\xi}_1^h\|_{0,\Omega_I}^2 + \|\boldsymbol{\xi}_1^h\|_{1,\Omega}^2 \leq C\left(\|\boldsymbol{\varepsilon}_0^h\|_{0,\Omega_I}^2 + \|\boldsymbol{\xi}_0^h\|_{1,\Omega}^2\right).$$

As we have $\boldsymbol{\xi}_0^h = \boldsymbol{\theta}^h(0)$, the last term is treated as in (6.17). The remaining term yields

$$\|\boldsymbol{\varepsilon}_0^h\|_{0,\Omega_I}^2 \leq C\left(\|\partial_{\Delta t}\boldsymbol{\theta}_1^h\|_{0,\Omega_I}^2 + \|\boldsymbol{\pi}_0^h\|_{0,\Omega_I}^2 + \frac{\Delta t^2}{4}\|\boldsymbol{\rho}_0^h\|_{0,\Omega_I}^2\right), \tag{6.44}$$

We have with Lemma 6.5

$$\begin{aligned}\|\partial_{\Delta t}\boldsymbol{\theta}_1^h\|_{0,\Omega_I}^2 &= \frac{1}{\Delta t}\left\|\int_0^{\Delta t}\dot{\boldsymbol{\theta}}^h(t)dt\right\|_{0,\Omega_I}^2 \leq \int_0^{\Delta t}\|\dot{\boldsymbol{\theta}}^h(t)\|_{0,\Omega_I}^2 dt \\ &\leq \|Z^h\dot{\mathbf{u}} - \dot{\mathbf{u}}\|_{0;0,\Omega_I}^2 + \|\dot{\mathbf{u}} - \dot{\mathbf{w}}^h\|_{0;0,\Omega_I}^2 \\ &\leq Ch^2\left(\|\dot{\mathbf{u}}\|_{0;2,\Omega}^2 + \left\|\frac{\partial^3\mathbf{u}}{\partial t^3}\right\|_{0;1,\Omega}^2\right).\end{aligned}$$

The second and third term of (6.44) are estimated by means of (6.5c), Lemma 6.2 and Taylor expansion:

$$\|\boldsymbol{\pi}_0^h\|_{0,\Omega_I}^2 + \frac{\Delta t^2}{4}\|\boldsymbol{\rho}_0^h\|_{0,\Omega_I}^2 \leq C\Delta t^4\sum_{s=3}^{4}\left\|\frac{\partial^s\mathbf{u}}{\partial t^s}\right\|_{0;0,\Omega}^2.$$

All together, the initial terms are bounded by

$$\begin{aligned}&\|\partial_{\Delta t}\boldsymbol{\xi}_1^h\|_{0,\Omega_I}^2 + \|\boldsymbol{\xi}_1^h\|_{1,\Omega}^2 \\ &\leq C\left(h^2\left(\sum_{s=0}^{1}\left\|\frac{\partial^s\mathbf{u}}{\partial t^s}\right\|_{0;2,\Omega}^2 + \sum_{s=2}^{3}\left\|\frac{\partial^s\mathbf{u}}{\partial t^s}\right\|_{0;1,\Omega}^2\right) + \Delta t^4\sum_{s=3}^{4}\left\|\frac{\partial^s\mathbf{u}}{\partial t^s}\right\|_{0;0,\Omega}^2\right).\end{aligned}$$

Combined with (6.43), this yields the desired inequality (6.29). □

The estimates in the $\mathbf{L}^2(\Omega_I)$-norm stated in Theorem 6.11 are proved along the lines of [7, Theorem 5.1], using (5.12) instead of the norm equivalence

$$c\|\boldsymbol{\chi}^h\|_{0,\Omega}^2 \leq m(\boldsymbol{\chi}^h, \boldsymbol{\chi}^h) \leq C\|\boldsymbol{\chi}^h\|_{0,\Omega}^2.$$

Some further details can be found in [73].

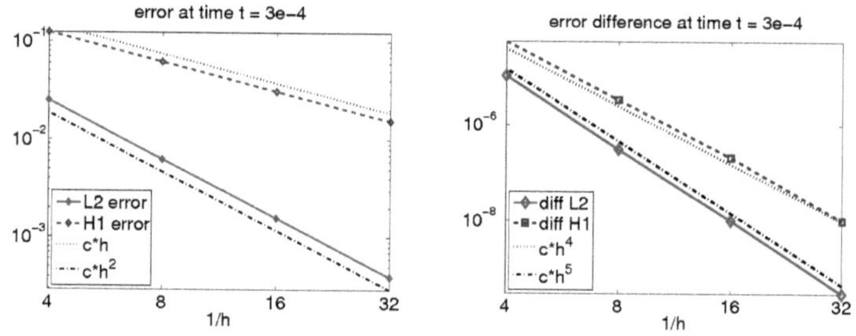

Figure 6.1: Error at time t_{30} with respect to h.

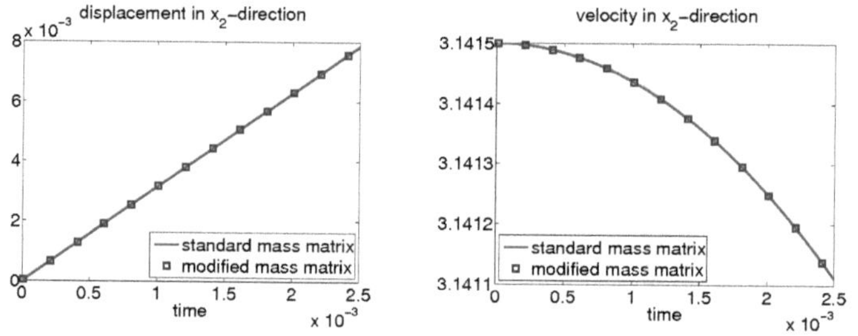

Figure 6.2: Vertical displacement and velocity at (5.0, 1.0) for $h = \frac{1}{8}$ and M^h, $\bar{M}_0^h$.

6.3 Numerical results

For the numerical validation of the error estimates proved in the previous sections, we consider the geometry and the material parameters of the elastic beam presented in Section 5.4.1. The right hand side $\mathbf{f}$ and the Neumann boundary conditions on $\partial\Omega$ are chosen such that the exact solution is given by

$$\mathbf{u}(\mathbf{x}, t) = \big(0,\ \ 0.02\, x_1 \cdot (10 - x_1) \cdot x_2^2 \cdot \sin(2\pi t)\big)^T.$$

This leads to homogeneous Neumann boundary conditions on the bottom boundary $[0, 10] \times \{0\}$ which we define as Γ_C. We use a uniform quadrilateral grid with mesh widths ranging from $h = \frac{1}{4}$ to $h = \frac{1}{32}$, and for the time discretization we compute 250 time steps with $\Delta t = 10^{-5}$.

The left picture of Figure 6.1 presents the exact error at time t_{30} with respect to the mesh size h. As the computation with $\bar{M}_1^h$ and $\bar{M}_0^h$ leads to almost exactly the same error, only one value is plotted. We obtain an error reduction of $\mathcal{O}(h^2)$ in the $\mathbf{L}^2(\Omega)$-norm and of $\mathcal{O}(h)$ in the $\mathbf{H}^1(\Omega)$-norm which matches the theoretical results of Sections 6.1 and 6.2. On the right picture of Figure 6.1, the absolute value of the difference between the error for the modified and the standard computation is plotted. Here, we observe numerically that the modified error converges to the standard error with a convergence order of 4 in the $\mathbf{H}^1(\Omega)$-norm and 5 in the $\mathbf{L}^2(\Omega)$-norm. Again, there is no visible difference between the results for $\bar{M}_0^h$ and $\bar{M}_1^h$.

In Figure 6.2, the vertical displacement and velocity of the node (5.0, 1.0) is depicted. One can see that the movement of the beam is accurately resolved.

Part IV

Iterative solvers for problems with different scales

7 Overlapping domain decomposition

In the previous two parts, we have dealt with efficient modeling and solution techniques for the numerical simulation of dynamic contact problems. However, characteristic applications like the forming of sheet metal or the rolling of a car tire are challenging tasks, as the problem features a complex threedimensional geometry as well as different nonlinear elastic materials. In addition, the contact zone is usually quite small compared to the size of the structure but needs to be resolved very accurately to get a good picture of the evolution of the contact stress during rolling contact. Hence, a fine triangulation is needed near the contact area; but discretizing the whole tire with such a fine grid yields a huge system that cannot be treated with the available computer architecture in the majority of cases.

In this part of the work, we are going to design a numerical scheme that combines a suitable multi-scale discretization for the geometry with the solution techniques for dynamic contact described in the previous chapters. The basic idea is to employ a decomposition of the original structure into several overlapping subdomains which have different grid spacing. The whole d-dimensional structure, with $d \in \{2, 3\}$, is discretized with a relatively coarse mesh that does not resolve the details along the contact boundary, whereas an overlapping patch with an independent fine triangulation is introduced at the contact area. An example of such a geometry is sketched in Figure 7.1. The transfer between the subdomains is only performed at the inner $(d-1)$-dimensional interface in order to avoid expensive volume coupling. This is especially important if the patch alters its position during the computation. For the transfer operators, we employ the variationally consistent mortar method (see, e.g., [11, 13]) with dual Lagrange multipliers [164] to enforce the weak continuity of the traces.

The subject of domain decomposition methods is already well-established in the literature; we refer to [51, 112, 122, 145, 158, 161] and the references therein for an overview of the topic. Overlapping domain decomposition techniques with independent grids have been considered in, e.g., [31, 61]. Further, the construction of domain decomposition schemes which are robust with respect to the mesh size as well as to jumps in the material parameters has been the topic of several papers (e.g., [18, 64, 107, 108, 109]). In this work, we make use of the overlapping decomposition in order to obtain an iterative solution scheme whose convergence rate is bounded independently of the mesh size or the material parameters.

For the efficient treatment of the contact inequality constraints, we use the NCP formulation and the resulting semismooth Newton iteration presented in Section 4.2. Here, the iterative subdomain coupling constitutes an efficient solver for the linear tangential system arising in each Newton step. In combination with appropriate stopping criteria for this inner linear iteration (see Theorem 3.10 or, e.g., [48, 49, 54]), the superlinear local convergence of the outer Newton method can be conserved.

The rubber material of a car tire usually shows an almost incompressible behaviour, but the numerical simulation of such materials using lowest order conforming Finite Elements often

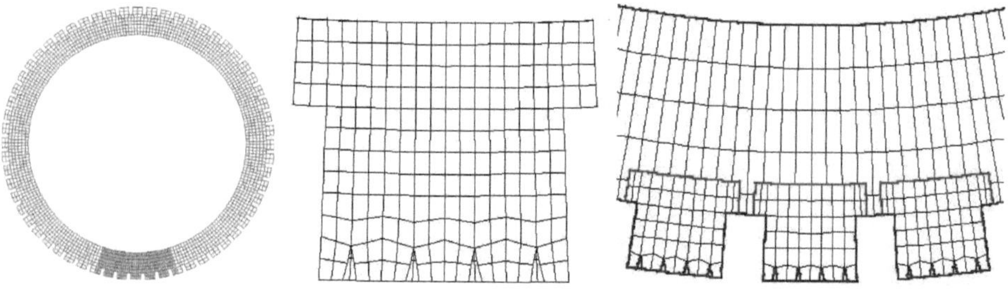

Figure 7.1: Sketch of overlapping domain decomposition; left: coarse domain; middle: fine patch resolving the details at the boundary; right: combined geometry with couping interface.

leads to volume locking (see, e.g., [20, 21, 33]). This can be avoided by using suitable mixed discretizations which introduce an additional variable for, e.g., the stress or the pressure [22, 23, 100, 101, 146]. In Section 7.3.7, we show numerically that our iterative algorithm can be extended to the case of nearly incompressible materials in a robust way. For the theoretical analysis using a generalized Hu–Washizu formulation [33, 115], we refer to [69].

Finally, we employ the decomposition into subdomains to define different time step sizes in each subdomains, allowing for a better resolution in time near the contact zone. Here, we present an appropriate definition of the interface conditions yielding an energy-conserving and thus stable time stepping algorithm.

Having introduced the main building blocks of our iterative algorithm, we turn to the structure of the rest of this part. The results therein have been obtained in cooperation with Manufacture Française des Pneumatiques Michelin.

Chapter 7 is concerned with the case of linear elastic problems. In Section 7.1, we introduce the notation, the governing equations and the fully coupled mortar system. Section 7.2 describes and analyses the iterative solution scheme that is used to solve the coupled system efficiently. Section 7.3 contains several numerical tests for linear elastic problems, confirming the theoretical results.

The two remaining chapters contain meaningful extensions of the scheme developed in Chapter 7. In Chapter 8, the case of nonlinear material laws and frictional contact is considered, whereas in Chapter 9, we examine the possibility of using different time step sizes in the subdomains.

7.1 Setting and problem formulation

This section contains the problem formulation as well as the basic notation for the rest of this part. In Subsection 7.1.1, the governing equations for the linear problem are stated in their strong and weak form, whereas Subsection 7.1.2 introduces the spatial discretization, including some properties of trace spaces and mortar operators which will be used in the sequel. The time discretization is sketched in Subsection 7.1.3, followed by the algebraic Schur complement formulation presented in Subsection 7.1.4.

7.1.1 Problem statement

As before, we consider a body $\Omega \subset \mathbb{R}^d$ with $d \in \{2,3\}$. But in this chapter, we assume that there exists a given subregion $\omega \subset \Omega$, with possibly different material parameters, where a more accurate local resolution is desired. The interface $\partial\omega \cap \Omega$ is denoted by Γ and the domain $\Omega \backslash \bar{\omega}$ by Ξ, as sketched on the left side of Figure 7.2. For simplicity, we assume $\text{meas}(\Gamma) > 0$, $\text{meas}(\Gamma_D) > 0$, $\bar{\Gamma}_D \cap \bar{\Gamma} = \emptyset$ as well as homogeneous Dirichlet boundary conditions on Γ_D.

Further, we suppose that there exist positive constants c, C independent of the material parameters E, ν, ϱ, as well as constant values ϱ_Θ, E_Θ, $\Theta \in \{\Xi, \omega\}$, such that

$$c\,E_\Theta \|\boldsymbol{\varepsilon}\|^2 \le \mathbb{C}^{\text{el}}(\mathbf{x})\boldsymbol{\varepsilon} : \boldsymbol{\varepsilon} \le C E_\Theta \|\boldsymbol{\varepsilon}\|^2, \quad \mathbf{x} \in \Theta, \tag{7.1a}$$

$$c\,\varrho_\Theta \le \varrho(\mathbf{x}) \le C \varrho_\Theta, \quad \mathbf{x} \in \Theta, \tag{7.1b}$$

with (7.1a) holding for all symmetric tensors $\boldsymbol{\varepsilon} \in \mathbb{R}^{d \times d}$. In other words, the parameters vary only moderately within the subdomains Ξ, ω but can exhibit a large jump along Γ. Furthermore, (7.1a) also implies that the material is compressible; the case of incompressible material is considered in [69] and Section 7.3.7.

Due to the decomposition into subdomains, we extend the definitions in (2.1) and (2.5) to functions which are not necessarily continuous along the interface Γ. For each of the (sub)domains $\Theta \in \{\Omega, \omega, \Xi\}$, we define

$$\mathbf{V}(\Theta) := [H^1(\Theta)]^d, \quad \mathbf{V}_0(\Theta) := \{\mathbf{v} \in \mathbf{V}(\Theta) : \mathbf{v}|_{\Gamma_D \cap \partial\Theta} = \mathbf{0}\}, \tag{7.2}$$

and

$$m_\Theta(\mathbf{u}, \mathbf{w}) := \int_\Theta \varrho\, \mathbf{u} \cdot \mathbf{w}\, d\mathbf{x}, \tag{7.3a}$$

$$a_\Theta(\mathbf{u}, \mathbf{w}) := \int_\Theta \mathbb{C}^{\text{el}} \boldsymbol{\varepsilon}(\mathbf{u}) : \boldsymbol{\varepsilon}(\mathbf{w})\, d\mathbf{x}, \tag{7.3b}$$

$$f_\Theta(\mathbf{w}) := \int_\Theta \mathbf{l} \cdot \mathbf{w}\, d\mathbf{x} + \int_{\Gamma_N \cap \partial\Theta} \mathbf{g}_N \cdot \mathbf{w}\, ds. \tag{7.3c}$$

Let $\mathbf{W}_\Gamma := \mathbf{H}^{1/2}(\Gamma)$ be the space of traces on Γ. With (7.3), the linear problem (6.3) can be formulated as two separate problems on the subdomains Ξ, ω, where the continuity of the displacements along Γ is enforced weakly by means of a Lagrange multiplier $\boldsymbol{\zeta}_\Gamma \in \mathbf{M}_\Gamma := \mathbf{W}_\Gamma'$. Thus, we obtain the continuous problem: find $(\mathbf{u}|_\Xi, \mathbf{u}|_\omega, \boldsymbol{\zeta}_\Gamma) \in \mathbf{V}_0(\Xi) \times \mathbf{V}_0(\omega) \times \mathbf{M}_\Gamma$ such that for all $t \in (0, T]$

$$\begin{aligned} m_\Xi(\ddot{\mathbf{u}}, \mathbf{w}) + a_\Xi(\mathbf{u}, \mathbf{w}) - \langle \mathbf{w}, \boldsymbol{\zeta}_\Gamma \rangle_\Gamma &= f_\Xi(\mathbf{w}), && \mathbf{w} \in \mathbf{V}_0(\Xi), \\ m_\omega(\ddot{\mathbf{u}}, \mathbf{w}) + a_\omega(\mathbf{u}, \mathbf{w}) + \langle \mathbf{w}, \boldsymbol{\zeta}_\Gamma \rangle_\Gamma &= f_\omega(\mathbf{w}), && \mathbf{w} \in \mathbf{V}_0(\omega), \\ \langle \mathbf{u}|_\omega, \boldsymbol{\mu}_\Gamma \rangle_\Gamma - \langle \mathbf{u}|_\Xi, \boldsymbol{\mu}_\Gamma \rangle_\Gamma &= 0, && \boldsymbol{\mu}_\Gamma \in \mathbf{M}_\Gamma, \end{aligned} \tag{7.4}$$

plus appropriate initial conditions

$$\mathbf{u}|_{t=0} = \mathbf{u}_0, \quad \dot{\mathbf{u}}|_{t=0} = \mathbf{v}_0. \tag{7.5}$$

Above, we have used the duality pairing $\langle \cdot, \cdot \rangle_\Gamma$ on Γ given by $\langle \mathbf{w}, \boldsymbol{\zeta} \rangle_\Gamma = \int_\Gamma \mathbf{w} \cdot \boldsymbol{\zeta} \, ds$. We remark that for the static case, i.e., $\varrho = 0$, (7.4) has the structure of a saddle point problem [27], whereas for $\varrho > 0$, the continuity conditions $(7.4)_3$ turn (7.4) into a DAE of index 3 (cf. Section 5.1). In the latter case, the problem is well posed if both initial conditions (7.5) satisfy condition $(7.4)_3$ [76]. For the static case, (7.4) is well posed if the bilinear form $a_\Xi(\cdot,\cdot) + a_\omega(\cdot,\cdot)$ is uniformly coercive for all functions in $\mathbf{V}_0(\Xi) \oplus \mathbf{V}_0(\omega)$ satisfying $(7.4)_3$ [26, 27]. To verify this, we consider the kernel of (7.3b), i.e., the space of rigid body modes on $\Theta \in \{\Xi, \omega\}$ which is denoted by

$$\mathcal{RB}_\Theta := \{ \mathbf{z} \in \mathbf{V}_0(\Theta) : a_\Theta(\mathbf{z}, \mathbf{z}) = 0 \}. \tag{7.6}$$

If $\text{meas}(\bar{\Gamma}_D \cap \partial\Theta) > 0$ holds, there are no free rigid body modes on the subdomain Θ and we have $\mathcal{RB}_\Theta = \{\mathbf{0}\}$. For $\text{meas}(\bar{\Gamma}_D \cap \partial\Theta) = 0$, we obtain $\dim(\mathcal{RB}_\Theta) \leq 3$ for $d = 2$ and $\dim(\mathcal{RB}_\Theta) \leq 6$ for $d = 3$. But in this case, the rigid body component on Θ is fixed by the continuity condition $(7.4)_3$, such that the well-posedness of (7.4) follows from the condition $\text{meas}(\Gamma_D) > 0$ and Korn's inequality [25].

7.1.2 Spatial discretization

For the spatial discretization of (7.4), we first triangulate the global domain Ω in terms of a quasi-uniform regular mesh $\mathcal{T}^H$ of simplicial or quadrilateral/ hexahedral elements of size H. We assume that the Dirichlet boundary Γ_D as well as the interface Γ are resolved by this triangulation. On $\mathcal{T}^H$, we consider lowest order conforming finite element basis functions ϕ_p^H, $p \in \mathcal{N}^H$, with $\mathcal{N}_\Gamma^H \subset \mathcal{N}^H$ denoting the subset of vertices of $\mathcal{T}^H$ on the interface Γ (see Figure 7.2). On the patch ω, we introduce a second regular and quasi-uniform discretization $\mathcal{T}^h$ with

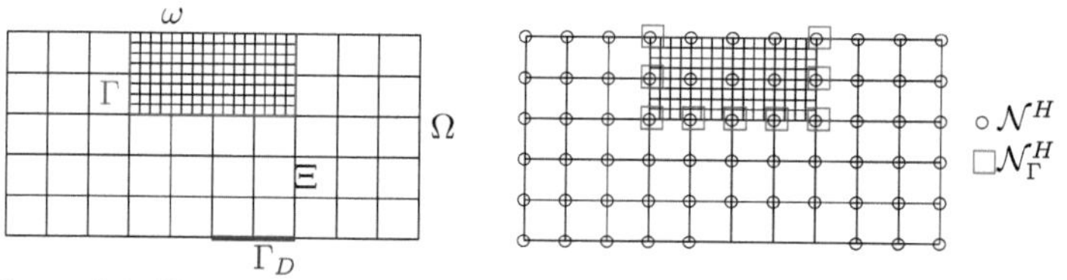

Figure 7.2: Illustration of the subdomains and the degrees of freedom

elements of size $h < H$, the basis functions ϕ_p^h, $p \in \mathcal{N}^h$, and the interface nodes $\mathcal{N}_\Gamma^h \subset \mathcal{N}^h$.

With these triangulations, we define the following vector valued finite element spaces for the coarse and the fine discretization:

$$\mathbf{V}_\Xi^H := \text{span}\{\phi_p^H|_\Xi\}_{p \in \mathcal{N}^H} \subset \mathbf{V}_0(\Xi), \qquad \mathbf{V}_\omega^H := \text{span}\{\phi_p^H|_\omega\}_{p \in \mathcal{N}^H} \subset \mathbf{V}_0(\omega), \tag{7.7a}$$

$$\mathbf{V}^H := \{ (\mathbf{v}^H, \mathbf{w}^H) \in \mathbf{V}_\Xi^H \oplus \mathbf{V}_\omega^H : \mathbf{v}^H|_\Gamma = \mathbf{w}^H|_\Gamma \} \subset \mathbf{V}_0(\Omega), \tag{7.7b}$$

$$\mathbf{V}^h = \mathbf{V}_\omega^h := \text{span}\{\phi_p^h\}_{p \in \mathcal{N}^h} \subset \mathbf{V}_0(\omega). \tag{7.7c}$$

We remark that the lower index $\cdot_0$ indicating the homogeneous Dirichlet conditions is omitted in (7.7) and in the definition of $\mathcal{N}^m$, $m \in \{h, H\}$, for ease of notation.

The triangulations $\mathcal{T}^m$, $m \in \{h, H\}$, induce discretizations $\mathcal{G}_\Gamma^m$ of the interface Γ, on which we define the trace spaces $\mathbf{W}_\Gamma^m := \mathbf{V}^m|_\Gamma$ spanned by the basis functions $\varphi_p^m := \phi_p^m|_\Gamma$, $p \in \mathcal{N}_\Gamma^m$, $m \in \{h, H\}$. With each trace space $\mathbf{W}_\Gamma^m \subset \mathbf{W}_\Gamma$, we associate a Lagrange multiplier space $\mathbf{M}_\Gamma^m \subset \mathbf{M}_\Gamma$ spanned by dual basis functions $\{\psi_p^m\}_{p\in\mathcal{N}_\Gamma^m}$ defined as in (2.8). We remark that this definition yields $\dim(\mathbf{M}_\Gamma^m) = \dim(\mathbf{W}_\Gamma^m)$, i.e., we do not modify the basis functions of $\mathbf{M}_\Gamma^m$ near the locations where Γ is nondifferentiable. Further, we emphasize that neither the finite element spaces $\mathbf{V}_\omega^h$, $\mathbf{V}_\omega^H$ nor the trace spaces $\mathbf{W}_\Gamma^h$, $\mathbf{W}_\Gamma^H$ are assumed to be nested. Even if $\mathbf{W}_\Gamma^H \subset \mathbf{W}_\Gamma^h$ holds, the use of the dual basis functions for the multiplier space generally yields $\mathbf{M}_\Gamma^H \not\subset \mathbf{M}_\Gamma^h$.

Inserting this space discretization into (7.4) and enforcing the weak continuity along Γ by means of the fine Lagrange multiplier space $\mathbf{M}_\Gamma^h$, we obtain the following problem: find $(\mathbf{u}^H, \mathbf{u}^h, \zeta_\Gamma^h) \in \mathbf{V}_\Xi^H \times \mathbf{V}^h \times \mathbf{M}_\Gamma^h$ such that for all $t \in (0, T)$

$$
\begin{aligned}
m_\Xi(\ddot{\mathbf{u}}^H, \mathbf{w}^H) + a_\Xi(\mathbf{u}^H, \mathbf{w}^H) - \langle \mathbf{w}^H, \zeta_\Gamma^h \rangle_\Gamma &= f_\Xi(\mathbf{w}^H), && \mathbf{w}^H \in \mathbf{V}_\Xi^H,\\
m_\omega(\ddot{\mathbf{u}}^h, \mathbf{w}^h) + a_\omega(\mathbf{u}^h, \mathbf{w}^h) + \langle \mathbf{w}^h, \zeta_\Gamma^h \rangle_\Gamma &= f_\omega(\mathbf{w}^h), && \mathbf{w}^h \in \mathbf{V}^h,\\
\langle \mathbf{u}^h - \mathbf{u}^H, \boldsymbol{\mu}_\Gamma^h \rangle_\Gamma &= 0, && \boldsymbol{\mu}_\Gamma^h \in \mathbf{M}_\Gamma^h,
\end{aligned}
\tag{7.8}
$$

together with suitable initial conditions

$$
(\mathbf{u}^H, \mathbf{u}^h)|_{t=0} = (\mathbf{u}_0^H, \mathbf{u}_0^h), \quad (\dot{\mathbf{u}}^H, \dot{\mathbf{u}}^h)|_{t=0} = (\mathbf{v}_0^H, \mathbf{v}_0^h). \tag{7.9}
$$

We remark that for the case $\mathbf{W}_\Gamma^H \subset \mathbf{W}_\Gamma^h$, the definition of $\mathbf{M}_\Gamma^h$ and $(7.8)_3$ imply that the finite element solution is continuous on the whole domain Ω.

Remark 7.1. As stated in [165], it is possible to define the Lagrange multiplier space with respect to a coarser triangulation than $\mathcal{G}_\Gamma^h$. If $\mathcal{G}_\Gamma^h$ is for example obtained from $\mathcal{G}_\Gamma^{2h}$ by one uniform refinement step, then the mortar problem (7.8) with $\mathbf{M}_\Gamma^h$ replaced by the multiplier space $\mathbf{M}_\Gamma^{2h}$ associated with $\mathcal{G}_\Gamma^{2h}$ still gives a well-defined problem.

Especially in the case that the trace spaces $\mathbf{W}_\Gamma^H \subset \mathbf{W}_\Gamma^h$ are nested, it is possible to define the Lagrange multiplier with respect to the coarser triangulation $\mathcal{G}^H$. However, the global continuity of the finite element solution of (7.8) is no longer guaranteed.

The case of a coarse multiplier space $\mathbf{M}_\Gamma^H$ is considered in more detail in [69]; some numerical results are shown in Section 7.3.6.

In the rest of this subsection, we formulate the algebraic version of (7.8) and state an assumption on the trace spaces $\mathbf{W}_\Gamma^h$, $\mathbf{W}_\Gamma^H$ that is used later on. For this, we introduce the mortar projection operator onto $\mathbf{W}_\Gamma^l$ given by

$$
P^l : \mathbf{W}_\Gamma \to \mathbf{W}_\Gamma^l : \; \langle P^l(\mathbf{w}_\Gamma), \boldsymbol{\mu}_\Gamma^l \rangle_\Gamma = \langle \mathbf{w}_\Gamma, \boldsymbol{\mu}_\Gamma^l \rangle_\Gamma, \quad \boldsymbol{\mu}_\Gamma^l \in \mathbf{M}_\Gamma^l, \quad l \in \{h, H\}, \tag{7.10}
$$

which satisfies the following Lemma proved in [165]:

Lemma 7.2. *The mortar projection P^l defined in (7.10) is uniformly continuous with respect to the $\mathbf{L}_2(\Gamma)$-, the $\mathbf{H}^1(\Gamma)$- and the $\mathbf{H}^{1/2}(\Gamma)$-norm. Furthermore, there exists a constant independent of l such that*

$$
\|\mathbf{w}_\Gamma - P^l \mathbf{w}_\Gamma\|_{0,\Gamma}^2 \le Cl |\mathbf{w}_\Gamma|_{1/2,\Gamma}^2, \quad \mathbf{w}_\Gamma \in \mathbf{W}_\Gamma, \quad l \in \{h, H\}. \tag{7.11}
$$

Defining the matrices $D_\Gamma^{lm} \in \mathbb{R}^{d|\mathcal{N}_\Gamma^l| \times d|\mathcal{N}_\Gamma^m|}$, $l, m \in \{h, H\}$, which are composed of the $d \times d$ submatrices

$$(D_\Gamma^{lm})_{pq} = \mathrm{Id}_d \cdot \int_\Gamma \psi_p^l \varphi_q^m \, ds, \quad p \in \mathcal{N}_\Gamma^l, q \in \mathcal{N}_\Gamma^m, \tag{7.12}$$

we can formulate the algebraic representation of the operator $P^l|_{\mathbf{W}_\Gamma^m} : \mathbf{W}_\Gamma^m \to \mathbf{W}_\Gamma^l$ given by

$$\Pi^{lm} = (D_\Gamma^{ll})^{-1} D_\Gamma^{lm} \in \mathbb{R}^{d|\mathcal{N}_\Gamma^l| \times d|\mathcal{N}_\Gamma^m|}. \tag{7.13}$$

The extension of the matrices D_Γ^{lh}, $l \in \{h, H\}$, by zero to the vector space $\mathbf{V}^h$ is named $D^{lh} \in \mathbb{R}^{d|\mathcal{N}_\Gamma^l| \times d|\mathcal{N}^h|}$; similarly, the extension of D_Γ^{lH} to the coarse upper space $\mathbf{V}_\Xi^H$ is denoted by D^{lH}.

Remark 7.3. Due to the use of the dual basis functions (2.8) for the Lagrange multiplier spaces $\mathbf{M}_\Gamma^l$, $l \in \{h, H\}$, the matrices D_Γ^{ll} in (7.12) are diagonal and can easily be inverted.

Using (7.12) as well as standard matrix notation for the mass and stiffness matrices as in (2.12), the algebraic version of the discrete mortar system (7.8) reads

$$M_\Xi^H \ddot{\mathbf{u}}^H + A_\Xi^H \mathbf{u}^H - (D^{hH})^T \zeta_\Gamma^h = \mathbf{f}^H, \tag{7.14a}$$

$$M_\omega^h \ddot{\mathbf{u}}^h + A_\omega^h \mathbf{u}^h + (D^{hh})^T \zeta_\Gamma^h = \mathbf{f}^h, \tag{7.14b}$$

$$D^{hh} \mathbf{u}^h - D^{hH} \mathbf{u}^H = \mathbf{0}. \tag{7.14c}$$

In Section 7.2, we make use of the following assumption relating the trace spaces $\mathbf{W}_\Gamma^H$, $\mathbf{W}_\Gamma^h$:

Assumption 7.4. *There exists a constant* $c_\Gamma > 0$ *independent of* h, H *such that*

$$\|P^h(\mathbf{w}_\Gamma^H)\|_{1/2,\Gamma} \geq c_\Gamma \|\mathbf{w}_\Gamma^H\|_{1/2,\Gamma}, \quad \mathbf{w}_\Gamma^H \in \mathbf{W}_\Gamma^H. \tag{7.15}$$

Remark 7.5. Clearly, Assumption 7.4 can only hold if $\dim(\mathbf{W}_\Gamma^h) \geq \dim(\mathbf{W}_\Gamma^H)$. In the case $\mathbf{W}_\Gamma^H \subseteq \mathbf{W}_\Gamma^h$, it is trivially satisfied with the constant $c_\Gamma = 1$, as P^h is the identity in (7.15). If the trace spaces are not nested, Assumption 7.4 is not as easy to verify; a possible numerical indicator if (7.15) is satisfied is to check whether the matrix

$$Q^{HH} := \Pi^{Hh} \Pi^{hH} \in \mathbb{R}^{d|\mathcal{N}_\Gamma^H| \times d|\mathcal{N}_\Gamma^H|} \tag{7.16}$$

is strictly diagonally dominant with a bound

$$\min_{1 \leq i \leq d|\mathcal{N}_\Gamma^H|} \left(|Q_{ii}^{HH}| - \sum_{j=1,\, j \neq i}^{d|\mathcal{N}_\Gamma^H|} |Q_{ij}^{HH}| \right) \geq c_0 > 0$$

and c_0 independent of h, H.

From [21, Lemma III.4.2], we obtain that Assumption 7.4 can be equivalently reformulated as follows:

Lemma 7.6. *Assumption 7.4 is equivalent to the uniform inf–sup condition of the spaces* $\mathbf{W}_\Gamma^H$ *and* $\mathbf{M}_\Gamma^h$, *i.e., there exists a constant* $\beta_\Gamma > 0$ *independent of* h, H *such that*

$$\inf_{\mathbf{w}_\Gamma^H \in \mathbf{W}_\Gamma^H} \sup_{\boldsymbol{\mu}_\Gamma^h \in \mathbf{M}_\Gamma^h} \frac{\langle \mathbf{w}_\Gamma^H, \boldsymbol{\mu}_\Gamma^h \rangle_\Gamma}{\|\mathbf{w}_\Gamma^H\|_{1/2,\Gamma} \|\boldsymbol{\mu}_\Gamma^h\|_{-1/2,\Gamma}} \geq \beta_\Gamma. \tag{7.17}$$

The constant β_Γ *is related to* c_Γ *by the bounds* $c_\Gamma \beta^h \leq \beta_\Gamma \leq c_\Gamma$, *where* β^h *is the inf–sup constant of the pairing* $\mathbf{W}_\Gamma^h$, $\mathbf{M}_\Gamma^h$.

Proof. Let Assumption 7.4 hold, which implies

$$\begin{aligned}
&\inf_{\mathbf{w}_\Gamma^H \in \mathbf{W}_\Gamma^H} \sup_{\boldsymbol{\mu}_\Gamma^h \in \mathbf{M}_\Gamma^h} \frac{\langle \mathbf{w}_\Gamma^H, \boldsymbol{\mu}_\Gamma^h \rangle_\Gamma}{\|\mathbf{w}_\Gamma^H\|_{1/2,\Gamma} \|\boldsymbol{\mu}_\Gamma^h\|_{-1/2,\Gamma}} \\
&\geq \inf_{\mathbf{w}_\Gamma^H \in \mathbf{W}_\Gamma^H} \sup_{\boldsymbol{\mu}_\Gamma^h \in \mathbf{M}_\Gamma^h} \frac{\langle P^h(\mathbf{w}_\Gamma^H), \boldsymbol{\mu}_\Gamma^h \rangle_\Gamma}{c_\Gamma^{-1} \|P^h(\mathbf{w}_\Gamma^H)\|_{1/2,\Gamma} \|\boldsymbol{\mu}_\Gamma^h\|_{-1/2,\Gamma}} \\
&\geq c_\Gamma \inf_{\mathbf{w}_\Gamma^h \in \mathbf{W}_\Gamma^h} \sup_{\boldsymbol{\mu}_\Gamma^h \in \mathbf{M}_\Gamma^h} \frac{\langle \mathbf{w}_\Gamma^h, \boldsymbol{\mu}_\Gamma^h \rangle_\Gamma}{\|\mathbf{w}_\Gamma^h\|_{1/2,\Gamma} \|\boldsymbol{\mu}_\Gamma^h\|_{-1/2,\Gamma}} \geq c_\Gamma \beta^h,
\end{aligned}$$

i.e., there exists $\beta_\Gamma \geq c_\Gamma \beta^h$ such that (7.17) is satisfied. Conversely, let (7.17) hold. Then, we get for any $\mathbf{w}_\Gamma^H \in \mathbf{W}_\Gamma^H$

$$\begin{aligned}
\beta_\Gamma \|\mathbf{w}_\Gamma^H\|_{1/2,\Gamma} &\leq \sup_{\boldsymbol{\mu}_\Gamma^h \in \mathbf{M}_\Gamma^h} \frac{\langle P^h(\mathbf{w}_\Gamma^H), \boldsymbol{\mu}_\Gamma^h \rangle_\Gamma}{\|\boldsymbol{\mu}_\Gamma^h\|_{-1/2,\Gamma}} \\
&\leq \sup_{\boldsymbol{\mu}_\Gamma^h \in \mathbf{M}_\Gamma^h} \frac{\|P^h(\mathbf{w}_\Gamma^H)\|_{1/2,\Gamma} \|\boldsymbol{\mu}_\Gamma^h\|_{-1/2,\Gamma}}{\|\boldsymbol{\mu}_\Gamma^h\|_{-1/2,\Gamma}} = \|P^h(\mathbf{w}_\Gamma^H)\|_{1/2,\Gamma},
\end{aligned}$$

i.e., (7.15) holds with some $c_\Gamma \geq \beta_\Gamma$. □

7.1.3 Time discretization

Next, we discretize (7.14) in time using the same scheme as in Section 2.3 with a fixed parameter $\theta \in \left[\frac{1}{2}, 1\right]$. Employing the notation given in (6.27), we obtain the following problem: find sequences of vectors $(\mathbf{u}_j^H, \mathbf{u}_j^h)_{j=0}^M$, $(\mathbf{v}_j^H, \mathbf{v}_j^h)_{j=0}^M$, $(\boldsymbol{\zeta}_{\Gamma,j+\theta}^H)_{j=0}^{M-1}$ satisfying the initial conditions (7.9) as well as

$$M_\Xi^H \partial_{\Delta t} \mathbf{v}_{j+1}^H + A_\Xi^H \mathbf{u}_{j+\theta}^H - (D^{hH})^T \boldsymbol{\zeta}_{\Gamma,j+\theta}^h = \mathbf{f}_{j+\theta}^H, \tag{7.18a}$$

$$\mathbf{v}_{j+1/2}^H - \partial_{\Delta t} \mathbf{u}_{j+1}^H = \mathbf{0}, \tag{7.18b}$$

$$M_\omega^h \partial_{\Delta t} \mathbf{v}_{j+1}^h + A_\omega^h \mathbf{u}_{j+\theta}^h + (D^{hh})^T \boldsymbol{\zeta}_{\Gamma,j+\theta}^h = \mathbf{f}_{j+\theta}^h, \tag{7.18c}$$

$$\mathbf{v}_{j+1/2}^h - \partial_{\Delta t} \mathbf{u}_{j+1}^h = \mathbf{0}, \tag{7.18d}$$

$$D^{hh} \mathbf{u}_{j+1}^h - D^{hH} \mathbf{u}_{j+1}^H = \mathbf{0}. \tag{7.18e}$$

We emphasize that the continuity of the displacements (7.18e) is enforced at t_{j+1}, similar to the contact conditions (2.18b).

Analogeously to (2.15), the discrete energy at time t_j is the sum of the contributions from the subdomains Ξ, ω given by

$$\begin{aligned} \mathbb{E}_j = \mathbb{E}^H_{\Xi,j} + \mathbb{E}^h_{\omega,j} := &\Big(\frac{1}{2}(\mathbf{v}^H_j)^T M^H_\Xi \mathbf{v}^H_j + \frac{1}{2}(\mathbf{u}^H_j)^T A^H_\Xi \mathbf{u}^H_j - (f^H_j)^T \mathbf{u}^H_j\Big) \\ &+ \Big(\frac{1}{2}(\mathbf{v}^h_j)^T M^h_\omega \mathbf{v}^h_j + \frac{1}{2}(\mathbf{u}^h_j)^T A^h_\omega \mathbf{u}^h_j - (f^h_j)^T \mathbf{u}^h_j\Big). \end{aligned} \tag{7.19}$$

For the case $\theta = \frac{1}{2}$ and time-independent outer loads, the energy is conserved for the discrete system (7.18):

$$\begin{aligned} \mathbb{E}_{j+1} - \mathbb{E}_j &= \Delta t\Big((\partial_{\Delta t}\mathbf{v}^H_{j+1})^T M^H_\Xi \mathbf{v}^H_{j+1/2} + (\partial_{\Delta t}\mathbf{u}^H_{j+1})^T A^H_\Xi \mathbf{u}^H_{j+1/2} - (f^H)^T \partial_{\Delta t}\mathbf{u}^H_{j+1}\Big) \\ &\quad + \Delta t\Big((\partial_{\Delta t}\mathbf{v}^h_{j+1})^T M^h_\omega \mathbf{v}^h_{j+1/2} + (\partial_{\Delta t}\mathbf{u}^h_{j+1})^T A^h_\omega \mathbf{u}^h_{j+1/2} - (f^h)^T \partial_{\Delta t}\mathbf{u}^h_{j+1}\Big) \\ &= \Delta t\Big((\partial_{\Delta t}\mathbf{u}^H_{j+1})^T (D^{hH})^T \boldsymbol{\zeta}^h_{\Gamma,j+1/2} - (\partial_{\Delta t}\mathbf{u}^h_{j+1})^T (D^{hh})^T \boldsymbol{\zeta}^h_{\Gamma,j+1/2}\Big) = 0, \end{aligned}$$

where we have used (7.18b), (7.18d) as well as (7.18e) for both times t_j and t_{j+1}.

7.1.4 Schur complement formulation

Condensing of the velocity at time t_{j+1} from (7.18) and using the notation (2.22), (2.23), the system to be solved for the displacement at time t_{j+1} becomes

$$\begin{pmatrix} K^H_\Xi & \mathbf{0} & -(D^{hH})^T \\ \mathbf{0} & K^h_\omega & (D^{hh})^T \\ -D^{hH} & D^{hh} & \mathbf{0} \end{pmatrix} \begin{pmatrix} \mathbf{u}^H_{j+1} \\ \mathbf{u}^h_{j+1} \\ \boldsymbol{\zeta}^h_{\Gamma,j+\theta} \end{pmatrix} = \begin{pmatrix} \boldsymbol{\varrho}^H_j, \\ \boldsymbol{\varrho}^h_j, \\ \mathbf{0} \end{pmatrix}. \tag{7.20}$$

As in Section 7.1, we shortly look at the well-posedness of the discrete problem (7.20) which is guaranteed if the symmetric matrix

$$\begin{pmatrix} K^H_\Xi & \mathbf{0} \\ \mathbf{0} & K^h_\omega \end{pmatrix} \tag{7.21}$$

is uniformly coercive on the space

$$\Big\{(\mathbf{w}^H, \mathbf{w}^h) \in (\mathbf{V}^H_\Xi \oplus \mathbf{V}^h_\omega) : D^{hH}\mathbf{w}^H - D^{hh}\mathbf{w}^h = \mathbf{0}\Big\}. \tag{7.22}$$

For $\varrho > 0$, this is always satisfied; otherwise, we need to consider the rigid body modes defined in (7.6). In contrast to Section 7.1, the functions in (7.22) are in general not globally continuous; however, as any rigid body mode $\mathbf{z} \in \mathcal{RB}_\Theta$ is linear, its trace $\mathrm{tr}_\Gamma(\mathbf{z})$ is contained in $\mathbf{W}^H_\Gamma$ as well as in $\mathbf{W}^h_\Gamma$. This implies that the operator P^h defined in (7.10) restricted to $\mathrm{tr}_\Gamma(\mathcal{RB}_\Xi)$ is the identity, and $\ker(D^{hH}) \cap \mathcal{RB}_\Xi = \ker(D^{hh}) \cap \mathcal{RB}_\Xi = \{\mathbf{0}\}$ holds. With $\mathrm{meas}(\Gamma_D) > 0$ and the Korn inequalities established in [25], we obtain that the matrix (7.21) is coercive on (7.22) with a constant independent of h, H or the diameters of the subdomains ω, Ξ.

Next, we partition the vectors $(\mathbf{u}^H, \mathbf{u}^h)$ and the matrix (7.21) into its components associated with the nodes on the interface Γ and the inner degrees of freedom by

$$\mathbf{u}^H = \begin{pmatrix} \mathbf{u}^H_\Xi \\ \mathbf{u}^H_\Gamma \end{pmatrix}, \quad K^H_\Xi = \begin{pmatrix} K^H_{\Xi\Xi} & K^H_{\Xi\Gamma} \\ K^H_{\Gamma\Xi} & K^H_{\Gamma\Gamma} \end{pmatrix}, \quad \mathbf{u}^h = \begin{pmatrix} \mathbf{u}^h_\Gamma \\ \mathbf{u}^h_\omega \end{pmatrix}, \quad K^h_\omega = \begin{pmatrix} K^h_{\Gamma\Gamma} & K^h_{\Gamma\omega} \\ K^h_{\omega\Gamma} & K^h_{\omega\omega} \end{pmatrix}.$$

Static condensation of the inner variables yields from (7.20)

$$\mathbf{u}^h_{\omega,j+1} = (K^h_{\omega\omega})^{-1}\left(\boldsymbol{\varrho}^h_{\omega,j} - K^h_{\omega\Gamma}\mathbf{u}^h_{\Gamma,j+1}\right), \tag{7.23a}$$
$$\mathbf{u}^H_{\Xi,j+1} = (K^H_{\Xi\Xi})^{-1}\left(\boldsymbol{\varrho}^H_{\Xi,j} - K^H_{\Xi\Gamma}\mathbf{u}^H_{\Gamma,j+1}\right). \tag{7.23b}$$

Hence, by introducing the symmetric positive semi-definite Schur complement matrices $S^H_\Xi \in \mathbb{R}^{d|\mathcal{N}^H_\Gamma|\times d|\mathcal{N}^H_\Gamma|}$, $S^h_\omega \in \mathbb{R}^{d|\mathcal{N}^h_\Gamma|\times d|\mathcal{N}^h_\Gamma|}$ via

$$S^H_\Xi := K^H_{\Gamma\Gamma} - K^H_{\Gamma\Xi}(K^H_{\Xi\Xi})^{-1}K^H_{\Xi\Gamma}, \tag{7.24a}$$
$$S^h_\omega := K^h_{\Gamma\Gamma} - K^h_{\Gamma\omega}(K^h_{\omega\omega})^{-1}K^h_{\omega\Gamma}, \tag{7.24b}$$

we can rewrite (7.20) as

$$\begin{pmatrix} S^H_\Xi & \mathbf{0} & -(D^{hH}_\Gamma)^T \\ \mathbf{0} & S^h_\omega & (D^{hh}_\Gamma)^T \\ -D^{hH}_\Gamma & D^{hh}_\Gamma & \mathbf{0} \end{pmatrix} \begin{pmatrix} \mathbf{u}^H_{\Gamma,j+1} \\ \mathbf{u}^h_{\Gamma,j+1} \\ \boldsymbol{\zeta}^h_{\Gamma,j+\theta} \end{pmatrix} = \begin{pmatrix} \bar{\boldsymbol{\varrho}}^H_{\Xi,j} \\ \bar{\boldsymbol{\varrho}}^h_{\omega,j} \\ \mathbf{0} \end{pmatrix}, \tag{7.25}$$

where the right hand side of (7.25) is given by

$$\bar{\boldsymbol{\varrho}}^H_{\Xi,j} := \boldsymbol{\varrho}^H_{\Gamma,j} - K^H_{\Gamma\Xi}(K^H_{\Xi\Xi})^{-1}\boldsymbol{\varrho}^H_{\Xi,j}, \qquad \bar{\boldsymbol{\varrho}}^h_{\omega,j} := \boldsymbol{\varrho}^h_{\Gamma,j} - K^h_{\Gamma\omega}(K^h_{\omega\omega})^{-1}\boldsymbol{\varrho}^h_{\omega,j}. \tag{7.26}$$

We remark that the Schur matrices (7.24) are the algebraic representations of discrete linear Dirichlet–to–Neumann maps $\mathcal{S}^H_\Xi : \mathbf{W}^H_\Gamma \to (\mathbf{W}^H_\Gamma)'$, $\mathcal{S}^h_\omega : \mathbf{W}^h_\Gamma \to (\mathbf{W}^h_\Gamma)'$ [145].

Instead of solving the fully coupled problem (7.25) directly, we are going to consider an iterative solution scheme that incorporates a coarse mortar problem on $\mathbf{V}^H$. In Section 7.2, a corresponding algorithm is derived and theoretically analysed, including a short discussion about estimating the corresponding algebraic error. Numerical results are presented in Section 7.3.

7.2 Iterative coupling algorithm

In this section, we derive and investigate an iterative solution scheme for the mortar system (7.25). In Subsection 7.2.1, the method is introduced and formulated as a block Gauß–Seidel iteration applied to an augmented problem. The error propagation and the convergence rate of the iterative scheme are analysed in Subsections 7.2.2 and 7.2.3, respectively, followed by a short discussion how its algebraic error can be measured.

7.2.1 Derivation

We construct an iterative solution algorithm for the mortar system (7.25) by considering an augmented version of (7.25), with one additional line containing the auxiliary coefficient vector $\boldsymbol{\mu}^H_\Gamma \in \mathbb{R}^{d|\mathcal{N}^H_\Gamma|}$:

$$\underbrace{\begin{pmatrix} S^H_\Xi + S^H_\omega & (D^{HH}_\Gamma)^T & \mathbf{0} & -(D^{hH}_\Gamma)^T \\ S^H_\omega & (D^{HH}_\Gamma)^T & \mathbf{0} & \mathbf{0} \\ -D^{hH}_\Gamma & \mathbf{0} & D^{hh}_\Gamma & \mathbf{0} \\ \mathbf{0} & \mathbf{0} & S^h_\omega & (D^{hh}_\Gamma)^T \end{pmatrix}}_{=:\widehat{G}_d} \underbrace{\begin{pmatrix} \mathbf{u}^H_{\Gamma,j+1} \\ \boldsymbol{\mu}^H_{\Gamma,j+\theta} \\ \mathbf{u}^h_{\Gamma,j+1} \\ \boldsymbol{\zeta}^h_{\Gamma,j+\theta} \end{pmatrix}}_{=:\widehat{\mathbf{z}}_d} = \underbrace{\begin{pmatrix} \bar{\boldsymbol{\varrho}}^H_{\Xi,j} + \bar{\boldsymbol{\varrho}}^H_{\omega,j} \\ \bar{\boldsymbol{\varrho}}^H_{\omega,j} \\ \mathbf{0} \\ \bar{\boldsymbol{\varrho}}^h_{\omega,j} \end{pmatrix}}_{=:\widehat{\mathbf{F}}_d}. \tag{7.27}$$

In (7.27), we have used the coarse Schur complement matrix $S_\omega^H \in \mathbb{R}^{d|\mathcal{N}_\Gamma^H|\times d|\mathcal{N}_\Gamma^H|}$ and the right hand side $\bar{\boldsymbol{\varrho}}_{\omega,j}^H$ which are assumed to be suitable coarse grid approximations of the fine grid quantities S_ω^h, $\bar{\boldsymbol{\varrho}}_{\omega,j}^h$ defined in (7.24b), (7.26). The exact definition of these approximations is not fixed; however, we require them to comply with the Dirichlet boundary conditions on $\Gamma_D \cap \partial\omega$ and with the rigid body modes in $\mathcal{RB}_\omega$, i.e., we assume that

$$S_\omega^H \Pi^{Hh} \mathbf{z}_\Gamma^h = \mathbf{0} \quad \text{and} \quad \left(\Pi^{Hh}\mathbf{z}_\Gamma^h, \bar{\boldsymbol{\varrho}}_{\omega,j}^H\right) = \left(\mathbf{z}_\Gamma^h, \bar{\boldsymbol{\varrho}}_{\omega,j}^h\right) \tag{7.28}$$

holds for every $\mathbf{z}_\Gamma^h \in \mathrm{tr}_\Gamma(\mathcal{RB}_\omega) \subset \mathbf{W}_\Gamma^h$, where $(\cdot,\cdot)$ denotes the Euclidean scalar product.

For the linear problem we investigate in this chapter, a natural definition of S_ω^H is

$$S_\omega^H := K_{\Gamma\Gamma}^H - K_{\Gamma\omega}^H (K_{\omega\omega}^H)^{-1} K_{\omega\Gamma}^H, \tag{7.29}$$

with K_ω^H defined as in (2.22) but assembled with respect to the coarse grid $\mathcal{T}^H|_\omega$ instead of $\mathcal{T}^h$. In Chapter 8, other ways of defining suitable coarse grid approximations will be discussed. We remark that the additional line $(7.27)_2$ can be interpreted as an auxiliary coarse grid problem on the patch ω with a fixed trace on Γ, with $\boldsymbol{\mu}_\Gamma^H$ being the corresponding coarse Lagrange multiplier in $\mathbf{M}_\Gamma^H$. A related algorithm with an augmented coarse problem has been investigated in [82].

As the components $(\mathbf{u}_{\Gamma,j+1}^H, \mathbf{u}_{\Gamma,j+1}^h, \boldsymbol{\zeta}_{\Gamma,j+\theta}^h)$ of the exact solution of the augmented system (7.27) also solve the mortar system (7.25), any convergent iterative method for (7.27) yields an approximation of the solution of (7.25). Hence, we consider the iterative solution of (7.27) by means of a block Gauß–Seidel scheme according to the splitting $\widehat{G}_d = (\widehat{G}_d - \widehat{K}_d) + \widehat{K}_d$, with $\widehat{K}_d$ being the upper triangular part of $\widehat{G}_d$, i.e., $\widehat{K}_d$ is defined to be zero except for its first row which reads

$$\begin{pmatrix} \mathbf{0} & (D_\Gamma^{HH})^T & \mathbf{0} & -(D_\Gamma^{hH})^T \end{pmatrix}.$$

Given some starting vector $\widehat{\mathbf{z}}_d^{(0)}$, the Gauß–Seidel iteration on (7.27) reads for $l \geq 0$:

$$(\widehat{G}_d - \widehat{K}_d)\delta\widehat{\mathbf{z}}_d^{(l)} = \widehat{\mathbf{F}}_d - \widehat{G}_d\widehat{\mathbf{z}}_d^{(l)}, \qquad \widehat{\mathbf{z}}_d^{(l+1)} = \widehat{\mathbf{z}}_d^{(l)} + \delta\widehat{\mathbf{z}}_d^{(l)}. \tag{7.30}$$

Equation (7.30) leads to the following system that has to be solved in the l-th iteration step

$$\begin{pmatrix} (S_\Xi^H + S_\omega^H) & \mathbf{0} & \mathbf{0} & \mathbf{0} \\ S_\omega^H & (D_\Gamma^{HH})^T & \mathbf{0} & \mathbf{0} \\ -D_\Gamma^{hH} & \mathbf{0} & D_\Gamma^{hh} & \mathbf{0} \\ \mathbf{0} & \mathbf{0} & S_\omega^h & (D_\Gamma^{hh})^T \end{pmatrix} \begin{pmatrix} \delta\mathbf{u}_{\Gamma,j+1}^{H,(l)} \\ \delta\boldsymbol{\mu}_{\Gamma,j+\theta}^{H,(l)} \\ \delta\mathbf{u}_{\Gamma,j+1}^{h,(l)} \\ \delta\boldsymbol{\zeta}_{\Gamma,j+\theta}^{h,(l)} \end{pmatrix} = \begin{pmatrix} \mathbf{r}_\Xi^{H,(l)} + \mathbf{r}_\omega^{H,(l)} \\ \mathbf{r}_\omega^{H,(l)} \\ \boldsymbol{\nu}_\Gamma^{h,(l)} \\ \mathbf{r}_\omega^{h,(l)} \end{pmatrix}, \tag{7.31}$$

where the residuals on the right hand side of (7.31) are defined in (7.35). One can see that (7.31) is a linear system with a lower block-triangular matrix and can thus be solved by two sequential subproblems on $\mathbf{V}^H$ and $\mathbf{V}^h$. The corresponding scheme is described in Algorithm 1.

Remark 7.7. After one iteration of Algorithm 1, i.e., for $l \geq 1$, the residuals (7.35b) to (7.35d) vanish.

Algorithm 1 Two-way coupling scheme with augmented coarse grid problem

Starting from some initial guess $\widehat{\mathbf{z}}_d^{(0)}$, compute sequentially for $l = 0, 1, \dots$

(i) Solve problem on coarse space $\mathbf{V}^H$ with interface load on Γ inherited from fine computation on ω:

$$\begin{pmatrix} (S_\Xi^H + S_\omega^H) & \mathbf{0} \\ S_\omega^H & (D_\Gamma^{HH})^T \end{pmatrix} \begin{pmatrix} \delta \mathbf{u}_{\Gamma,j+1}^{H,(l)} \\ \delta \boldsymbol{\mu}_{\Gamma,j+\theta}^{H,(l)} \end{pmatrix} = \begin{pmatrix} \mathbf{r}_\Xi^{H,(l)} + \mathbf{r}_\omega^{H,(l)} \\ \mathbf{r}_\omega^{H,(l)} \end{pmatrix}. \tag{7.32}$$

(ii) Solve problem on fine space $\mathbf{V}^h$ with weakly imposed trace on Γ inherited from coarse computation on Ω:

$$\begin{pmatrix} S_\omega^h & (D_\Gamma^{hh})^T \\ D_\Gamma^{hh} & \mathbf{0} \end{pmatrix} \begin{pmatrix} \delta \mathbf{u}_{\Gamma,j+1}^{h,(l)} \\ \delta \boldsymbol{\zeta}_{\Gamma,j+\theta}^{h,(l)} \end{pmatrix} = \begin{pmatrix} \mathbf{r}_\omega^{h,(l)} \\ \boldsymbol{\nu}_\Gamma^{h,(l)} \end{pmatrix} + \begin{pmatrix} \mathbf{0} \\ D_\Gamma^{hH} \delta \mathbf{u}_{\Gamma,j+1}^{H,(l)} \end{pmatrix}. \tag{7.33}$$

(iii) Update the solution vector:

$$\widehat{\mathbf{z}}_d^{(l+1)} := \widehat{\mathbf{z}}_d^{(l)} + \delta \widehat{\mathbf{z}}_d^{(l)}. \tag{7.34}$$

The residuals of (7.32), (7.33) are given by

$$\mathbf{r}_\Xi^{H,(l)} = \bar{\boldsymbol{\varrho}}_{\Xi,j}^H - S_\Xi^H \mathbf{u}_{\Gamma,j+1}^{H,(l)} + (D_\Gamma^{hH})^T \boldsymbol{\zeta}_{\Gamma,j+\theta}^{h,(l)}, \tag{7.35a}$$

$$\mathbf{r}_\omega^{H,(l)} = \bar{\boldsymbol{\varrho}}_{\omega,j}^H - S_\omega^H \mathbf{u}_{\Gamma,j+1}^{H,(l)} - (D_\Gamma^{HH})^T \boldsymbol{\mu}_{\Gamma,j+\theta}^{H,(l)}, \tag{7.35b}$$

$$\mathbf{r}_\omega^{h,(l)} = \bar{\boldsymbol{\varrho}}_{\omega,j}^h - S_\omega^h \mathbf{u}_{\Gamma,j+1}^{h,(l)} - (D_\Gamma^{hh})^T \boldsymbol{\zeta}_{\Gamma,j+\theta}^{h,(l)}, \tag{7.35c}$$

$$\boldsymbol{\nu}_\Gamma^{h,(l)} = D_\Gamma^{hH} \mathbf{u}_{\Gamma,j+1}^{H,(l)} - D_\Gamma^{hh} \mathbf{u}_{\Gamma,j+1}^{h,(l)}. \tag{7.35d}$$

7.2.2 Error propagation

In order to analyse the error propagation of Algorithm 1, we are going to reformulate both the mortar system (7.25) and the iteration (7.30) as equations with the only unknown $\mathbf{u}_{\Gamma,j+1}^H \in \mathbf{W}_\Gamma^H$. Introducing the Schur matrix $S_\omega^{HhH} \in \mathbb{R}^{d|\mathcal{N}_\Gamma^H| \times d|\mathcal{N}_\Gamma^H|}$ with

$$S_\omega^{HhH} := -\begin{pmatrix} \mathbf{0} & (D_\Gamma^{hH})^T \end{pmatrix} \begin{pmatrix} S_\omega^h & (D_\Gamma^{hh})^T \\ D_\Gamma^{hh} & \mathbf{0} \end{pmatrix}^{-1} \begin{pmatrix} \mathbf{0} \\ D_\Gamma^{hH} \end{pmatrix} = (\Pi^{hH})^T S_\omega^h \Pi^{hH}, \tag{7.36}$$

the mortar system (7.25) can be rewritten as a Schur complement system for the coarse trace $\mathbf{u}_{\Gamma,j+1}^H$:

$$(S_\Xi^H + S_\omega^{HhH}) \mathbf{u}_{\Gamma,j+1}^H = \mathbf{g}_{\Gamma,j}^H := (\Pi^{hH})^T \bar{\boldsymbol{\varrho}}_{\omega,j}^h + \bar{\boldsymbol{\varrho}}_{\Xi,j}^H. \tag{7.37}$$

Remark 7.8. If the Lagrange multiplier $\boldsymbol{\zeta}_{\Gamma,j+\theta}^h$ is not associated with $\mathcal{G}_\Gamma^h$ but with a coarser triangulation (cf. Remark 7.1), the second equality of (7.36) does not hold, and the Schur complement matrix S_ω^{HhH} is defined by the first formula. In the special case that $\mathbf{W}_\Gamma^H \subset \mathbf{W}_\Gamma^h$ and the multiplier is chosen from $\mathbf{M}_\Gamma^H$, (7.36) becomes

$$S_\omega^{HhH} := -\begin{pmatrix} \mathbf{0} & (D_\Gamma^{HH})^T \end{pmatrix} \begin{pmatrix} S_\omega^h & (D_\Gamma^{Hh})^T \\ D_\Gamma^{Hh} & \mathbf{0} \end{pmatrix}^{-1} \begin{pmatrix} \mathbf{0} \\ D_\Gamma^{HH} \end{pmatrix}.$$

The following lemma shows that the iterative scheme (7.30), which is realized by Algorithm 1, can be reformulated as a fixed point iteration on $\mathbf{u}^H_{\Gamma,j+1}$:

Lemma 7.9. *Let $\hat{\mathbf{z}}_d$ denote the exact solution of the augmented mortar system (7.27), and let $\hat{\mathbf{z}}_d^{(l)}$, $l = 0, 1, \ldots$, be the sequence of vectors obtained from Algorithm 1. For $l \geq 1$, the error $\mathbf{e}_\Gamma^{H,(l)} := (\mathbf{u}^{H,(l)}_{\Gamma,j+1} - \mathbf{u}^H_{\Gamma,j+1})$ satisfies the relation*

$$\mathbf{e}_\Gamma^{H,(l+1)} = \Big(Id - \big(S_\Xi^H + S_\omega^H\big)^{-1}\big(S_\Xi^H + S_\omega^{HhH}\big)\Big)\mathbf{e}_\Gamma^{H,(l)}. \tag{7.38}$$

Proof. With (7.30) and $\widehat{\mathbf{F}}_d = \widehat{G}_d\hat{\mathbf{z}}_d$ from (7.27), we get

$$\begin{aligned}(\widehat{G}_d - \widehat{K}_d)\hat{\mathbf{z}}_d^{(l+1)} &= \widehat{\mathbf{F}}_d - \widehat{K}_d\hat{\mathbf{z}}_d^{(l)},\\ (\widehat{G}_d - \widehat{K}_d)\big(\hat{\mathbf{z}}_d^{(l+1)} - \hat{\mathbf{z}}_d\big) &= \widehat{G}_d\hat{\mathbf{z}}_d - (\widehat{G}_d - \widehat{K}_d)\hat{\mathbf{z}}_d - \widehat{K}_d\hat{\mathbf{z}}_d^{(l)} = -\widehat{K}_d\big(\hat{\mathbf{z}}_d^{(l)} - \hat{\mathbf{z}}_d\big).\end{aligned} \tag{7.39}$$

From the definition of $\widehat{K}_d$, we obtain that the right hand side of (7.39) is zero except for its first component which reads

$$\Big(-\widehat{K}_d\big(\hat{\mathbf{z}}_d^{(l)} - \hat{\mathbf{z}}_d\big)\Big)_1 = (D_\Gamma^{hH})^T\big(\boldsymbol{\zeta}^{h,(l)}_{\Gamma,j+\theta} - \boldsymbol{\zeta}^h_{\Gamma,j+\theta}\big) - (D_\Gamma^{HH})^T\big(\boldsymbol{\mu}^{H,(l)}_{\Gamma,j+\theta} - \boldsymbol{\mu}^H_{\Gamma,j+\theta}\big).$$

Using the Schur matrix (7.36), we get

$$S_\omega^H\mathbf{e}_\Gamma^{H,(l+1)} = -(D_\Gamma^{HH})^T\big(\boldsymbol{\mu}^{H,(l+1)}_{\Gamma,j+\theta} - \boldsymbol{\mu}^H_{\Gamma,j+\theta}\big) \quad \text{from} \quad (7.39)_2, \tag{7.40a}$$

$$S_\omega^{HhH}\mathbf{e}_\Gamma^{H,(l+1)} = -(D_\Gamma^{hH})^T\big(\boldsymbol{\zeta}^{h,(l+1)}_{\Gamma,j+\theta} - \boldsymbol{\zeta}^h_{\Gamma,j+\theta}\big) \quad \text{from} \quad (7.39)_{3-4}. \tag{7.40b}$$

For $l \geq 1$, the relations (7.40) also hold for the previous iteration with (l) instead of $(l+1)$. Hence, $(7.39)_1$ yields

$$\begin{aligned}(S_\Xi^H + S_\omega^H)\mathbf{e}_\Gamma^{H,(l+1)} &= (D_\Gamma^{hH})^T\big(\boldsymbol{\zeta}^{h,(l)}_{\Gamma,j+\theta} - \boldsymbol{\zeta}^h_{\Gamma,j+\theta}\big) - (D_\Gamma^{HH})^T\big(\boldsymbol{\mu}^{H,(l)}_{\Gamma,j+\theta} - \boldsymbol{\mu}^H_{\Gamma,j+\theta}\big)\\ &= \big(S_\omega^H - S_\omega^{HhH}\big)\mathbf{e}_\Gamma^{H,(l)}.\end{aligned}$$

Using the relation

$$\big(S_\Xi^H + S_\omega^H\big)^{-1}\big(S_\omega^H - S_\omega^{HhH}\big) = \mathrm{Id} - \big(S_\Xi^H + S_\omega^H\big)^{-1}\big(S_\Xi^H + S_\omega^{HhH}\big),$$

we finally obtain (7.38). □

Remark 7.10. If the coarse and the fine grid on ω coincide, i.e., $\mathbf{V}^h = \mathbf{V}^H_\omega$ holds with the definitions from (7.7), then we have $S_\omega^{HhH} = S_\omega^H$, and Lemma 7.9 implies that the exact coarse trace $\mathbf{u}^H_{\Gamma,j+1}$ is obtained after at most two steps. In this case, (7.30) yields that the other components of the solution vector $\hat{\mathbf{z}}_d^{(2)}$ coincide with the exact solution $\hat{\mathbf{z}}_d$ of (7.27), too. If the starting vector $\hat{\mathbf{z}}_d^{(0)}$ satisfies $\widehat{K}_d\hat{\mathbf{z}}_d^{(0)} = \widehat{K}_d\hat{\mathbf{z}}_d$, then the exact solution of (7.27) is already obtained after the first iteration.

A damped version of Algorithm 1 with a suitable damping parameter $\alpha_l > 0$ yields the error propagation

$$\mathbf{e}_\Gamma^{H,(l+1)} = \Big(\mathrm{Id} - \alpha_l\big(S_\Xi^H + S_\omega^H\big)^{-1}\big(S_\Xi^H + S_\omega^{HhH}\big)\Big)\mathbf{e}_\Gamma^{H,(l)}. \tag{7.41}$$

If $\alpha_l = \alpha$ does not depend on l, (7.41) is equivalent to a preconditioned Richardson iteration for the Schur complement system (7.37). But in general, it is more effective to choose α_l in each step according to a conjugate gradient algorithm (see, e.g., [18, 149, 161]).

Remark 7.11. From (7.38), one can see that the difference between Algorithm 1 and the classical Dirichlet–Neumann coupling with Ξ as the Neumann subdomain is the additional term of S_ω^H in the factor $\left(S_\Xi^H + S_\omega^H\right)^{-1}$. The benefit of this term can be seen in the next subsection, where we analyse the robustness of Algorithm 1 with respect to jumps in the material parameters.

7.2.3 Condition number analysis

The convergence properties of the iteration (7.38) or its damped version (7.41) are determined by the condition number of the iteration matrix in (7.38). On this account, we investigate the spectral properties of the matrices S_ω^H and S_ω^{HhH} in Theorem 7.12 below. For its proof, we use the zero extension operator $R_\omega^m : \mathbf{W}_\Gamma^m \to \mathbf{V}_\omega^m$, $m \in \{h, H\}$, given by

$$R_\omega^m \mathbf{w}_\Gamma^m(p) = \begin{cases} \mathbf{w}_\Gamma^m(p), & p \in \mathcal{N}_\Gamma^m, \\ \mathbf{0}, & \text{else}, \end{cases} \tag{7.42}$$

as well as a projection operator $Z_\omega^m : \mathbf{V}_0(\omega) \to \mathbf{V}_\omega^m$, $m \in \{h, H\}$, which is constructed similar to the Scott–Zhang operator Z^h defined in [153] and used in Chapter 6. The only difference is that we change the original definition of [153] near the interface Γ such that the relation $(Z_\omega^m \mathbf{v})|_\Gamma = P^m(\mathbf{v}|_\Gamma)$ holds for $\mathbf{v} \in \mathbf{V}_0(\omega)$, where the mortar projection P^m has been defined in (7.10). This modified operator Z_ω^m can be constructed similarly as in [153, Section 5] and thus satisfies the approximation properties (6.5a) as well as the $\mathbf{H}^1(\omega)$-stability estimate (6.5b). However, the uniform $\mathbf{L}_2(\omega)$-stability (6.5c) cannot be valid in general. A counterexample is given by the function $\mathbf{v}^h = R_\omega^h \mathbf{w}_\Gamma^h$ with $\mathbf{w}_\Gamma^h \in \mathbf{W}_\Gamma^h$, where we obtain

$$\|Z_\omega^H \mathbf{v}^h\|_{0,\omega}^2 \le CH \|P^H(\mathbf{w}_\Gamma^h)\|_{0,\omega}^2 \le CH \|\mathbf{w}_\Gamma^h\|_{0,\omega}^2 \le C\frac{H}{h}\|\mathbf{v}^h\|_{0,\omega}^2. \tag{7.43}$$

Nevertheless, using (6.5a) and an inverse estimate, the projection of $\mathbf{v}^H \in \mathbf{V}_\omega^H$ onto $\mathbf{V}^h$ gives the stability estimate

$$\begin{aligned} \|Z_\omega^h \mathbf{v}^H\|_{0,\omega} &\le \|Z_\omega^h \mathbf{v}^H - \mathbf{v}^H\|_{0,\omega} + \|\mathbf{v}^H\|_{0,\omega} \le Ch|\mathbf{v}^H|_{1,\omega} + \|\mathbf{v}^H\|_{0,\omega} \\ &\le \left(1 + C\frac{h}{H}\right)\|\mathbf{v}^H\|_{0,\omega} \le C\|\mathbf{v}^H\|_{0,\omega}. \end{aligned} \tag{7.44}$$

Theorem 7.12. *Assume that the material parameters E, ν, and ϱ satisfy the estimates (7.1) on ω as well as that there exists a constant C_{mass} with*

$$\frac{\varrho_\omega}{\Delta t^2} \le C_{mass}\frac{E_\omega}{H^2}. \tag{7.45}$$

Let Assumption 7.4 hold. Then, there exist constants c^*, C^* independent of the diameter of ω, h, H, Δt and the values E_ω, ϱ_ω, such that the following estimates are satisfied for any function $\mathbf{w}_\Gamma^H \in \mathbf{W}_\Gamma^H$:

$$c^*(1 + c_\Gamma^{-2} + C_{mass})^{-1}\left(\mathbf{w}_\Gamma^H, S_\omega^H \mathbf{w}_\Gamma^H\right) \le \left(\mathbf{w}_\Gamma^H, S_\omega^{HhH}\mathbf{w}_\Gamma^H\right) \le C^*\left(\mathbf{w}_\Gamma^H, S_\omega^H \mathbf{w}_\Gamma^H\right). \tag{7.46}$$

If the finite element spaces (7.7) satisfy $\mathbf{V}_\omega^H \subset \mathbf{V}^h$, then the constant C^* in (7.46) can be replaced by one.

Proof. According to (2.22), we define the bilinear forms

$$k_\Theta(\mathbf{v},\mathbf{w}) := \frac{2}{\Delta t^2} m_\Theta(\mathbf{v},\mathbf{w}) + \theta a_\Theta(\mathbf{v},\mathbf{w}), \quad \mathbf{v},\mathbf{w} \in \mathbf{V}_0(\Theta),\ \Theta \in \{\Xi,\omega\}.$$

Due to the fact that the Schur complement matrices S_ω^H, S_ω^{HhH} defined in (7.29), (7.36) correspond to discrete harmonic extensions onto ω with respect to $k_\omega(\cdot,\cdot)$, we obtain

$$\left(\mathbf{w}_\Gamma^H, S_\omega^{HhH}\mathbf{w}_\Gamma^H\right) = \inf_{\substack{\mathbf{v}^h\in\mathbf{V}^h\\ \mathbf{v}^h|_\Gamma = P^h(\mathbf{w}_\Gamma^H)}} k_\omega(\mathbf{v}^h,\mathbf{v}^h), \quad \left(\mathbf{w}_\Gamma^H, S_\omega^{H}\mathbf{w}_\Gamma^H\right) = \inf_{\substack{\mathbf{v}^H\in\mathbf{V}_\omega^H\\ \mathbf{v}^H|_\Gamma = \mathbf{w}_\Gamma^H}} k_\omega(\mathbf{v}^H,\mathbf{v}^H). \tag{7.47}$$

As the spaces $\mathbf{V}^h$, $\mathbf{V}_\omega^H$ are finite dimensional, we can define the values

$$\widehat{\mathbf{v}}^h := \arg \inf_{\substack{\mathbf{v}^h\in\mathbf{V}^h\\ \mathbf{v}^h|_\Gamma = P^h(\mathbf{w}_\Gamma^H)}} k_\omega(\mathbf{v}^h,\mathbf{v}^h), \qquad \widehat{\mathbf{v}}^H := \arg \inf_{\substack{\mathbf{v}^H\in\mathbf{V}_\omega^H\\ \mathbf{v}^H|_\Gamma = \mathbf{w}_\Gamma^H}} k_\omega(\mathbf{v}^H,\mathbf{v}^H). \tag{7.48}$$

Further, recalling that $\mathbf{V}(\Theta) = \mathbf{H}^1(\Theta)$, $\Theta \in \{\Xi,\omega\}$, and $\mathbf{W}_\Gamma = \mathbf{H}^{1/2}(\Gamma)$, we introduce the seminorms

$$|\mathbf{v}|_{\widetilde{\mathbf{V}}(\Theta)} := \inf_{\mathbf{z}\in\mathcal{RB}_\Theta} \|\mathbf{v}+\mathbf{z}\|_{1,\Theta}, \quad \mathbf{v}\in\mathbf{V}(\Theta), \tag{7.49a}$$

$$|\mathbf{w}_\Gamma|_{\widetilde{\mathbf{W}}_\Gamma(\Theta)} := \inf_{\mathbf{z}\in\mathcal{RB}_\Theta} \|\mathbf{w}_\Gamma + \operatorname{tr}_\Gamma \mathbf{z}\|_{1/2,\Gamma}, \quad \mathbf{w}_\Gamma \in \mathbf{W}_\Gamma. \tag{7.49b}$$

Then, we get with the stability estimates (6.5b), (7.44), the equation $(Z_\omega^h\widehat{\mathbf{v}}^H)|_\Gamma = P^h(\mathbf{w}_\Gamma^H)$ as well as the fact that $Z_\omega^h\mathbf{z} = \mathbf{z}$ holds for $\mathbf{z}\in\mathcal{RB}_\omega \subset (\mathbf{V}^h \cap \mathbf{V}_\omega^H)$:

$$\begin{aligned}
k_\omega(\widehat{\mathbf{v}}^h,\widehat{\mathbf{v}}^h) &\le k_\omega\left(Z_\omega^h\widehat{\mathbf{v}}^H, Z_\omega^h\widehat{\mathbf{v}}^H\right)\\
&\le C\left(\theta E_\omega |Z_\omega^h\widehat{\mathbf{v}}^H|^2_{\widetilde{\mathbf{V}}(\omega)} + \frac{\varrho_\omega}{\Delta t^2}\|Z_\omega^h\widehat{\mathbf{v}}^H\|^2_{0,\omega}\right)\\
&\le C\left(\theta E_\omega |\widehat{\mathbf{v}}^H|^2_{\widetilde{\mathbf{V}}(\omega)} + \frac{\varrho_\omega}{\Delta t^2}\|\widehat{\mathbf{v}}^H\|^2_{0,\omega}\right)\\
&\le C^* k_\omega\left(\widehat{\mathbf{v}}^H,\widehat{\mathbf{v}}^H\right),
\end{aligned} \tag{7.50}$$

implying the second inequality of (7.46).

For the other estimate, we proceed similar but need to correct the boundary values

$$(Z_\omega^H\widehat{\mathbf{v}}^h)|_\Gamma = P^H P^h(\mathbf{w}_\Gamma^H),$$

which we do by means of the zero extension operator R_ω^H defined in (7.42). With this, we obtain

$$\begin{aligned}
k_\omega(\widehat{\mathbf{v}}^H,\widehat{\mathbf{v}}^H) &\le k_\omega\Big(Z_\omega^H\widehat{\mathbf{v}}^h + R_\omega^H(\mathrm{Id} - P^HP^h)\mathbf{w}_\Gamma^H, Z_\omega^H\widehat{\mathbf{v}}^h + R_\omega^H(\mathrm{Id} - P^HP^h)\mathbf{w}_\Gamma^H\Big)\\
&\le C\theta E_\omega\left(|Z_\omega^H\widehat{\mathbf{v}}^h|^2_{\widetilde{\mathbf{V}}(\omega)} + \left|R_\omega^H(\mathrm{Id} - P^HP^h)\mathbf{w}_\Gamma^H\right|^2_{1,\omega}\right)\\
&\quad + C\frac{\varrho_\omega}{\Delta t^2}\left(\|Z_\omega^H\widehat{\mathbf{v}}^h\|^2_{0,\omega} + \left\|R_\omega^H(\mathrm{Id} - P^HP^h)\mathbf{w}_\Gamma^H\right\|^2_{0,\omega}\right).
\end{aligned} \tag{7.51}$$

First, we deal with the mass terms in (7.51). Choosing any fixed rigid body motion $\mathbf{z}\in\mathcal{RB}_\omega$, the first term gives

$$\begin{aligned}
\|Z_\omega^H\widehat{\mathbf{v}}^h\|^2_{0,\omega} &\le C\left(\|(Z_\omega^H - \mathrm{Id})\left(\widehat{\mathbf{v}}^h + \mathbf{z}\right)\|^2_{0,\omega} + \|\widehat{\mathbf{v}}^h\|^2_{0,\omega}\right)\\
&\le C\left(H^2|\widehat{\mathbf{v}}^h + \mathbf{z}|^2_{1,\omega} + \|\widehat{\mathbf{v}}^h\|^2_{0,\omega}\right),
\end{aligned} \tag{7.52}$$

where we have used the approximation property of Z_ω^H (6.5a). For the boundary mass term, we make use of the approximation property of P^h (7.11) to arrive at

$$\begin{aligned}\left\|R_\omega^H(\mathrm{Id}-P^HP^h)\mathbf{w}_\Gamma^H\right\|_{0,\omega}^2 &\le CH\|(\mathrm{Id}-P^HP^h)(\mathbf{w}_\Gamma^H+\mathrm{tr}_\Gamma\,\mathbf{z})\|_{0,\Gamma}^2\\ &\le CH\|(\mathrm{Id}-P^H+P^H-P^HP^h)(\mathbf{w}_\Gamma^H+\mathrm{tr}_\Gamma\,\mathbf{z})\|_{0,\Gamma}^2\\ &\le CH\|P^H(\mathrm{Id}-P^h)(\mathbf{w}_\Gamma^H+\mathrm{tr}_\Gamma\,\mathbf{z})H\|_{0,\Gamma}^2\\ &\le CHh\|\mathbf{w}_\Gamma^H+\mathrm{tr}_\Gamma\,\mathbf{z}\|_{1/2,\Gamma}^2.\end{aligned}\tag{7.53}$$

Using Assumption 7.4 and an inverse estimate, we obtain

$$\begin{aligned}\left\|R_\omega^H(\mathrm{Id}-P^HP^h)(\mathbf{w}_\Gamma^H+\mathrm{tr}_\Gamma\,\mathbf{z})\right\|_{0,\omega}^2 &\le Cc_\Gamma^{-2}Hh\|P^h(\mathbf{w}_\Gamma^H+\mathrm{tr}_\Gamma\,\mathbf{z})\|_{1/2,\Gamma}^2\\ &\le Cc_\Gamma^{-2}Hh\|\hat{\mathbf{v}}_\Gamma^h+\mathbf{z}\|_{0,\omega}\|\hat{\mathbf{v}}_\Gamma^h+\mathbf{z}\|_{1,\omega}\\ &\le Cc_\Gamma^{-2}H\|\hat{\mathbf{v}}_\Gamma^h+\mathbf{z}\|_{0,\omega}^2.\end{aligned}\tag{7.54}$$

Adding (7.52) and (7.54) and minimizing over $\mathbf{z}\in\mathcal{RB}_\omega$ finally gives

$$\|Z_\omega^H\hat{\mathbf{v}}^h\|_{0,\omega}^2+\left\|R_\omega^H(\mathrm{Id}-P^HP^h)\mathbf{w}_\Gamma^H\right\|_{0,\omega}^2\le C\left(H^2|\hat{\mathbf{v}}^h|_{\mathbf{V}(\omega)}^2+(1+c_\Gamma^{-2}H)\|\hat{\mathbf{v}}^h\|_{0,\omega}^2\right).\tag{7.55}$$

Next, we consider the stiffness terms in (7.51). The first one can directly be bounded using (6.5b). For the second one, we obtain with the discrete inequalities stated in, e.g., [171], and Assumption 7.4

$$\begin{aligned}&\left|R_\omega^H(\mathrm{Id}-P^HP^h)\mathbf{w}_\Gamma^H\right|_{1,\omega}^2\\ &\le CH^{d-2}\sum_{e\in\mathcal{G}_\Gamma^H}\sum_{p,q\in\mathcal{N}_\Gamma^H\cap\bar{e}}\left(\left((\mathrm{Id}-P^HP^h)\mathbf{w}_\Gamma^H\right)(p)-\left((\mathrm{Id}-P^HP^h)\mathbf{w}_\Gamma^H\right)(q)\right)^2\\ &\le CH^{d-2}\sum_{p\in\mathcal{N}_\Gamma^H}\left((\mathrm{Id}-P^HP^h)\mathbf{w}_\Gamma^H\right)^2(p)\\ &\le CH^{-1}\|(\mathrm{Id}-P^HP^h)\mathbf{w}_\Gamma^H\|_{0,\Gamma}^2\le C|\mathbf{w}_\Gamma^H|_{\widetilde{\mathbf{W}}_\Gamma(\omega)}^2\\ &\le Cc_\Gamma^{-2}|P^h(\mathbf{w}_\Gamma^H)|_{\widetilde{\mathbf{W}}_\Gamma(\omega)}^2\le Cc_\Gamma^{-2}|\hat{\mathbf{v}}^h|_{\mathbf{V}(\omega)}^2.\end{aligned}\tag{7.56}$$

The inequality (7.56) can be shown analogeously to the estimate (7.53).

Combining the above results with (7.55), we obtain from (7.51)

$$k_\omega(\hat{\mathbf{v}}^H,\hat{\mathbf{v}}^H)\le C(1+c_\Gamma^{-2})\theta E_\omega|\hat{\mathbf{v}}^h|_{\mathbf{V}(\omega)}^2+C\frac{\varrho_\omega}{\Delta t^2}\left(H^2|\hat{\mathbf{v}}^h|_{\mathbf{V}(\omega)}^2+(1+c_\Gamma^{-2}H)\|\hat{\mathbf{v}}^h\|_{0,\omega}^2\right).$$

Using Assumption (7.45), we finally get the upper bound

$$\begin{aligned}k_\omega(\hat{\mathbf{v}}^H,\hat{\mathbf{v}}^H)&\le C(1+c_\Gamma^{-2}+C_{\mathrm{mass}})\theta E_\omega|\hat{\mathbf{v}}^h|_{\mathbf{V}(\omega)}^2+C\frac{\varrho_\omega}{\Delta t^2}(1+c_\Gamma^{-2}H)\|\hat{\mathbf{v}}^h\|_{0,\omega}^2\\ &\le (c^*)^{-1}(1+c_\Gamma^{-2}+C_{\mathrm{mass}})k_\omega(\hat{\mathbf{v}}^h,\hat{\mathbf{v}}^h),\end{aligned}\tag{7.57}$$

implying (7.46).

Finally, we prove that the constant C^* in (7.46) can be replaced by one if the finite element spaces $\mathbf{V}_\omega^H\subset\mathbf{V}^h$ are nested. For this, we need to show the inequality

$$\left(\mathbf{w}_\Gamma^H,(S_\omega^H-S_\omega^{HhH})\mathbf{w}_\Gamma^H\right)\ge 0.\tag{7.58}$$

The nestedness of $\mathbf{W}_\Gamma^H \subset \mathbf{W}_\Gamma^h$ implies that P^h is the identity. Observing that

$$\left\{\mathbf{w}^H \in \mathbf{V}_\omega^H : \mathbf{w}^H|_\Gamma = \mathbf{w}_\Gamma^H\right\} \subset \left\{\mathbf{w}^h \in \mathbf{V}^h : \mathbf{w}^h|_\Gamma = P^h(\mathbf{w}_\Gamma^H)\right\},$$

we find

$$\left(\mathbf{w}_\Gamma^H, (S_\omega^H - S_\omega^{HhH})\mathbf{w}_\Gamma^H\right) = \inf_{\substack{\mathbf{w}^H \in \mathbf{V}_\omega^H \\ \mathbf{w}^H|_\Gamma = \mathbf{w}_\Gamma^H}} k_\omega(\mathbf{w}^H, \mathbf{w}^H) - \inf_{\substack{\mathbf{w}^h \in \mathbf{V}^h \\ \mathbf{w}^h|_\Gamma = P^h(\mathbf{w}_\Gamma^H)}} k_\omega(\mathbf{w}^h, \mathbf{w}^h) \geq 0.$$

□

Thus, if the assumptions of Theorem 7.12 are satisfied, the condition number of the iteration matrix in (7.38) is bounded by

$$\kappa\left(\left(S_\Xi^H + S_\omega^H\right)^{-1}\left(S_\Xi^H + S_\omega^{HhH}\right)\right) \leq \frac{\max(1, C^*)}{\min\left(1, c^*(1 + c_\Gamma^{-2} + C_{\text{mass}})^{-1}\right)}. \tag{7.59}$$

with the constants c^*, C^* from Theorem 7.12. Furthermore, the iterates of (7.41) converge to the solution $\mathbf{u}_{\Gamma,j+1}^H$ of (7.37) if the damping parameter α_l is chosen within the set $0 < \alpha_{\min} \leq \alpha_l < \frac{2}{C^*}$ for some fixed value of $\alpha_{\min}$ (see, e.g., [161]). Hence, if the finite element spaces $\mathbf{V}_\omega^H \subset \mathbf{V}^h$ are nested, Theorem 7.12 implies the convergence of (7.41) for $\alpha_l = 1$, i.e., the convergence of Algorithm 1.

Remark 7.13. In the case that the mass term on ω is dominant and Assumption (7.45) is only satisfied with a large constant C_{mass}, the lower bound in (7.46) degenerates. The proof of Theorem 7.12 shows that this assumption is only needed to bound the H^1-term in (7.52). Instead, we can bound this term by means of (6.5a) and an inverse estimate, leading (up to logarithmic terms) to an upper bound of

$$k_\omega(\hat{\mathbf{v}}^H, \hat{\mathbf{v}}^H) \leq (c^*)^{-1} \max\left(1 + c_\Gamma^{-2}, 1 + \frac{H}{h} + c_\Gamma^{-2} H\right) k_\omega(\hat{\mathbf{v}}^h, \hat{\mathbf{v}}^h)$$

instead of (7.57). The resulting dependence of the convergence rate of Algorithm 1 on the ratio H/h can be observed from the numerical results in Section 7.3.3; furthermore, it is in agreeement with the theoretical estimates of the spectrum of discrete Dirichlet–to–Neumann operators recently presented in [171].

Remark 7.14. Algorithm 1 can also be defined with a Lagrange multiplier space $\mathbf{M}_\Gamma^{h*}$ associated with a coarser triangulation $\mathcal{G}^{h*} \subset \mathcal{G}^h$ (cf. Remarks 7.1 and 7.8). The results of Lemma 7.9 and Theorem 7.12 can be transferred onto this case, provided that the spaces $\mathbf{W}_\Gamma^H$ and $\mathbf{M}_\Gamma^{h*}$ satisfy a uniform inf–sup condition according to Lemma 7.6. The latter is trivially fulfilled for the special case $\mathbf{W}_\Gamma^H \subset \mathbf{W}_\Gamma^h$ and $\mathbf{M}_\Gamma^{h*} = \mathbf{M}_\Gamma^H$.

The disadvantage of Theorem 7.12 is the need for Assumption 7.4 if the trace spaces are not nested. In the following, we state a variant of Theorem 7.12 that replaces Assumption 7.4 by a stronger restriction of the material parameters on the subdomains.

Lemma 7.15. *Assume that the material parameters E, ν, ϱ satisfy the estimate (7.1) and in addition*

$$E_\omega \le C_{par} E_\Xi, \qquad \varrho_\omega \le C_{par} \varrho_\Xi, \tag{7.60}$$

for some given value of C_{par}. Then, there exist a constant c^* independent of the diameter of ω, h, H, Δt or the material parameters such that the following estimate is satisfied for any function $\mathbf{w}_\Gamma^H \in \mathbf{W}_\Gamma^H$:

$$c^*(1 + C_{par})^{-1} \left(\mathbf{w}_\Gamma^H, (S_\Xi^H + S_\omega^H)\mathbf{w}_\Gamma^H\right) \le \left(\mathbf{w}_\Gamma^H, (S_\Xi^H + S_\omega^{HhH})\mathbf{w}_\Gamma^H\right). \tag{7.61}$$

Proof. In the proof of Theorem 7.12, we have used Assumption 7.4 for the upper bound of the two boundary terms in (7.51). If this assumption is not satisfied, we can bound these terms by their counterparts on Ξ. For this, we define

$$\check{\mathbf{v}}^H := \arg \inf_{\substack{\mathbf{v}^H \in \mathbf{V}_\Xi^H \\ \mathbf{v}^H|_\Gamma = \mathbf{w}_\Gamma^H}} k_\Xi(\mathbf{v}^H, \mathbf{v}^H).$$

For the mass term in (7.51), we obtain

$$\left\| R_\omega^H (\mathrm{Id} - P^H P^h)\mathbf{w}_\Gamma^H \right\|_{0,\omega}^2 \le CH \|\mathbf{w}_\Gamma^H\|_{0,\Gamma}^2 \le C \|R_\Xi^H \mathbf{w}_\Gamma^H\|_{0,\Xi}^2 \le C \|\check{\mathbf{v}}^H\|_{0,\Xi}^2.$$

For the stiffness term, we obtain similarly as in (7.56)

$$\left| R^H (\mathrm{Id} - P^H P^h)\mathbf{w}_\Gamma^H \right|_{1,\omega}^2 \le C |\mathbf{w}_\Gamma^H|_{\widetilde{\mathbf{W}}_\Gamma(\Xi)}^2.$$

Finally, the second-to-last term in (7.52) can be bounded by

$$\begin{aligned} H^2 |\hat{\mathbf{v}}^h|_{\widetilde{\mathbf{V}}(\omega)}^2 &\le CH^2 |P^h(\mathbf{w}_\Gamma^H)|_{1/2,\Gamma}^2 \le CH^2 |\mathbf{w}_\Gamma^H|_{1/2,\Gamma}^2 \\ &\le CH^2 |\check{\mathbf{v}}^H|_{1,\Xi}^2 \le C \|\check{\mathbf{v}}^H\|_{0,\Xi}^2. \end{aligned}$$

With (7.51), the above estimates and Assumption (7.60), we obtain

$$\begin{aligned} \inf_{\substack{\mathbf{v}^H \in \mathbf{V}_\omega^H \\ \mathbf{v}^H|_\Gamma = \mathbf{w}_\Gamma^H}} k_\omega(\mathbf{v}^H, \mathbf{v}^H) &\le C k_\omega(\hat{\mathbf{v}}^h, \hat{\mathbf{v}}^h) + C \left(\theta E_\omega |\mathbf{w}_\Gamma^H|_{\widetilde{\mathbf{W}}_\Gamma(\Xi)}^2 + \frac{\varrho_\omega}{\Delta t^2} \|\check{\mathbf{v}}^H\|_{0,\Xi}^2\right) \\ &\le C \left(k_\omega(\hat{\mathbf{v}}^h, \hat{\mathbf{v}}^h) + C_{\mathrm{par}} k_\omega(\check{\mathbf{v}}^H, \check{\mathbf{v}}^H)\right), \end{aligned}$$

implying (7.61). □

However, in the unfavorable case that the material parameters satisfy $E_\omega \gg E_\Xi$ or $\varrho_\omega \gg \varrho_\Xi$ and Assumption 7.4 is not satisfied, both Theorem 7.12 and Lemma 7.15 do not provide a uniform bound for the condition number of (7.59). Concerning the trace spaces $\mathbf{W}_\Gamma^H$, $\mathbf{W}_\Gamma^h$, this means that there exist some "high frequency" components $\mathbf{w}_\Gamma^H \in \mathbf{W}_\Gamma^H$ that are strongly damped or even annihilated by the discrete mortar operator Π^{hH}. As Π^{hH} is present in the exact mortar system (7.37) but not in the preconditioning matrix $(S_\Xi^H + S_\omega^H)$, these frequencies can spoil the good convergence of Algorithm 1, as a numerical example in Section 7.3.5 shows.

A possible remedy in the above sketched situation is to define the auxiliary coarse grid operator S_ω^H in a different way such that these spurious frequencies are damped. This can for example be done by replacing S_ω^H in (7.27), (7.38) and (7.41) by the nonsymmetric matrix $S_\omega^H Q^{HH}$ with $Q^{HH} = \Pi^{Hh}\Pi^{hH}$ (cf. (7.16)). This additional factor is likely to have a damping effect on the high frequency components of $\mathbf{W}_\Gamma^H$. Furthermore, if the trace spaces $\mathbf{W}_\Gamma^H \subset \mathbf{W}_\Gamma^h$ are nested and Theorem 7.12 can be applied, Q^{HH} is the identity on $\mathbf{W}_\Gamma^H$. The numerical performance of the modified nonsymmetric version of Algorithm 1 is investigated in Section 7.3.5.

7.2.4 Stopping criteria

In this subsection, we state a measure of the algebraic error introduced by solving (7.27) by means of the iterative scheme (7.30).

As the residuals (7.35b) to (7.35d) vanish for $l \geq 1$, the only nonzero component of the residual vector is (7.35a) which can be written as

$$\mathbf{r}_\Xi^{H,(l)} = \left(-\widehat{K}_d \delta \widehat{\mathbf{z}}_d^{(l-1)}\right)_1 = (D_\Gamma^{hH})^T \delta \boldsymbol{\zeta}_{\Gamma,j+\theta}^{h,(l-1)} - (D_\Gamma^{HH})^T \delta \boldsymbol{\mu}_{\Gamma,j+\theta}^{H,(l-1)}. \tag{7.62}$$

As the energy norm of this residual given by

$$\left((S_\Xi^H + S_\omega^{HhH})^{-1}\mathbf{r}_\Xi^{H,(l)}, \mathbf{r}_\Xi^{H,(l)}\right)$$

is too expensive to compute, it is approximated by the value

$$\left((S_\Xi^H + S_\omega^H)^{-1}\mathbf{r}_\Xi^{H,(l)}, \mathbf{r}_\Xi^{H,(l)}\right) = \left(\delta \mathbf{u}_{\Gamma,j+1}^{H,(l)}, \mathbf{r}_\Xi^{H,(l)}\right), \tag{7.63}$$

where the last equality follows from (7.32) for $l \geq 1$. Thus, we propose to use the following relative algebraic error estimator for $l \geq 1$:

$$\left(\eta_{\text{alg}}^{(l)}\right)^2 := \frac{\left(\delta \mathbf{u}_{\Gamma,j+1}^{H,(l)}, \mathbf{r}_\Xi^{H,(l)}\right)}{\left(\mathbf{u}_{\Gamma,j+1}^{H,(l)}, (S_\Xi^H + S_\omega^H)\mathbf{u}_{\Gamma,j+1}^{H,(l)}\right)}. \tag{7.64}$$

As (7.64) is a norm of the error between the the exact mortar solution and the approximate solution of (7.27) by means of Algorithm 1, it can be used to define a stopping criterion for the iterative process (see also Section 8.3).

Remark 7.16. In order to compute the denominator of (7.64) without solving a Schur complement problem in each step, one can use the relation

$$(S_\Xi^H + S_\omega^H)\mathbf{u}_{\Gamma,j+1}^{H,(l)} = (S_\Xi^H + S_\omega^H)\mathbf{u}_{\Gamma,j+1}^{H,(0)} + \mathbf{r}_\omega^{H,(0)} + \sum_{k=0}^{l-1} \mathbf{r}_\Xi^{H,(k)}.$$

7.3 Numerical results

7.3.1 Geometry and parameters

For the first numerical tests, we consider the domain $\Omega = [0, 2] \times [0, 1]$ which is split into the patch $\omega = [0.5, 1.5] \times [0, 0.5]$ and the upper domain $\Xi = \Omega \backslash \bar{\omega}$. Both subdomains are initially discretized by quadrilaterals of size $H = h = 0.125$; afterwards, we perform L additional refinements of the fine grid on ω. The resulting grid for $L = 2$ is depicted on the left side of Figure 7.3. Hence, we have nested finite element spaces $\mathbf{V}_\omega^H \subset \mathbf{V}^h$ such that Assumption 7.4 is satisfied with $c_\Gamma = 1$.

Furthermore, the nestedness of the triangulations $\mathcal{G}_\Gamma^H$, $\mathcal{G}_\Gamma^h$ implies that Algorithm 1 is still well-defined if the Lagrange multiplier $\boldsymbol{\zeta}_\Gamma$ is chosen with respect to $\mathcal{G}_\Gamma^H$ (cf. Remarks 7.1, 7.8 and 7.14). In [69], we have tested both possibilities $\mathbf{M}_\Gamma^h$ and $\mathbf{M}_\Gamma^H$ as spaces for the Lagrange multiplier; however, as the convergence rates for the coarse multiplier space are slightly worse than for the fine space, we only present the results for the space $\mathbf{M}_\Gamma^h$.

Remark 7.17. As stated in Section 7.1, each node of the interface Γ is associated with a degree of freedom for the Lagrange multiplier, i.e., there are no modifications due to cross points.

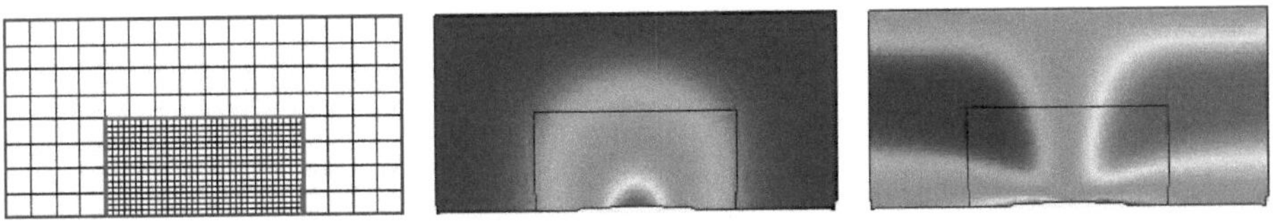

Figure 7.3: First example; left: grid for $L = 2$; middle: effective stress; right: pressure.

On the discretized domain Ω, we solve the equations of linear elasticity with the constant elasticity module $E_\Xi = 100$ and the Poisson ratio $\nu = 0.3$. The elasticity module on ω is chosen as $E_\omega = 10^{\text{par}} E_\Xi$, $\text{par} \in \mathbb{Z}$. The coarse matrix S_ω^H and the right hand side $\bar{\boldsymbol{\varrho}}_j^H$ are assembled as in (7.29) and (2.23), (7.26) with h replaced by H, respectively, and the starting vector is taken to be $\mathbf{0}$. Thus, all computations with $L = 0$ yield the exact mortar solution after one iteration and are not presented in the following.

7.3.2 Algebraic error for static case

For the first set of tests, we set $\varrho = 0$, $\theta = 1$ and $\mathbf{l} = \mathbf{0}$; further, we enforce homogeneous Dirichlet conditions on the upper boundary, homogeneous Neumann boundary conditions on the left and right side and a surface load of $\mathbf{g}_N = 10^6 \cdot \max(0.25 - |x_1 - 1|, 0)$ at the bottom. The effective stress and the pressure of the corresponding mortar solution are depicted in Figure 7.3.

In order to investigate the decrease of the algebraic error for Algorithm 1, the difference between the approximations $(\mathbf{u}^{H,(l)}, \mathbf{u}^{h,(l)})$ obtained by Algorithm 1 and the solution $(\mathbf{u}^H, \mathbf{u}^h)$ of the discrete mortar system (7.25) is measured with respect to the relative energy norm by

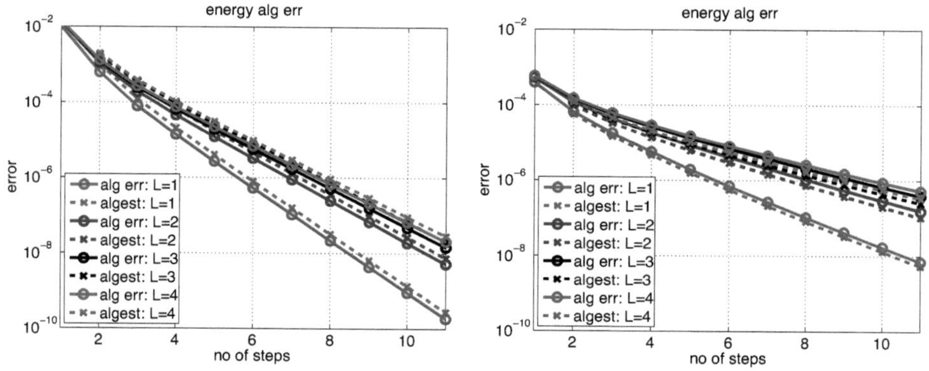

Figure 7.4: True $e_{\text{alg}}^{(l)}$ and estimated $\eta_{\text{alg}}^{(l)}$ relative algebraic error for Algorithm 1 with $L \in \{1,2,3,4\}$ with respect to l; left: $E_\omega = E_\Xi$; right: $E_\omega = 10^5 E_\Xi$.

computing

$$\left(e_{\text{alg}}^{(l)}\right)^2 := \frac{(\mathbf{u}^{H,(l)} - \mathbf{u}^H)^T A_\Xi^H (\mathbf{u}^{H,(l)} - \mathbf{u}^H) + (\mathbf{u}^{h,(l)} - \mathbf{u}^h)^T A_\omega^h (\mathbf{u}^{h,(l)} - \mathbf{u}^h)}{(\mathbf{u}^H)^T A_\Xi^H \mathbf{u}^H + (\mathbf{u}^h)^T A_\omega^h \mathbf{u}^h}. \tag{7.65}$$

The results for this relative algebraic error and the corresponding estimator (7.64) are depicted in Figure 7.4 for different values of $L \in \{1,2,3,4\}$. The left picture shows the results for equal material parameters $E_\omega = E_\Xi$, whereas the right picture displays the convergence for discontinuous parameters $E_\omega = 10^5 E_\Xi$. One can see that the decay rate with respect to the number of iterations is the same for the true and the estimated algebraic error and that the difference between them is very small. This indicates that η_{alg} is well suited to measure the algebraic error due to the iterative solution of the coupled system.

The dependence of the algebraic error reduction on the ratio $H/h = 2^L$ as well as on the jump in the material parameters $E_\omega = 10^{\text{par}} E_\Xi$ is further investigated in Figures 7.5 and 7.6. The former displays the effectivity index $\eta_{\text{alg}}^{(l)}/e_{\text{alg}}^{(l)}$ of the estimated and the true relative algebraic error for par $= 0$ and varying L on the left side and for $L = 2$ and different values of par on the right side. One can observe that the ratio is always in the interval $[0.6, 1.65]$, decreases for increasing L and par and converges to a fixed value for $l \to \infty$. This again demonstrates the adequacy of the error measure (7.64).

Figure 7.6 features the reduction factor $e_{\text{alg}}^{(l+1)}/e_{\text{alg}}^{(l)}$ of the relative algebraic error in the l-th step of Algorithm 1. The results show that the algorithm converges best for small values of L and par, but the reduction factor is limited from above by around 0.35 for par $= 0$ independently of L and by around 0.55 for $L = 2$ and par $\to \infty$. Hence, in the static case, Algorithm 1 is stable with respect to jumps in the coefficients, in accordance with the theoretical results of Theorem 7.12 for the case $\varrho_\omega = 0$.

Remark 7.18. From (7.38), one can see that for $l \to \infty$, the error in the coarse trace $\mathbf{e}_\Gamma^{H,(l)} = (\mathbf{u}_\Gamma^{H,(l)} - \mathbf{u}_\Gamma^H)$ converges to an eigenvector $\mathbf{v}_\Gamma^H$ of the iteration matrix in (7.38) corresponding to

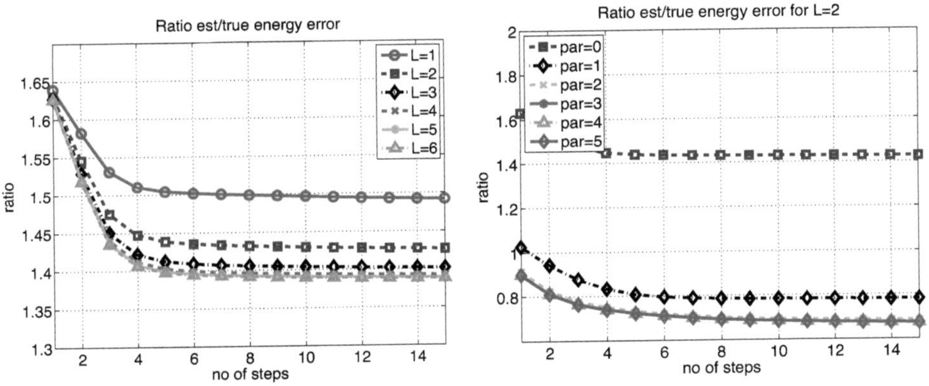

Figure 7.5: Ratio estimated/true relative algebraic error $\eta_{\text{alg}}^{(l)}/e_{\text{alg}}^{(l)}$ for $E_\omega = 10^{\text{par}} E_\Xi$ with respect to l; left: par $= 0$, $L \in \{1, \dots, 6\}$; right: $L = 2$, par $\in \{0, \dots, 5\}$.

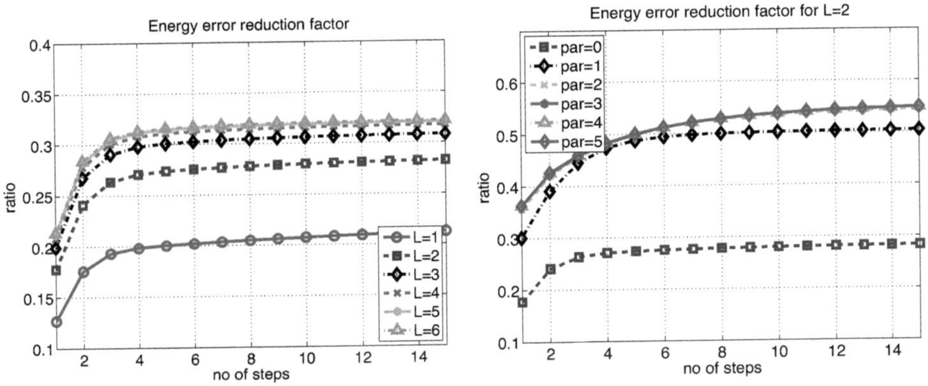

Figure 7.6: Algebraic error reduction factor $e_{\text{alg}}^{(l+1)}/e_{\text{alg}}^{(l)}$ for $E_\omega = 10^{\text{par}} E_\Xi$ with respect to l; left: par $= 0$, $L \in \{1, \dots, 6\}$; right: $L = 2$, par $\in \{0, \dots, 5\}$.

the largest eigenvalue $\lambda_{\max}$ such that $\mathbf{v}_\Gamma^H \cdot \mathbf{e}_\Gamma^{H,(0)} \neq 0$. Thus, the error reduction factor $e_{\text{alg}}^{(l+1)}/e_{\text{alg}}^{(l)}$ becomes

$$\frac{e_{\text{alg}}^{(l+1)}}{e_{\text{alg}}^{(l)}} = \frac{\left(\mathbf{e}_\Gamma^{H,(l+1)}, (S_\Xi^H + S_\omega^{HhH})\mathbf{e}_\Gamma^{H,(l+1)}\right)}{\left(\mathbf{e}_\Gamma^{H,(l)}, (S_\Xi^H + S_\omega^{HhH})\mathbf{e}_\Gamma^{H,(l)}\right)} \longrightarrow \lambda_{\max}^2.$$

Hence, the results depicted in Figures 7.6 allow for an estimate of the largest eigenvalue of the iteration matrix in (7.38).

7.3.3 Algebraic error for dynamic case

Next, we consider a dynamic setting by choosing $\theta = \frac{1}{2}$, $\Delta t = 10^{-3}$ and piecewise constant material parameters $E_\Xi = 100$, $\varrho_\Xi = 0.01$ and $E_\omega = 10^{\mathrm{par}_E} \cdot E_\Xi$, $\varrho_\omega = 10^{\mathrm{par}_\varrho} \cdot \varrho_\Xi$. The parameters are chosen such that the mass and the stiffness part on Ξ are approximately equilibrated, i.e., $\frac{E_\Xi}{H^2} \approx 10^4 = \frac{\varrho_\Xi}{\Delta t^2}$. We impose homogeneous Dirichlet boundary conditions on the top, the parameter-dependent Neumann compression forces

$$\mathbf{g}_N = f(\mathrm{par}_\varrho) \cdot \max(0, 0.25 - r), \quad \text{with} \quad r^2 = (0.5 - x_1)^2,$$
$$f(\mathrm{par}_\varrho) = 10^{2+\min(-2,\mathrm{par}_\varrho)+\delta_{\mathrm{par}_E>0}+\max(3,\min(0,0.5(\mathrm{par}_\varrho-\mathrm{par}_E)))},$$

on the bottom and homogeneous Neumann conditions elsewhere. Both the volume load and the initial velocity are set to zero.

$\mathrm{par}_E \setminus \mathrm{par}_\varrho$	-4	-2	0	2	4	6
-4	0.0001	0.0037	0.2572	0.7357	0.7498	0.7500
-2	0.0034	0.0052	0.2449	0.7352	0.7498	0.7500
0	0.2311	0.2315	0.2792	0.6933	0.7498	0.7500
2	0.5441	0.5442	0.5473	0.5568	0.7065	0.7495
4	0.5462	0.5463	0.5516	0.5568	0.5624	0.7066
6	0.5462	0.5464	0.5516	0.5560	0.5558	0.5624

Table 7.1: Asymptotic error reduction rates $e_{\mathrm{alg}}^{(l+1)}/e_{\mathrm{alg}}^{(l)}$ for $L = 2$ and different values of $E_\omega = 10^{\mathrm{par}_E} E_\Xi$, $\varrho_\omega = 10^{\mathrm{par}_\varrho} \varrho_\Xi$ with $\mathrm{par}_\varrho, \mathrm{par}_E \in \{-4, -2, 0, 2, 4, 6\}$.

$\mathrm{par}_E \setminus \mathrm{par}_\varrho$	-4	-2	0	2	4	6
-4	0.0001	0.0043	0.3210	0.9196	0.9373	0.9375
-2	0.0039	0.0059	0.2887	0.9177	0.9373	0.9375
0	0.2617	0.2621	0.3159	0.8144	0.9373	0.9375
2	0.6063	0.6063	0.6095	0.6240	0.8298	0.9355
4	0.6090	0.6091	0.6139	0.6193	0.6302	0.8300
6	0.6091	0.6091	0.6138	0.6195	0.6194	0.6303

Table 7.2: Asymptotic error reduction rates $e_{\mathrm{alg}}^{(l+1)}/e_{\mathrm{alg}}^{(l)}$ for $L = 4$ and different values of $E_\omega = 10^{\mathrm{par}_E} E_\Xi$, $\varrho_\omega = 10^{\mathrm{par}_\varrho} \varrho_\Xi$ with $\mathrm{par}_\varrho, \mathrm{par}_E \in \{-4, -2, 0, 2, 4, 6\}$.

In Tables 7.1 and 7.2, the algebraic error reduction rates of Algorithm 1 are summarized for $L \in \{2, 4\}$, $\mathrm{par}_E, \mathrm{par}_\varrho \in \{-4, -2, 0, 2, 4, 6\}$. One can see that for fixed $L = \log_2(H/h)$, the error reduction factor is bounded from above by $(1 - h/H)$ independently of the ratio of the grid sizes or the material parameters E_ω and ϱ_ω. However, for $\varrho_\omega \gg E_\omega$, this L-dependent convergence rate is actually reached, in accordance with the theoretical results of Theorem 7.12 and Remark 7.13.

Algorithm 2 Dirichlet–Neumann coupling

Starting from some initial guess $\widehat{\mathbf{z}}^{(0)} := \left(\mathbf{u}^{H,(0)}_{\Gamma,j+1}, \mathbf{u}^{h,(0)}_{\Gamma,j+1}, \boldsymbol{\zeta}^{h,(0)}_{\Gamma,j+\theta}\right)$, compute sequentially for $l = 0, 1, \dots$

(i) Solve problem on coarse space $\mathbf{V}^H|_\Xi$ with boundary load on Γ inherited from fine computation on ω:

$$S^H_\Xi \delta\mathbf{u}^{H,(l)}_{\Gamma,j+1} = \bar{\boldsymbol{\varrho}}^H_{\Xi,j} - S^H_\Xi \mathbf{u}^{H,(l)}_{\Gamma,j+1} + (D^{hH}_\Gamma)^T \boldsymbol{\zeta}^{h,(l)}_{\Gamma,j+\theta},$$

(ii) Solve problem on fine space $\mathbf{V}^h$ with weakly imposed trace on Γ inherited from coarse computation on Ω:

$$\begin{pmatrix} S^h_\omega & (D^{hh}_\Gamma)^T \\ D^{hh}_\Gamma & \mathbf{0} \end{pmatrix} \begin{pmatrix} \delta\mathbf{u}^{h,(l)}_{\Gamma,j+1} \\ \delta\boldsymbol{\zeta}^{h,(l)}_{\Gamma,j+\theta} \end{pmatrix} = \begin{pmatrix} \mathbf{r}^{h,(l)}_\omega \\ \boldsymbol{\nu}^{h,(l)}_\Gamma \end{pmatrix} + \begin{pmatrix} \mathbf{0} \\ D^{hH}_\Gamma \delta\mathbf{u}^{H,(l)}_{\Gamma,j+1} \end{pmatrix},$$

(iii) Update the solution vector:

$$\widehat{\mathbf{z}}^{(l+1)} := \widehat{\mathbf{z}}^{(l)} + \delta\widehat{\mathbf{z}}^{(l)}.$$

The residuals $\mathbf{r}^{h,(l)}_\omega$, $\boldsymbol{\nu}^{h,(l)}_\Gamma$ are defined in (7.35c), (7.35d).

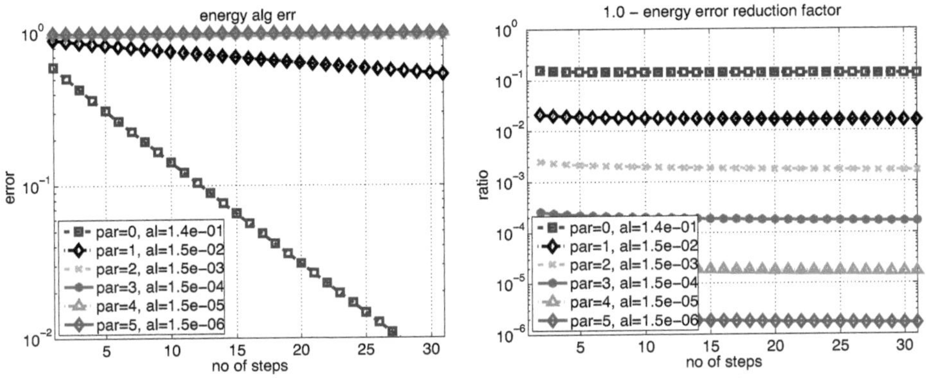

Figure 7.7: Rel. algebraic error $e^{(l)}_{\text{alg}}$ for Dirichlet–Neumann coupling with $\varrho = 0$, $E_\omega = 10^{\text{par}} E_\Xi$, $\text{par} \in \{0, \dots, 5\}$ with respect to l; left: $e^{(l)}_{\text{alg}}$; right: $\left(1 - e^{(l+1)}_{\text{alg}} / e^{(l)}_{\text{alg}}\right)$.

7.3.4 Comparison with Dirichlet–Neumann algorithm

In this subsection, we compare the performance of Algorithm 1 with that of a Dirichlet–Neumann coupling on the subdomains ω (Dirichlet) and Ξ (Neumann). This well-known algorithm (see, e.g., [18, 24, 127, 145]) can be implemented as stated in Algorithm 2. The left picture of Figure 7.7 shows the decay of the relative algebraic energy error $e^{(l)}_{\text{alg}}$ for the static test setting from Subsection 7.3.2 with $L = 0$, different material coefficients $E_\omega = 10^{\text{par}} E_\Xi$, par $\in \{0, \ldots, 5\}$, and suitable damping parameters α such that the algorithm converges. The right picture displays the quantity $\left(1 - e_{\text{alg}}^{(l+1)} / e_{\text{alg}}^{(l)}\right)$, again for different material parameters. We remark that the results for $L \in \{0, 1, 2, 3, 4\}$ are overall independent of L and are hence omitted in Figure 7.7. One can observe that a strong damping is necessary to ensure the convergence of the Dirichlet–Neumann algorithm, that its convergence rate is strongly dependent on the jump in the material parameters, and that its performance is much worse than that of Algorithm 1 even for the case par $= 0$.

7.3.5 Algebraic error for nonnested trace spaces

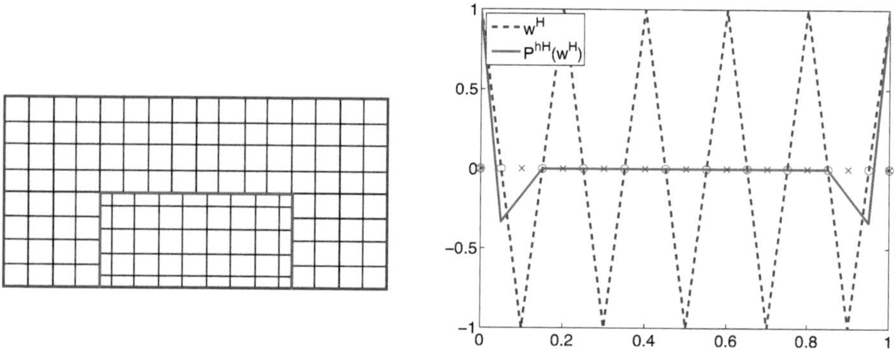

Figure 7.8: Triangulation of Ω with nonnested trace spaces; left: $\mathcal{T}^H$ and $\mathcal{T}^h$; right: trace function $\mathbf{w}^H \in \mathbf{W}_\Gamma^H$ and corresponding image $P^h(\mathbf{w}^H) \in \mathbf{W}_\Gamma^h$ with $\|P^h(\mathbf{w}^H)\|_{1/2,\Gamma} = c_\Gamma(H)\|\mathbf{w}^H\|_{1/2,\Gamma}$ and $c_\Gamma(H) \to 0$ for $H \to 0$.

Next, we investigate the case that the finite element and trace spaces are not nested. For this, we consider the same domains Ω, ω as described in Section 7.3.1 but now choose a different triangulation $\mathcal{T}^h$ on ω as depicted on the left of Figure 7.8. One can see that $\mathbf{W}_\Gamma^H \not\subset \mathbf{W}_\Gamma^h$; furthermore, there exist functions $\mathbf{w}^H \in \mathbf{W}_\Gamma^H$, like the one sketched on the right of Figure 7.8, such that (7.15) only holds for $c_\Gamma(H)$ depending on H and $c_\Gamma(H) \to 0$ for $H \to 0$. Hence, Assumption 7.4 is not satisfied for this discretization.

To begin with, we test the performance of Algorithm 1 for these nonnested trace spaces in the static case, i.e., we set $\varrho = 0$ and take the problem setting from Subsection 7.3.2. In Figure 7.9, the error decay is shown for a high jump in the elasticity module $E_\omega = 10^5 E_\Xi$ and different grid sizes $h = H = 2^{-3-R_H}$, $R_H \in \{0, 1, 2, 3\}$. One can see that the error reduction is considerably slower compared to the case of nested trace spaces and converges for $H \to 0$ to a value close to one; furthermore, the left picture shows that the algebraic error estimator (7.64) strongly underestimates the true value.

Hence, we try to achieve a better convergence rate by using the nonsymmetric modification of Algorithm 1 stated on page 122, i.e., we replace the Schur matrix S_ω^H by the matrix $S_\omega^H Q^{HH}$ with $Q^{HH} = \Pi^{Hh}\Pi^{hH}$. For the example considered here, this modification indeed improves the

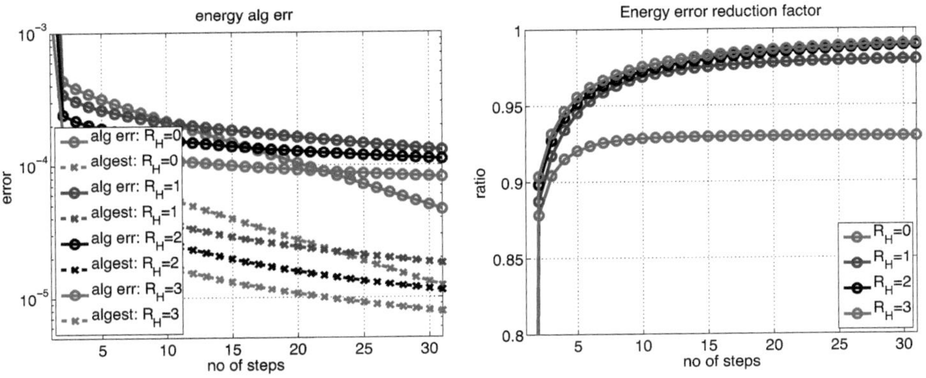

Figure 7.9: Performance of Algorithm 1 for $E_\omega = 10^5 E_\Xi$ and $R_H \in \{0,1,2,3\}$; left: true $e_{\text{alg}}^{(l)}$ and estimated $\eta_{\text{alg}}^{(l)}$ rel. alg. error; right: error reduction factor $e_{\text{alg}}^{(l+1)}/e_{\text{alg}}^{(l)}$.

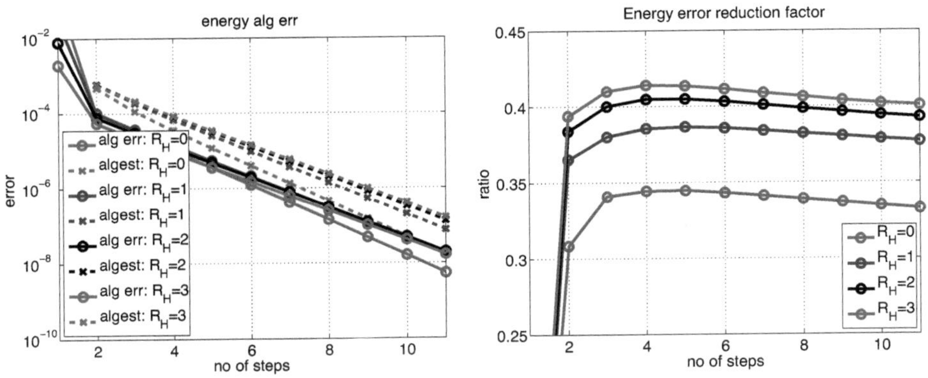

Figure 7.10: Performance of modified Algorithm 1 for $E_\omega = 10^5 E_\Xi$ and $R_H \in \{0,1,2,3\}$; left: $e_{\text{alg}}^{(l)}$ and $\eta_{\text{alg}}^{(l)}$; right: error reduction factor $e_{\text{alg}}^{(l+1)}/e_{\text{alg}}^{(l)}$.

convergence, as the results of Figures 7.10 and 7.11 show. The former figure demonstrates that the error decay of the modified version is comparable to the results of Algorithm 1 for nested spaces (cf. Figures 7.4 and 7.6). In Figure 7.11, we have fixed $R_H = 3$ and compare the error reduction factors for the original and modified algorithm for different material parameters. One can observe that the nonsymmetric version of Algorithm 1 shown on the right side gives much better results than the unmodified algorithm depicted on the left side.

Next, we apply the algorithm to the dynamic problem setting of Subsection 7.3.3 and test different combinations of the material parameters $E_\omega = 10^{\text{par}_E} E_\Xi$, $\varrho_\omega = 10^{\text{par}_\varrho} \varrho_\Xi$ with $\text{par}_\varrho, \text{par}_E \in \{0,2,4,6\}$. The asymptotic error reduction factors for $R_H \in \{1,3\}$ are given in

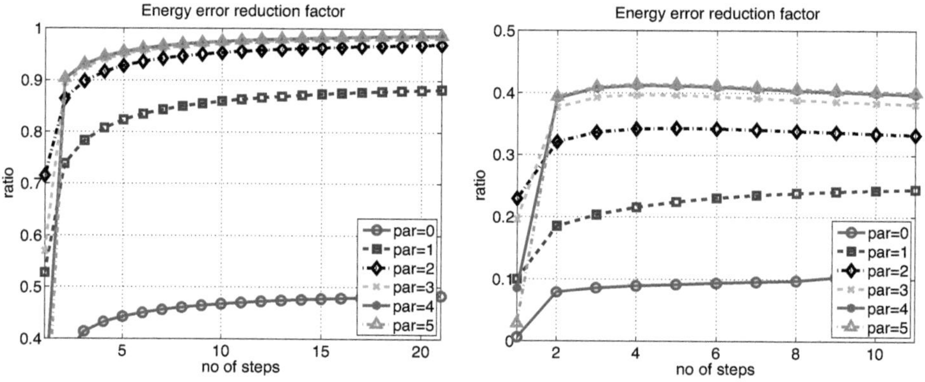

Figure 7.11: Error reduction factor of Algorithm 1 for $R_H = 3$ and $E_\omega = 10^{\text{par}} \cdot E_\Xi$, par $\in \{0, \dots, 5\}$; left: unmodified version; right: modified version.

$\text{par}_E \setminus \text{par}_\varrho$	0	2	4	6	0	2	4	6
0	0.4704	0.8705	0.9700	0.9716	0.1260	0.4101	0.5773	0.5952
2	0.9680	0.9360	0.9651	0.9695	0.3302	0.3455	0.5249	0.5934
4	0.9803	0.9472	0.9457	0.9660	0.3723	0.3737	0.3819	0.5369
6	0.9804	0.9437	0.9459	0.9458	0.4026	0.3301	0.3742	0.3824

Table 7.3: Asymptotic error reduction rates for $R_H = 1$ and different values of $E_\omega = 10^{\text{par}_E} E_\Xi$, $\varrho_\omega = 10^{\text{par}_\varrho} \varrho_\Xi$ with $\text{par}_\varrho, \text{par}_E \in \{0, 2, 4, 6\}$; left values: unmodified version; right values: modified version.

Tables 7.3 and 7.4; the results for the unmodified algorithm are summarized on the left side of the table, and those for the modified version are given on the right side. In accordance with Lemma 7.15, the performance of Algorithm 1 degrades if at least one of the two ratios E_ω / E_Ξ, $\varrho_\omega / \varrho_\Xi$ is large. But the nonsymmetric modification improves the convergence for either case.

$\text{par}_E \setminus \text{par}_\varrho$	0	2	4	6	0	2	4	6
0	0.4721	0.8589	0.9677	0.9950	0.1118	0.3347	0.6001	0.5840
2	0.9722	0.9690	0.9801	0.9907	0.3246	0.3185	0.5036	0.5896
4	0.9880	0.9881	0.9853	0.9894	0.3917	0.3861	0.3643	0.4260
6	0.9884	0.9893	0.9874	0.9855	0.4131	0.4020	0.3920	0.3696

Table 7.4: Asymptotic error reduction rates for $R_H = 3$ and different values of $E_\omega = 10^{\text{par}_E} E_\Xi$, $\varrho_\omega = 10^{\text{par}_\varrho} \varrho_\Xi$ with $\text{par}_\varrho, \text{par}_E \in \{0, 2, 4, 6\}$; left values: unmodified version; right values: modified version.

7.3.6 Alternative coupling algorithm

In this subsection, we present some numerical results for a different iterative algorithm which can be used for the solution of the mortar system (7.25) if the Lagrange multiplier is associated with the coarse side $\mathcal{G}^H$. The idea of the algorithm is to obtain an approximation of the coarse Lagrange multiplier ζ_Γ^H from an auxiliary coarse problem, solve a Neumann problem on ω with ζ_Γ^H applied as surface load, and perform the backcoupling via the fine trace $\mathbf{u}_\Gamma^h$. The detailed derivation of the algorithm as well as the theoretical analysis of its convergence rate can be found in [69]; here, we briefly present some numerical results illustrating the convergence behaviour of the scheme.

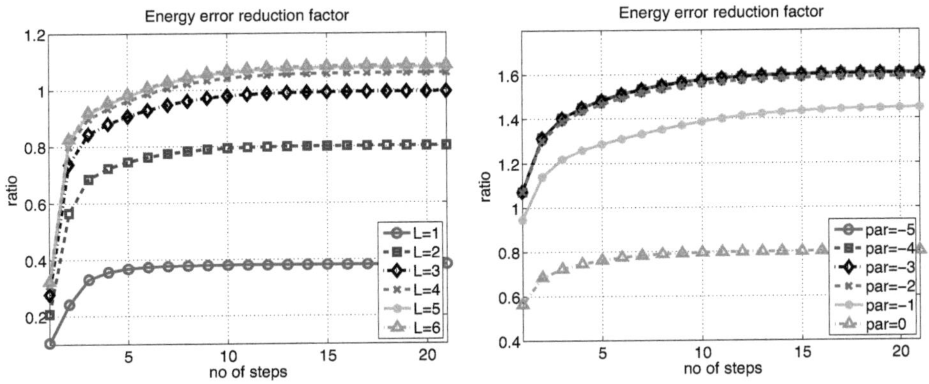

Figure 7.12: Error reduction factor $e_{\text{alg}}^{(l+1)}/e_{\text{alg}}^{(l)}$ for alternative algorithm with respect to l; left: $E_\omega = E_\Xi$ and $L \in \{1, \ldots, 6\}$; right: $L = 2$ and par $\in \{-5, \ldots, 0\}$.

First, we apply the alternative scheme to the static test setting described in Section 7.3.1. In Figure 7.12, the error quotient $e_{\text{alg}}^{(l+1)}/e_{\text{alg}}^{(l)}$ of the alternative algorithm is shown for $E_\omega = E_\Xi$, $L \in \{1, \ldots, 6\}$, on the left side and for $L = 2$ and different material parameters $E_\omega = 10^{\text{par}} E_\Xi$, par $\in \{-5, \ldots, 0\}$, on the right side. These results indicate that the largest eigenvalue of the iteration matrix is bounded from above by a value independent of L and par. However, in contrast to Algorithm 1, the alternative algorithm does not always converge if the finite element spaces are nested, as can be seen for the case $L \geq 4$. The results for par > 0 are omitted in Figure 7.12 because the algorithm converges quite fast in this case (see also Tables 7.5 and 7.6 below). This is in contrast to Algorithm 1, where the speed of convergence is generally better for smaller values of E_ω, corresponding to par being negative.

Next, we consider the dynamic test setting of Subsection 7.3.3 and vary the material parameters $E_\omega = 10^{\text{par}_E} E_\Xi$, $\varrho_\omega = 10^{\text{par}_\varrho} \varrho_\Xi$. The corresponding error quotients for $L \in \{2, 4\}$ are summarized in Tables 7.5 and 7.6. One can observe that the ratio becomes worst for $\text{par}_E < \text{par}_\varrho \leq 0$, i.e., if the mass contribution on ω is dominating. Especially for the case $\text{par}_E = -4$, $\text{par}_\varrho = 0$, the error quotient seems to be proportional to $\frac{H}{h} = 2^L$ which agrees with the theoretical results of [69]. Hence, in general, the alternative algorithm seems to be less efficient than Algorithm 1, which is why we focus on the latter one in this work.

$\mathrm{par}_E \setminus \mathrm{par}_\varrho$	−4	−2	0	2	4
−4	1.5344	3.3235	2.6359	0.1350	0.0006
−2	1.5914	1.5192	2.3174	0.1349	0.0006
0	0.9387	0.9350	0.7832	0.1286	0.0006
2	0.0565	0.0566	0.0561	0.0174	0.0005
4	0.0007	0.0007	0.0007	0.0005	0.0001

Table 7.5: Asymptotic error reduction rates $e_{\mathrm{alg}}^{(l+1)}/e_{\mathrm{alg}}^{(l)}$ for $L = 2$ and different values of $E_\omega = 10^{\mathrm{par}_E} E_\Xi$, $\varrho_\omega = 10^{\mathrm{par}_\varrho}\varrho_\Xi$ with $\mathrm{par}_\varrho, \mathrm{par}_E \in \{-4, -2, 0, 2, 4\}$.

$\mathrm{par}_E \setminus \mathrm{par}_\varrho$	−4	−2	0	2	4
−4	2.0880	10.8152	12.4981	0.6472	0.0035
−2	2.1075	2.0672	7.4568	0.6366	0.0035
0	1.2478	1.2429	1.0654	0.3368	0.0035
2	0.0686	0.0687	0.0686	0.0234	0.0011
4	0.0008	0.0008	0.0009	0.0006	0.0002

Table 7.6: Asymptotic error reduction rates $e_{\mathrm{alg}}^{(l+1)}/e_{\mathrm{alg}}^{(l)}$ for $L = 4$ and different values of $E_\omega = 10^{\mathrm{par}_E} E_\Xi$, $\varrho_\omega = 10^{\mathrm{par}_\varrho}\varrho_\Xi$ with $\mathrm{par}_\varrho, \mathrm{par}_E \in \{-4, -2, 0, 2, 4\}$.

7.3.7 Nearly incompressible material

Finally, we illustrate the fact that Algorithm 1 can be generalized to the case of nearly incompressible elastic material, which corresponds to $\lambda \to \infty$ or equivalently $\nu \to 0.5$. As the numerical simulation of such materials using lowest order conforming Finite Elements usually leads to volume locking, we employ a numerical implementation based on a modified version of the Hu–Washizu formulation [33, 115]. With this, we obtain a spectral equivalence result similar to Theorem 7.12, provided that a stable mixed discretization is employed. For details of the proof, we refer to [69].

In order to investigate the convergence behaviour of Algorithm 1 applied to the nearly incompressible case, we consider a simple 3D test setting with the domain $\Omega = [0,1] \times [0,1] \times [0,2]$ and $\omega = [0,1]^3$. We impose homogeneous Dirichlet boundary conditions on the top ($\Gamma_D = [0,1]^2 \times \{2\}$), the Neumann compression forces

$$\mathbf{g}_N = 1000 \cdot \max(0, 0.25 - r), \quad \text{with} \quad r^2 = (0.5 - x_1)^2 + (0.5 - x_2)^2$$

on the bottom ($[0,1]^2 \times \{0\}$) and homogeneous Neumann conditions elsewhere. We use uniform hexahedral grids of mesh size $H = \frac{1}{4}$ and $h = \frac{1}{16}$, and the parameters for the static linear elastic material are chosen according to $E_\omega = E_\Xi = 100$, $\nu_\Xi = 0.33$ and $\nu_\omega \to 0.5$. The computation is done using the method of mixed enhanced strains [100, 101].

In Figure 7.13, the error decay of Algorithm 1 is depicted for different values of ν_ω. One can see that the convergence rate is completely independent of ν_ω and hence also stable for $\nu_\omega \to 0.5$.

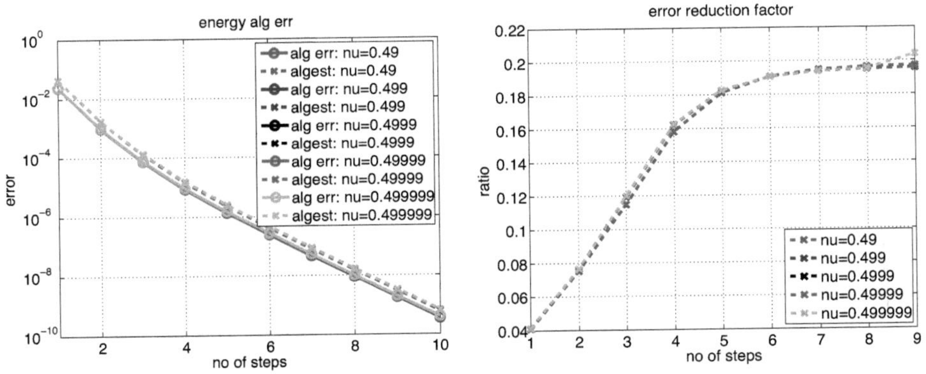

Figure 7.13: Performance of Algorithm 1 for $L = 2$, $E_\omega = E_\Xi$ and $\nu_\omega \to 0.5$; left: true $e_{\text{alg}}^{(l)}$ and estimated $\eta_{\text{alg}}^{(l)}$ rel. alg. error; right: error reduction factor $e_{\text{alg}}^{(l+1)}/e_{\text{alg}}^{(l)}$.

8 ODDM for nonlinear problems

Motivated by the challenge of simulating the rolling of a car tire, we have presented and analysed a flexible iterative scheme that allows for a locally improved spatial resolution. But so far, we have assumed that the underlying problem is linear, which usually does not apply for the tire problem. Hence, in this chapter, we extend the considerations of Chapter 7 to the nonlinear case by incorporating some of the effects presented in Chapter 1, namely, geometrical or material nonlinearity as well as frictional contact. After sketching the general setting in Section 8.1, possible approximate solution schemes are presented in Subsection 8.2. The following Section 8.3 contains several numerical examples including complex geometries, nonlinear material laws and contact.

8.1 Nonlinear setting

In comparison with the mortar system (7.20), the matrices K^m_Θ, $\Theta \in \{\Xi, \omega\}$, $m \in \{h, H\}$, are now replaced by nonlinear operators $\mathcal{K}^m_\Theta : \mathbf{V}^m_\Theta \to (\mathbf{V}^m_\Theta)'$ induced either by a nonlinear definition of the stress tensor $\mathbf{S}(\mathbf{u})$ as in (1.3) or by incorporation of nondifferentiable effects like plasticity and contact. The nonlinear version of the mortar coupled discrete problem (7.20) reads

$$\begin{pmatrix} \mathcal{K}^H_\Xi(\mathbf{u}^H_{j+1}) \\ \mathcal{K}^h_\omega(\mathbf{u}^h_{j+1}) \\ D^{hh}\mathbf{u}^h_{j+1} - D^{hH}\mathbf{u}^H_{j+1} \end{pmatrix} + \begin{pmatrix} -(D^{hH})^T \\ (D^{hh})^T \\ \mathbf{0} \end{pmatrix} \zeta^h_{\Gamma,j+\theta} = \mathbf{0}. \tag{8.1}$$

Above, we have included the volume and surface forces as well as the terms from the last time step (2.23) in the definition of the maps $\mathcal{K}^m_\Theta$, i.e., $\mathcal{K}^m_\Theta(\mathbf{u}^m_{j+1}) = \mathcal{K}^m_\Theta(\mathbf{u}^m_{j+1}, \mathbf{u}^m_j, \mathbf{v}^m_j)$, but we omit the dependence on the latter two arguments for ease of notation. Further, $\mathcal{K}^m_\Theta$ depends on the kind of nonlinearity considered. If for example plasticity or contact is included, $\mathcal{K}^m_\Theta$ also depends on the inner plastic or dual contact variables introduced in Chapter 2. There, we have seen that the corresponding constraints can be reformulated as a set of semismooth equations. Hence, we assume in the following that (8.1) is a system of semismooth equations, such that the framework of Chapter 3 can be applied.

Similar to (7.25), (8.1) can also be formulated with respect to the interface variables only. For this, we introduce nonlinear Dirichlet–to–Neumann operators $\mathcal{S}^m_\Theta : \mathbf{W}^m_\Gamma \to (\mathbf{W}^m_\Gamma)'$ defined by

$$\langle \mathcal{S}^m_\Theta \mathbf{u}^m_\Gamma, \mathbf{w}^m|_\Gamma \rangle_\Gamma = (\mathcal{K}^m_\Theta(\mathbf{u}^m), \mathbf{w}^m), \quad \mathbf{w}^m \in \mathbf{V}^m_\Theta, \tag{8.2}$$

where $\mathbf{u}^m \in \mathbf{V}^m_\Theta$ is such that $\mathbf{u}^m|_\Gamma = \mathbf{u}^m_\Gamma$ and

$$(\mathcal{K}^m_\Theta(\mathbf{u}^m), \widehat{\mathbf{w}}^m) = 0, \quad \widehat{\mathbf{w}}^m \in \mathbf{V}^m_\Theta,\ \widehat{\mathbf{w}}^m|_\Gamma = \mathbf{0}.$$

With this, (8.1) can be written as

$$\begin{pmatrix} \mathcal{S}_{\Xi}^{H}(\mathbf{u}_{\Gamma,j+1}^{H}) \\ \mathcal{S}_{\omega}^{h}(\mathbf{u}_{\Gamma,j+1}^{h}) \\ D_{\Gamma}^{hh}\mathbf{u}_{\Gamma,j+1}^{h} - D_{\Gamma}^{hH}\mathbf{u}_{\Gamma,j+1}^{H} \end{pmatrix} + \begin{pmatrix} -(D_{\Gamma}^{hH})^{T} \\ (D_{\Gamma}^{hh})^{T} \\ \mathbf{0} \end{pmatrix} \zeta_{\Gamma,j+\theta}^{h} = \mathbf{0}. \tag{8.3}$$

8.2 Approximate solution schemes

In this section, we transfer the idea of Algorithm 1 to the semismooth problem (8.1).

8.2.1 Nested iterations

As sketched in Chapter 3, the nonlinear system (8.1) can be solved by a semismooth Newton method. This solution procedure is now combined with the subdomain iteration presented in Section 7.2, such that we have to combine two iterative processes: on the one hand, the semismooth Newton loop with iteration index k, on the other hand, the subdomain coupling with iteration index l. These loops can be nested in two different ways:

(a) kl version: Linearize (8.1) and solve the resulting tangential problem using Algorithm 1.

(b) lk version: Approximate (8.1) by two separate nonlinear subproblems coupled at the interface Γ and solve each subproblem by an inner semismooth Newton loop.

Both (a) and (b) result in an algorithm with an inner and an outer iteration. The efficiency of these methods can possibly be increased by replacing the inner iteration with an approximate solver; one can for example perform only a fixed small number of inner iteration steps or use an adaptive stopping criterion as in, e.g., [48].

First, we consider possibility (a) in more detail. The application of the semismooth Newton scheme to the nonlinear problem (8.1) leads to the following linear system to be solved for the next Newton iterate $(\mathbf{u}_{j+1}^{H,(k+1)}, \mathbf{u}_{j+1}^{h,(k+1)}, \zeta_{\Gamma,j+\theta}^{h,(k+1)})$:

$$\begin{pmatrix} K_{\Xi}^{H,(k)} & \mathbf{0} & -(D^{hH})^{T} \\ \mathbf{0} & K_{\omega}^{h,(k)} & (D^{hh})^{T} \\ -D^{hH} & D^{hh} & \mathbf{0} \end{pmatrix} \begin{pmatrix} \mathbf{u}_{j+1}^{H,(k+1)} \\ \mathbf{u}_{j+1}^{h,(k+1)} \\ \zeta_{\Gamma,j+\theta}^{h,(k+1)} \end{pmatrix} = \begin{pmatrix} \mathbf{q}^{H,(k)} \\ \mathbf{q}^{h,(k)} \\ \mathbf{0} \end{pmatrix}, \tag{8.4}$$

with the tangential matrices

$$K_{\Theta}^{m,(k)} := \left(\partial_{\mathbf{u}_{j+1}^{m}} \mathcal{K}_{\Theta}^{m}\right)(\mathbf{u}_{j+1}^{m,(k)}), \quad \Theta \in \{\Xi, \omega\}, \quad m \in \{h, H\}, \tag{8.5}$$

and the right hand side vectors

$$\mathbf{q}^{H,(k)} := K_{\Xi}^{H,(k)}\mathbf{u}_{j+1}^{H,(k)} - \mathcal{K}_{\Xi}^{H}(\mathbf{u}_{j+1}^{H,(k)}), \tag{8.6a}$$

$$\mathbf{q}^{h,(k)} := K_{\omega}^{h,(k)}\mathbf{u}_{j+1}^{h,(k)} - \mathcal{K}_{\omega}^{h}(\mathbf{u}_{j+1}^{h,(k)}). \tag{8.6b}$$

This linear system has exactly the same structure as (7.20) and can thus be solved inexactly by means of Algorithm 1. For convenience, the resulting nested iteration is summarized in

Algorithm 3 Inexact nonlinear two-way coupling scheme

Start from some initial guess $\widehat{\mathbf{z}}_d^{(0,l_{\max})}$.

Newton loop: Compute sequentially for $k = 0, 1, \ldots$

(1) Initialize the solution $\widehat{\mathbf{z}}_d^{(k+1,0)} = \widehat{\mathbf{z}}_d^{(k,l_{\max})}$. Compute the stiffness matrices $S_\Xi^{H,(k)}$, $S_\omega^{h,(k)}$ and the right hand side vectors $\bar{\mathbf{q}}_\Xi^{H,(k)}$, $\bar{\mathbf{q}}_\omega^{h,(k)}$.
Define coarse grid approximations $S_\omega^{H,(k)}$, $\bar{\mathbf{q}}_\omega^{H,(k)}$ of $S_\omega^{h,(k)}$, $\bar{\mathbf{q}}_\omega^{h,(k)}$.

(2) Gauß–Seidel loop: Compute sequentially for $l = 0, 1, \ldots,$

(*i*) Solve problem on coarse space $\mathbf{V}^H$ with interface load on Γ inherited from fine computation on ω:

$$\begin{pmatrix} (S_\Xi^{H,(k)} + S_\omega^{H,(k)}) & \mathbf{0} \\ S_\omega^{H,(k)} & (D_\Gamma^{HH})^T \end{pmatrix} \begin{pmatrix} \delta\mathbf{u}_{\Gamma,j+1}^{H,(k+1,l)} \\ \delta\boldsymbol{\mu}_{\Gamma,j+\theta}^{H,(k+1,l)} \end{pmatrix} = \begin{pmatrix} \mathbf{r}_\Xi^{H,(k+1,l)} + \mathbf{r}_\omega^{H,(k+1,l)} \\ \mathbf{r}_\omega^{H,(k+1,l)} \end{pmatrix}, \tag{8.7}$$

(*ii*) Solve problem on fine space $\mathbf{V}^h$ with weakly imposed trace on Γ inherited from coarse computation on Ω:

$$\begin{pmatrix} S_\omega^{h,(k)} & (D_\Gamma^{hh})^T \\ D_\Gamma^{hh} & \mathbf{0} \end{pmatrix} \begin{pmatrix} \delta\mathbf{u}_{\Gamma,j+1}^{h,(k+1,l)} \\ \delta\boldsymbol{\zeta}_{\Gamma,j+\theta}^{h,(k+1,l)} \end{pmatrix} = \begin{pmatrix} \mathbf{r}_\omega^{h,(k+1,l)} \\ \boldsymbol{\nu}_\Gamma^{h,(k+1,l)} \end{pmatrix} + \begin{pmatrix} \mathbf{0} \\ D_\Gamma^{hH}\delta\mathbf{u}_{\Gamma,j+1}^{H,(k+1,l)} \end{pmatrix}, \tag{8.8}$$

(*iii*) Update the solution vector:

$$\widehat{\mathbf{z}}_d^{(k+1,l+1)} := \widehat{\mathbf{z}}_d^{(k+1,l)} + \delta\widehat{\mathbf{z}}_d^{(k+1,l)}, \tag{8.9}$$

The residuals are given by

$$\mathbf{r}_\Xi^{H,(k+1,l)} = \bar{\mathbf{q}}_\Xi^{H,(k)} - S_\Xi^{H,(k)}\mathbf{u}_{\Gamma,j+1}^{H,(k+1,l)} + (D_\Gamma^{hH})^T\boldsymbol{\zeta}_{\Gamma,j+\theta}^{h,(k+1,l)}, \tag{8.10a}$$
$$\mathbf{r}_\omega^{H,(k+1,l)} = \bar{\mathbf{q}}_\omega^{H,(k)} - S_\omega^{H,(k)}\mathbf{u}_{\Gamma,j+1}^{H,(k+1,l)} - (D_\Gamma^{HH})^T\boldsymbol{\mu}_{\Gamma,j+\theta}^{H,(k+1,l)}, \tag{8.10b}$$
$$\mathbf{r}_\omega^{h,(k+1,l)} = \bar{\mathbf{q}}_\omega^{h,(k)} - S_\omega^{h,(k)}\mathbf{u}_{\Gamma,j+1}^{h,(k+1,l)} - (D_\Gamma^{hh})^T\boldsymbol{\zeta}_{\Gamma,j+\theta}^{h,(k+1,l)}, \tag{8.10c}$$
$$\boldsymbol{\nu}_\Gamma^{h,(k+1,l)} = D_\Gamma^{hH}\mathbf{u}_{\Gamma,j+1}^{H,(k+1,l)} - D_\Gamma^{hh}\mathbf{u}_{\Gamma,j+1}^{h,(k+1,l)}. \tag{8.10d}$$

(*iv*) If $l+1 = l_{\max}$ or (8.11) is satisfied, set $\widehat{\mathbf{z}}_d^{(k+1,l_{\max})} = \widehat{\mathbf{z}}_d^{(k+1,l+1)}$ and stop.

(3) Check convergence of Newton iteration.

Algorithm 3, using the notation $\widehat{\mathbf{z}}_d$ as in (7.27) for the union of all unknowns on the interface Γ and $S_\Theta^{m,(k)}$ for the Schur complement of the matrix (8.5).

The convergence rate of the inner Gauß–Seidel iteration can be analysed with the results from Section 7.2. Furthermore, according to Theorem 3.10, the superlinear local convergence of the outer Newton iteration can be preserved if the norm of the residual of the inner iteration is bounded in terms of the norm of the Newton residual

$$\mathbf{R}^{(k)} := \begin{pmatrix} \mathcal{K}_\Xi^H(\mathbf{u}_{j+1}^{H,(k)}) - (D^{hH})^T\boldsymbol{\zeta}_{\Gamma,j+\theta}^{h,(k)} \\ \mathcal{K}_\omega^h(\mathbf{u}_{j+1}^{h,(k)}) + (D^{hh})^T\boldsymbol{\zeta}_{\Gamma,j+\theta}^{h,(k)} \end{pmatrix}.$$

The residual of the inner loop for $l \geq 1$ is given by the vector (8.10a), such that the Gauß–Seidel iteration has to be solved until the condition

$$\left\|\mathbf{r}_{\Xi}^{H,(k+1,l)}\right\| = o\left(\left\|\mathbf{R}^{(k)}\right\|\right) \tag{8.11}$$

is satisfied.

Second, we look at possibility (b) which can easier be formulated using the trace formulation (8.3). We introduce a coarse nonlinear Dirichlet–to–Neumann map $\mathcal{S}_{\omega}^{H} : \mathbf{W}_{\Gamma}^{H} \to (\mathbf{W}_{\Gamma}^{H})'$ and augment (8.3) with an additional equation:

$$\widehat{\mathbf{F}}(\widehat{\mathbf{z}}_d) + \widehat{D}_d\widehat{\mathbf{z}}_d := \begin{pmatrix} \mathcal{S}_{\Xi}^{H}(\mathbf{u}_{\Gamma,j+1}^{H}) + \mathcal{S}_{\omega}^{H}(\mathbf{u}_{\Gamma,j+1}^{H}) \\ \mathcal{S}_{\omega}^{H}(\mathbf{u}_{\Gamma,j+1}^{H}) \\ \mathcal{S}_{\omega}^{h}(\mathbf{u}_{\Gamma,j+1}^{h}) \\ D_{\Gamma}^{hh}\mathbf{u}_{\Gamma,j+1}^{h} - D_{\Gamma}^{hH}\mathbf{u}_{\Gamma,j+1}^{H} \end{pmatrix} + \begin{pmatrix} (D_{\Gamma}^{HH})^T & -(D_{\Gamma}^{hH})^T \\ (D_{\Gamma}^{HH})^T & \mathbf{0} \\ \mathbf{0} & (D_{\Gamma}^{hh})^T \\ \mathbf{0} & \mathbf{0} \end{pmatrix} \begin{pmatrix} \boldsymbol{\mu}_{\Gamma,j+\theta}^{H} \\ \boldsymbol{\zeta}_{\Gamma,j+\theta}^{h} \end{pmatrix} = \mathbf{0}. \tag{8.12}$$

Then, a Gauß–Seidel iteration similar to (7.30) is applied to the augmented nonlinear system (8.12). Using the matrix $\widehat{K}_d$ already introduced on page 114, the resulting nonlinear fixed point iteration can be written as

$$\widehat{\mathbf{F}}\left(\widehat{\mathbf{z}}_d^{(l+1)}\right) + \left(\widehat{D}_d - \widehat{K}_d\right)\widehat{\mathbf{z}}_d^{(l+1)} = -\widehat{K}_d\widehat{\mathbf{z}}_d^{(l)}. \tag{8.13}$$

The nonlinear part of (8.13) naturally decouples into two nonlinear problems on the subdomains which can be solved separately by inexact semismooth Newton methods.

In summary, both coupling schemes can be interpreted as the combination of a fixed point iteration with index l and a semismooth Newton iteration with index k; possibility (a) is the kl version with the Newton method as outer loop, whereas (b) corresponds to the lk version with the linear iteration as outer loop. However, as a Newton step is, in general, more expensive than a linear fixed point step due to the reassembly of the stiffness matrix, the former version is likely to be more efficient than the latter one. This conjecture has been confirmed by some simple numerical tests [69], such that the numerical results presented in Section 8.3 focus on the variant (a).

Remark 8.1. In the special case that the inner iterations of (a) and (b) are both stopped after only one step, the schemes collapse to the same inexact iterative scheme which can be obtained by setting $l_{\max} = 1$ in Algorithm 3 or by solving the nonlinear equation (8.13) inexactly by means of a single semismooth Newton iteration. The result can be written as

$$\left(\widehat{V}_k + \widehat{D}_d - \widehat{K}_d\right)\delta\mathbf{z}_d^{(k)} = -\left(\widehat{\mathbf{F}}\left(\widehat{\mathbf{z}}_d^{(k)}\right) + \widehat{D}_d\widehat{\mathbf{z}}_d^{(k)}\right), \quad \widehat{V}_k \in \boldsymbol{\partial}_B\widehat{\mathbf{F}}\left(\widehat{\mathbf{z}}_d^{(k)}\right), \tag{8.14}$$

where the notation $\boldsymbol{\partial}_B\widehat{\mathbf{F}}$ refers to the Schur system of the tangential matrices (8.5). Equation (8.14) can be interpreted as a semismooth quasi-Newton iteration to solve the nonlinear equation (8.12), where the tangential stiffness matrix $(\widehat{V}_k + \widehat{D}_d)$ is approximated by $(\widehat{V}_k + \widehat{D}_d - \widehat{K}_d)$.

8.2.2 Coarse grid approximations

The convergence rate of the inner iteration and thus the efficiency of Algorithm 3 strongly depends on the coarse grid matrix $S_\omega^{H,(k)}$. In this subsection, we present suitable definitions of this quantity for the case of nonlinear material or frictional contact.

In the former case, we define an auxiliary coarse displacement vector $\mathbf{u}_\omega^H \in \mathbf{V}_\omega^H$ as the extension of the solution of the global coarse grid problem (8.7) onto ω. Then, we construct both $S_\omega^{h,(k)}$, $\bar{\mathbf{q}}_\omega^{h,(k)}$ and $S_\omega^{H,(k)}$, $\bar{\mathbf{q}}_\omega^{H,(k)}$ during the same nonlinear assembly routine, based on the value of the vectors $\mathbf{u}_\omega^h$, $\mathbf{u}_\omega^H$ at the previous iteration.

For the case of dynamic frictional contact, the approximation is more involved. Using the results from Section 2.4, the corresponding fine Dirichlet–to–Neumann operator $\mathcal{S}_\omega^h : \mathbf{W}_\Gamma^h \to (\mathbf{W}_\Gamma^h)'$ defined in (8.2) is given by

$$\mathcal{S}_\omega^h(\mathbf{u}_{\Gamma,j+1}^h) = K_{\Gamma\Gamma}^h \mathbf{u}_{\Gamma,j+1}^h + K_{\Gamma\omega}^h \mathbf{u}_{\omega,j+1}^h - \boldsymbol{\varrho}_{\Gamma,j}^h, \tag{8.15}$$

with $\mathbf{u}_{\omega,j+1}^h$ being the first component of the solution $(\mathbf{u}_{\omega,j+1}^h, \boldsymbol{\lambda}_{j+\theta}^h)$ of (cf. (2.21))

$$K_{\omega\omega}^h \mathbf{u}_{\omega,j+1}^h + B_{\text{co}}^h \boldsymbol{\lambda}_{j+\theta}^h = \boldsymbol{\varrho}_{\omega,j}^h - K_{\omega\Gamma}^h \mathbf{u}_{\Gamma,j+1}^h, \tag{8.16a}$$

$$\mathbf{C}^{\text{co}}(\mathbf{u}_{\omega,j+1}^h, \boldsymbol{\lambda}_{j+\theta}^h) = \mathbf{0}. \tag{8.16b}$$

Let $\mathcal{N}_{\text{co}}^m \subset \mathcal{N}^m$, $m \in \{h, H\}$, denote the set of contact degrees of freedom of $\mathbf{V}_\omega^m$. In Section 4.2, we have seen that the structure of the tangential stiffness matrix depends on the composition of the (in)active sets (4.19). Numerical tests have shown that the convergence rate of the inner iteration in Algorithm 3 degrades if the Dirichlet contact conditions, i.e., the active nodes $\mathcal{A}_n^{h,(k)}$ in normal direction and the sticky nodes $\mathcal{I}_t^{h,(k)}$ in tangential direction, are not respected within the coarse approximation. Hence, we propose to construct $S_\omega^{H,(k)}$, $\bar{\mathbf{q}}_\omega^{H,(k)}$ by defining a set $\mathcal{A}_n^{H,(k)} \subset \mathcal{N}_{\text{co}}^H$ of active coarse grid nodes approximating the fine contact zone. This coarse active set is constructed in terms of the finite element function $\chi_n^{h,(k)} \in \mathbf{W}^h$ given by

$$\chi_n^{h,(k)}(p) = \begin{cases} \mathbf{1}, & p \in \mathcal{A}_n^{h,(k)}, \\ \mathbf{0}, & \text{otherwise.} \end{cases}$$

Introducing the mortar operator $P_C^H : \mathbf{W}^h \to \mathbf{W}^H$ on Γ_C defined analogeously to (7.10), we set

$$\mathcal{A}_n^{H,(k)}(\tau_n) := \left\{ p \in \mathcal{N}_{\text{co}}^H : \left(P_C^H \chi_n^{h,(k)}\right)(p) > \tau_n \right\} \tag{8.17}$$

for an appropriate threshold value $\tau_n \geq 0$. The rest of the coarse nodes is set inactive, i.e., $\mathcal{I}_n^{H,(k)} := \mathcal{N}_{\text{co}}^H \backslash \mathcal{A}_n^{H,(k)}$.

For the tangential part, we proceed similarly and define the set $\mathcal{I}_t^{H,(k)}$ of coarse sticky nodes by

$$\mathcal{I}_t^{H,(k)}(\tau_t) := \left\{ p \in \mathcal{N}_{\text{co}}^H : \left(P_C^H \chi_t^{h,(k)}\right)(p) > \tau_t \right\}, \quad \text{with} \quad \chi_t^{h,(k)}(p) = \begin{cases} \mathbf{1}, & p \in \mathcal{I}_t^{h,(k)}, \\ \mathbf{0}, & \text{otherwise.} \end{cases} \tag{8.18}$$

As the Robin conditions for the slippy nodes are very difficult to model on the coarse grid, we free the remaining nodes $\mathcal{F}_t^{H,(k)} := \mathcal{N}_{\text{co}}^H \backslash \mathcal{I}_t^{H,(k)}$ in tangential direction, leading to the Neumann condition $\boldsymbol{\lambda}_t^{H,(k+1)} = \mathbf{0}$ at these nodes (cf. page 52).

Using the sets (8.17), (8.18), we construct $S_\omega^{H,(k)}$ and $\mathbf{q}_\omega^{H,(k)}$ in the same way as the fine grid counterparts. We remark that smaller values of τ_n, τ_t generally lead to larger coarse contact sets $\mathcal{A}_n^{H,(k)}$, $\mathcal{I}_t^{H,(k)}$.

Remark 8.2. We have also tried to map the fine values $\mathbf{u}^h|_{\Gamma_C}$, $\boldsymbol{\lambda}^h$ onto the coarse discretization and then compute the coarse active sets from these projections. However, the coarse active sets are much smaller than those obtained from (8.17) for small threshold values, generally resulting in worse convergence rates.

8.3 Numerical tests

8.3.1 Geometrically conforming setting in 2D

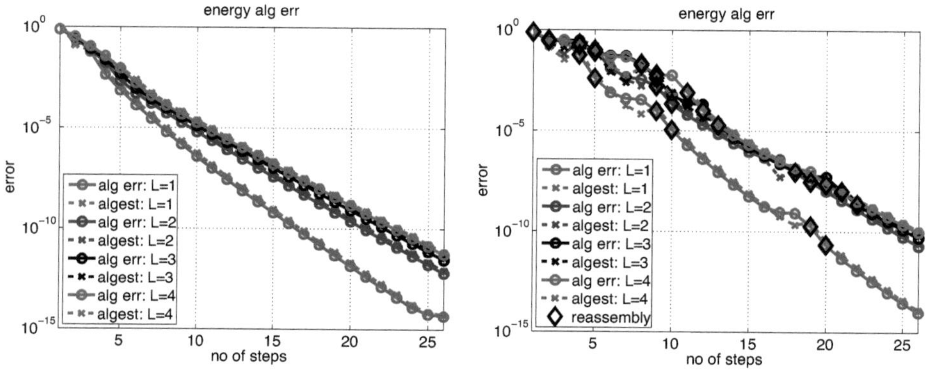

Figure 8.1: True e_{alg} and estimated η_{alg} relative algebraic error for Algorithm 3 with $L \in \{1,2,3,4\}$ and $E_\Xi = E_\omega$ with respect to number of total iterations; left: $l_{\max} = 1$; right: inner iteration stopped according to (8.19a).

For the first set of nonlinear computations, we use a similar test setting as in Subsections 7.3.1 and 7.3.2 with nested finite element spaces $\mathbf{V}_\omega^H \subset \mathbf{V}^h$. There are only two differences to the setting described there; firstly, the Neumann boundary conditions on the lower boundary for $x_2 = 0$ are replaced by unilateral frictionless contact with the fixed obstacle $\Gamma_{\text{obs}}(x_1) = (x_1, 0.3 \cdot \max(0.25 - |x_1 - 1|, 0))$, and secondly, we employ the nonlinear static Mooney–Rivlin elasticity law (1.9) with the parameters $c_m = 0.5$, $E_\Xi = E_\omega = 100$, $\nu = 0.3$ and $\varrho = 0$. The coarse grid approximation of the active set $\mathcal{A}_n^h$ is chosen according to (8.17) with $\tau_n = 0.15$.

As before, we test the performance of Algorithm 3 by computing the relative algebraic error measure η_{alg} defined in (7.64) and comparing it with the energy norm of the algebraic error e_{alg} given in (7.65). The results for different grid sizes $H = 2^{-3}$, $h = 2^{-L} \cdot H$, and different strategies of stopping the inner iteration are shown in Figures 8.1 and 8.2. For the quasi-Newton scheme with $l_{\max} = 1$ depicted on the left of Figure 8.1, the error decays linearly with respect to the number of iterations, and the error indicator η_{alg} accurately predicts the actual algebraic error

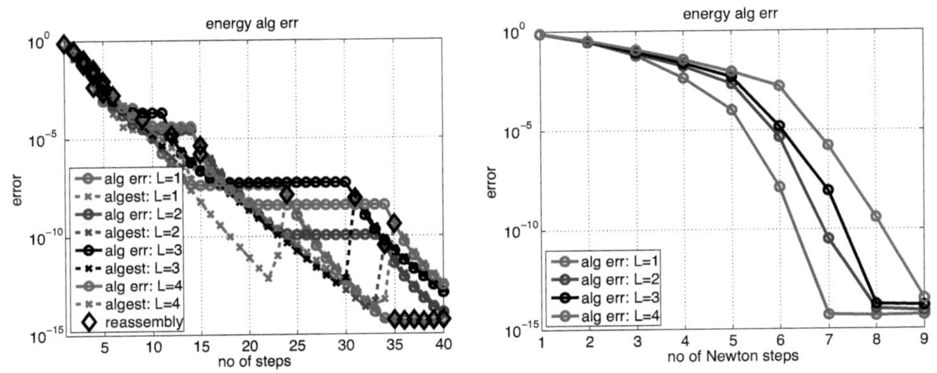

Figure 8.2: True e_{alg} and estimated η_{alg} rel. alg. error for Algorithm 3 for $L \in \{1,2,3,4\}$ and $E_{\Xi} = E_{\omega}$ with inner iteration stopped according to (8.19b); left: wrt. total steps; right: wrt. Newton steps.

e_{alg}. On the right side of Figure 8.1 and in Figure 8.2, we do not choose a fixed value of $l_{\max}$ but solve the inner iteration on l until the inner residual is bounded in terms of the Newton residual. For this, we employ the following conditions (cf. (7.63) and (8.11)):

$$\left(\eta_{\mathrm{abs}}^{(k+1,l)}\right)^2 := \left((S_{\Xi}^H + S_{\omega}^H)^{-1}\mathbf{r}_{\Xi}^{H,(k+1,l)}, \mathbf{r}_{\Xi}^{H,(k+1,l)}\right) \le E_{\omega}^{-1}\|\mathbf{R}^{(k)}\|^2, \tag{8.19a}$$

$$\left(\eta_{\mathrm{abs}}^{(k+1,l)}\right)^2 \le E_{\omega}\|\mathbf{R}^{(k)}\|^4. \tag{8.19b}$$

The iterations where the stiffness matrix is reassembled are marked with the symbol $\diamond$. The picture on the right side of Figure 8.1 shows the computations obtained with the linear stopping condition (8.19a). Comparing the results with those on the left picture where the stiffness matrix is reassembled in each step, one can observe that many of these assemblies can be avoided with only a slight loss of accuracy.

Figure 8.2 displays the error decay if the inner fixed point loop in Algorithm 3 is stopped according to (8.19b). One can observe that at times during the iteration, the estimated algebraic error η_{alg} is decreased by the additional inner steps but e_{alg} stays constant until the Newton matrix is updated, a phenomenon known as "oversolving" [54]. However, with the criterion (8.19b), we can guarantee that the local superlinear convergence of the outer Newton iteration is kept, as can be seen on the right side of Figure 8.2 where the error decay after each Newton step is plotted. Furthermore, for some of the more challenging problems like the one presented in Subsection 8.3.4, additional inner steps can increase the robustness of the scheme.

As in Subsection 7.3.2, we shortly test the performance of Algorithm 3 for varying values of L as well as for discontinuous material parameters $E_{\omega} = 10^{\mathrm{par}} E_{\Xi}$, $\mathrm{par} \ge 0$. In Figure 8.3, the reduction factor of the energy error is depicted with respect to the number of total iterations; the left picture shows the results for $\mathrm{par} = 0$, $l_{\max} = 1$ and different values of $L \in \{1, \dots, 4\}$, whereas the right picture displays the error reduction for $L = 2$, $l_{\max} = 2$ and $\mathrm{par} \in \{0, \dots, 6\}$. Especially in the latter case, one can observe that the error quotient is strongly varying in the

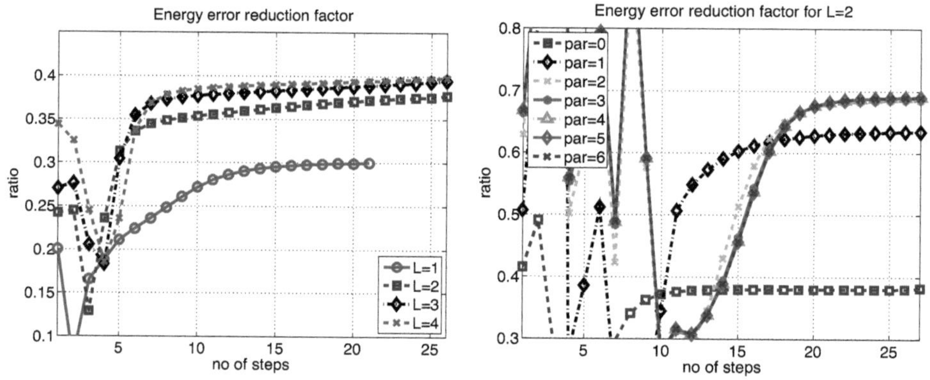

Figure 8.3: Mean algebraic error reduction factor for Algorithm 3 with respect to number of total iterations; left: $l_{\max} = 1$, $E_\Xi = E_\omega$ and $L \in \{1, 2, 3, 4\}$; right: $l_{\max} = 2$, $L = 2$ and $E_\omega = 10^{\text{par}} \cdot E_\Xi$, par $\in \{0, \ldots, 6\}$.

pre-asymptotic range due to the changes in the active sets. However, the asymptotic behaviour confirms that Algorithm 3 is robust with respect to the ratio E_ω / E_Ξ as well as the ratio H/h. We remark that the value $l_{\max} = 2$ has been chosen for the computations with par ≥ 1 because the algorithm with $\alpha = 1$ and $l_{\max} = 1$ has not converged.

8.3.2 Geometrically conforming setting in 3D

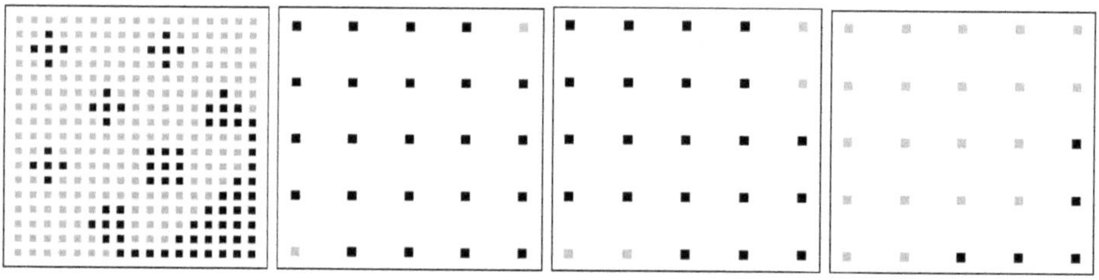

Figure 8.4: Fine normal active set $\mathcal{A}_n^h$ (black nodes) for the frictionless test and approximations $\mathcal{A}_n^H(\tau_n)$ for $\tau_n \in \{0,\ 0.1,\ 0.4\}$.

Before turning to applications with more complex geometries, we investigate the influence of the coarse grid approximations $\mathcal{A}_n^H(\tau_n)$, $\mathcal{I}_t^H(\tau_t)$ defined in (8.17), (8.18) on the convergence rate. For this, we consider the simple three-dimensional domain $\Omega = [0, 1] \times [0, 1] \times [0, 2]$ with the patch $\omega = [0, 1]^3$. We impose a fixed displacement of $(0, 0, -0.03)^T$ on the top, unilateral contact with the obstacle

$$\Gamma_{\text{obs}}(x_1, x_2) = \Big(x_1,\ x_2,\ 0.25 x_1 \cdot \sin(4\pi x_1) \cdot x_2 \cdot \sin(4\pi x_2)\Big)$$

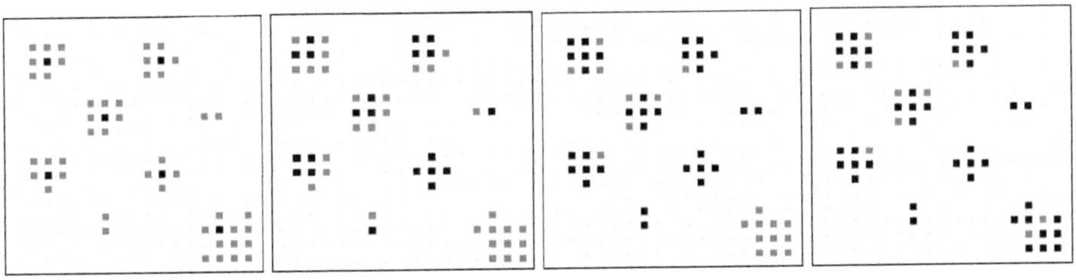

Figure 8.5: Fine tangential active sets $\mathcal{I}_t^h$ (black nodes) and $\mathcal{A}_t^h$ (grey nodes) for the frictional test with $\mathfrak{F} \in \{0.2,\ 0.35,\ 0.5,\ 1\}$.

on the bottom and homogeneous Neumann conditions elsewhere. The material parameters for the linear elastic body are $E_\omega = E_\Xi = 100$, $\nu = 0.33$ and $\varrho = 0$, and we use uniform hexahedral grids of mesh size $H = \frac{1}{4}$ and $h = \frac{1}{16}$, corresponding to $L = 2$.

We consider frictionless as well as frictional contact with $\mathfrak{F} \in \{0, 0.02, 0.2, 0.35, 0.5, 1, 5\}$ and $g_t = 0$. In Figures 8.4 and 8.5, the fine active sets $\mathcal{A}_n^h$, $\mathcal{I}_t^h$ of the converged solution are depicted, as well as the coarse grid approximations $\mathcal{A}_n^H(\tau_n)$ for $\tau_n \in \{0,\ 0.1,\ 0.4\}$.

$\mathfrak{F} \setminus \tau_n$	-1	0	0.1	0.2	0.4	2	$\lvert\mathcal{A}_n^h\rvert$	
0	0.3041	0.3041	0.3041	0.3063	0.6245	0.8299	77	−−
$\mathfrak{F} \setminus \tau_t$	-1	0	0.1	0.2	0.4	2	$\lvert\mathcal{A}_n^h\rvert$	$\lvert\mathcal{I}_t^h\rvert$
0.02	0.7411	0.3045	0.3045	0.3045	0.3045	0.3045	75	0
0.2	0.3712	0.3640	div	div	div	div	52	6
0.35	0.3046	0.3018	div	0.6988	div	div	51	21
0.5	0.3029	0.2991	0.3026	0.7299	div	div	51	32
1	0.3006	0.2994	0.2990	osc	osc	div	51	41
5	0.3018	0.3000	0.3010	0.3432	0.9429	div	49	46

Table 8.1: Asymptotic error reduction rates for Algorithm 3 and sizes of fine active sets for $L = 2$, $l_{\max} = 1$ and different values of $\mathfrak{F}$; first row: $\mathfrak{F} = 0$, $\tau_n \in \{-1, 0, 0.1, 0.2, 0.4, 2\}$; other rows: $\mathfrak{F} > 0$, $\tau_n = 0$, $\tau_t \in \{-1, 0, 0.1, 0.2, 0.4, 2\}$.

Table 8.1 summarizes the error reduction rates for Algorithm 3 with $l_{\max} = 1$ and different values of $\mathfrak{F}$, τ_n and τ_t. The upper subtable contains the frictionless case with $\tau_n \in \{-1, 0, 0.1, 0.2, 0.4, 2\}$, where the threshold values $\tau_n = -1$ and $\tau_n = 2$ correspond to the cases $\mathcal{A}_n^H = \emptyset$ and $\mathcal{I}_n^H = \emptyset$, respectively. We obtain a good convergence rate as long as $\tau_n \leq 0.2$, but for larger values of τ_n, the inner iterations of Algorithm 3 become less effective.

The remaining rows of Table 8.1 refer to the frictional case where we approximate the normal active set by (8.17) with $\tau_n = 0$ and compare the error reduction rates for different values of τ_t. The notation div indicates that the method has diverged, whereas osc refers to an oscillation of the active sets. One can observe that the convergence rate is best for the threshold value $\tau_t = 0$ whose error reduction is rather independent of the value of the friction coefficient. In contrast, smaller coarse sets $\mathcal{I}_t^H$ obtained by larger values of τ_t yield a more unstable convergence behaviour, and the extreme case $\mathcal{I}_t^H = \emptyset$ obtained for $\tau_t = 2$ hardly converges at all.

Remark 8.3. The convergence rate is less sensitive to the approximation of the contact boundary conditions if a mass term is included in the computation. For the frictionless contact problem with $\varrho = 10^{-2}$, $\Delta t = 10^{-3}$, we have obtained a convergence rate of 0.29 independent of the threshold value τ_n.

8.3.3 Tire application in 2D

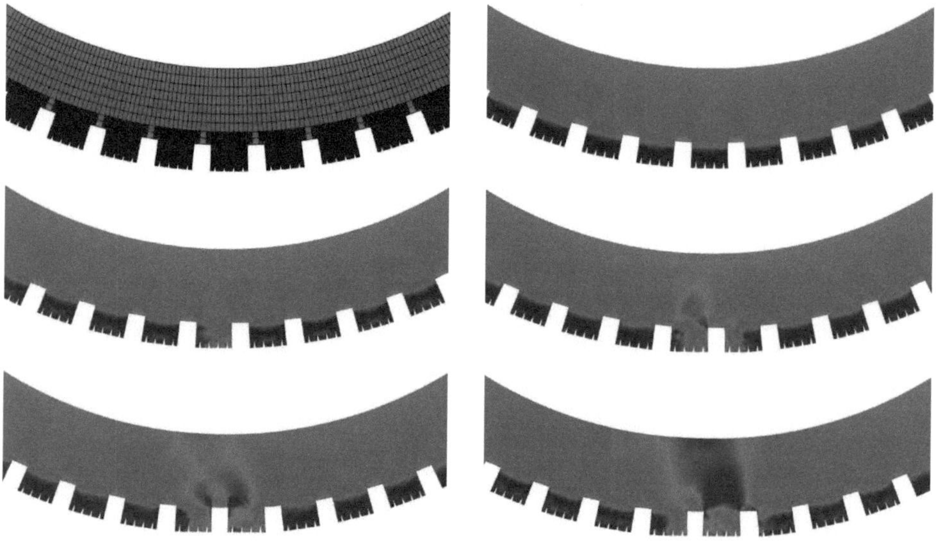

Figure 8.6: 2D tire example; upper row: initial grid for $H/h = 4$ and effective stress at time t_0; middle and lower row: effective stress at times t_{50}, t_{100}, t_{150} and t_{200}.

Next, we consider the simplified two-dimensional geometry of a car tire sketched in Figures 7.1 and 8.6. The domain Ω consists of a circular ring centered in the origin with the diameters $r_{\text{inner}} = 1.6$, $r_{\text{outer}} = 1.95$ and 60 additional salients of height 0.05. The fine domain ω is built by 60 separate patches which are associated with the salients but have an extended T-like shape in order to include the corner singularities in the fine triangulation. We remark that these patches can be computed in parallel as they do not overlap.

On the potential contact boundary Γ_C, each salient features four additional small sipes which are resolved by the fine grid $\mathcal{T}^h$ but are not respected by the coarse grid $\mathcal{T}^H$ (cf. Figure 7.1). Hence, we have a geometrically nonconforming situation where the coarse triangulation defines an approximate domain $\Omega^H \neq \Omega$, leading to $\mathbf{V}_\omega^H \not\subset \mathbf{V}_\omega^h$. On the upper left of Figure 8.6, the initial triangulation for $H/h = 4$ is shown which is constructed such that the trace spaces $\mathbf{W}_\Gamma^H \subset \mathbf{W}_\Gamma^h$ are nested.

For the dynamic computation with $\Delta t = 2.5 \cdot 10^{-6}$, we use the Mooney–Rivlin material law (1.9) with the parameters $E_\Xi = E_\omega = 4.4 \cdot 10^6$, $\varrho_\Xi = \varrho_\omega = 1$, $\nu = 0.33$ and $c_m = 0.5$. The tire

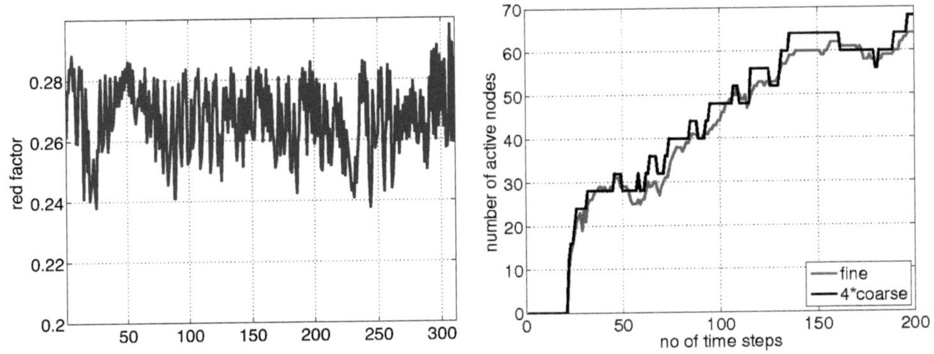

Figure 8.7: 2D tire example; left: error reduction factor for intermediate steps of Algorithm 3 with $l_{\max} = 1$; right: evolution of active sets $|\mathcal{A}^h_{n,j}|$ and $|\mathcal{A}^H_{n,j}(0)|$.

has an initial velocity of

$$\mathbf{v}_0(\mathbf{x}) = 150 \begin{pmatrix} x_2 \\ -x_1 \end{pmatrix} - \begin{pmatrix} 0 \\ 20 \end{pmatrix}$$

and an initial displacement corresponding to stationary rolling [83]. We apply a volume load of $(0, \ -19620)$ and homogeneous Neumann boundary conditions everywhere except for the potential contact boundary Γ_C where frictionless contact with a flat obstacle at $x_2 = -2.052$ is enforced. The linearized contact conditions are assembled in an updated Lagrangian manner, and the coarse contact set is defined according to (8.17) with $\tau_n = 0$.

In order to obtain stable and energy-conserving results, we use the persistency contact condition (2.19) combined with a modified mass matrix according to Chapter 5. Further, we employ a discrete gradient for the time discretization of the stress tensor $\mathbf{u}$ based on the approach presented in [67]. The resulting effective stress of the mortar solution at different time steps is depicted in Figure 8.6. The left picture of Figure 8.7 illustrates the error reduction rate of Algorithm 3 for $l_{\max} = 1$ and $k \geq 1$. One can observe that the quotients are about the same size as for the linear geometrically conforming setting in Subsection 7.3.3. On the right of Figure 8.7, the evolution of the number of fine active contact nodes and its coarse approximation (8.17) is shown.

8.3.4 Tire application in 3D

Finally, we apply our algorithm to a complex 3D geometry consisting of the lower half of a car tire centering on the x_2-axis and with an approximate radius of 318. In Figure 8.8, the domain Ω as well as the multiply connected fine patch ω are sketched, with a difference in the mesh sizes of about $4 \leq H/h \leq 8$. As in Subsection 8.3.3, the geometry is nonconforming because the fine triangulation $\mathcal{T}^h$ features some details at the potential contact boundary Γ_C which are not resolved by the coarse grid $\mathcal{T}^H$ (see Figure 8.9).

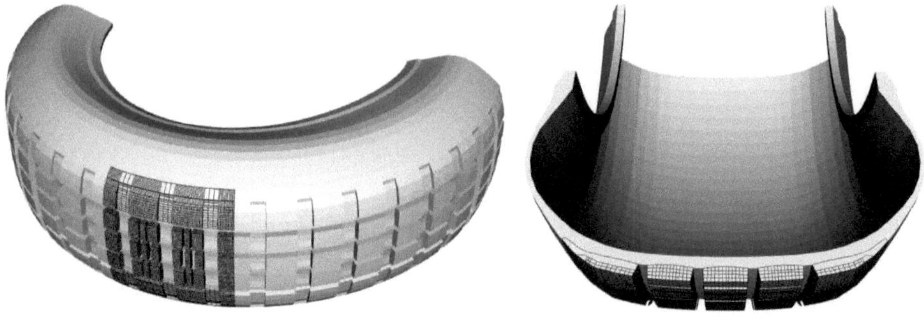

Figure 8.8: Geometry of the 3D tire example.

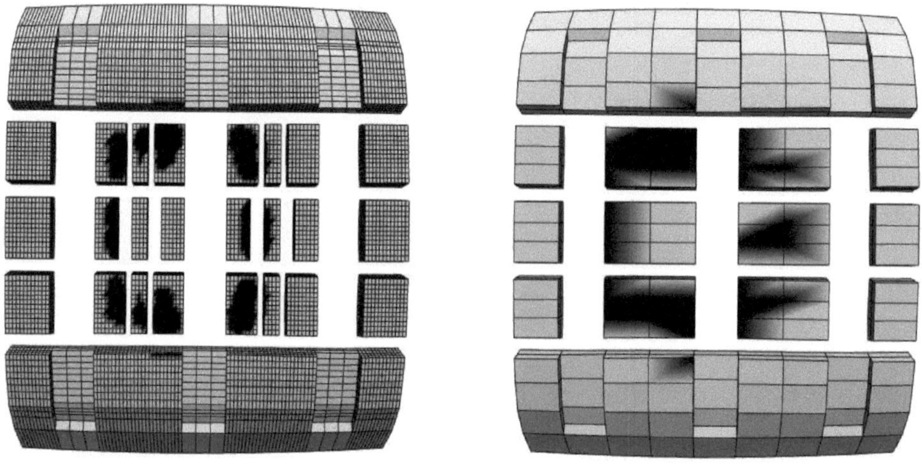

Figure 8.9: 3D tire example: Fine contact set $\mathcal{A}_n^h$ of exact solution with 555 nodes and approximation $\mathcal{A}_n^H(0.15)$ with 36 nodes.

First, we restrict ourselves to the case of linear static elasticity with frictionless contact. The contact plane is located at $x_3 = -320$, and the coarse approximation for the normal active set is done using (8.17). On the cutting faces of the tire (for $x_3 = 0$), a fixed displacement of $(0,\ 0,\ -1)$ is prescribed, whereas all other boundaries are free. Further, a volume force of $(0,\ 0,\ -1000)$ is applied to the tire, and the material parameters read $E_\Xi = E_\omega = 2.5 \cdot 10^7$ and $\nu = 0.33$. The active set $\mathcal{A}_n^h$ of the corresponding mortar solution is sketched on the left of Figure 8.9, whereas its coarse grid approximation $\mathcal{A}_n^H(0.15)$ can be seen on the right side.

Figure 8.10 displays the error decay and the mean error reduction factor of the damped version of Algorithm 3 with $\alpha = 0.6$, $l_{\max} = 4$ and different threshold values τ_n. One can observe that on the one hand, a too small threshold value leads to oscillations in the coarse

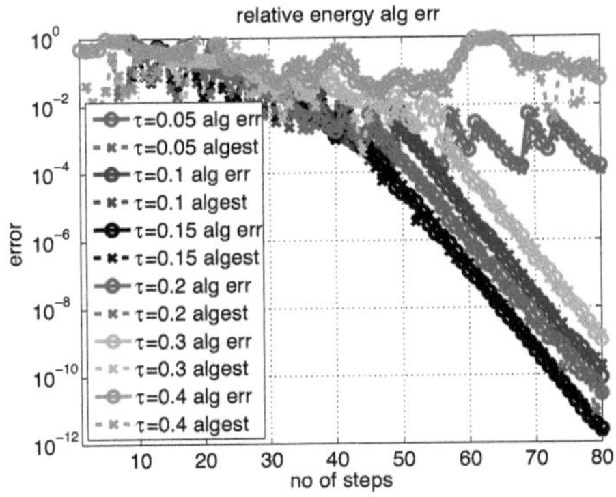

τ_n	$\lvert\mathcal{A}_n^H(\tau_n)\rvert$	err. red. factor
0.05	40/41	–
0.1	38	0.5826
0.15	36	0.5774
0.2	34	0.6014
0.3	24	0.5776

Figure 8.10: 3D tire example for $\alpha = 0.6$, $l_{\max} = 4$, $\tau_n \in \{0.05, 0.1, 0.15, 0.2, 0.3, 0.4\}$; left: True e_{alg} and estimated η_{alg} rel. alg. error; right: Size of $\mathcal{A}_n^H(\tau_n)$ and mean error reduction factor after correct active set has been found.

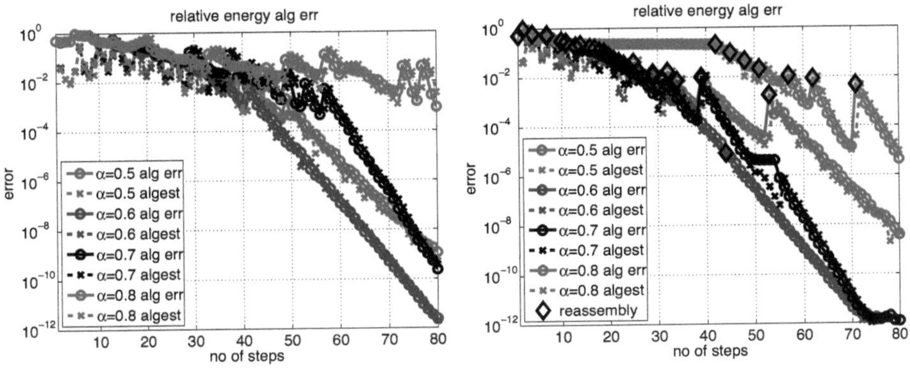

Figure 8.11: True e_{alg} and estimated η_{alg} relative alg. error with $\tau_n = 0.15$ for $\alpha \in \{0.5, 0.6, 0.7, 0.8\}$; left: $l_{\max} = 4$; right: inner iteration stopped according to (8.19a).

active set such that the algorithm does not converge. On the other hand, if τ_n is too large, the coarse contact set becomes smaller which degrades or even disables the convergence, as already observed in Subsection 8.3.2.

From now on, we fix $\tau_n = 0.15$ and investigate the performance of Algorithm 3 for different damping parameters α. On the left side of Figure 8.11, the error decay for $l_{\max} = 4$ and $\alpha \in \{0.5, 0.6, 0.7, 0.8\}$ is shown. For larger values of α, more Newton steps are necessary in order to detect the correct active set. But as soon as the correct active set has been found, the mean error reduction factor is better for larger values of α, varying between 0.49 for $\alpha = 0.7$ and 0.66 for $\alpha = 0.5$. This suggests an adaptive choice of α with respect to a change in the active sets.

In order to decrease the number of total iterations further, we replace the condition $l \le l_{\max}$ by the stopping criterion (8.19a), leading to

$$\left(\eta_{\mathrm{abs}}^{(k+1,l)}\right)^2 \le E_\omega^{-1}\left\|\mathbf{R}^{(k)}\right\|^2 = E_\omega^{-1}\left\|\mathbf{C}^{\mathrm{co}}(\mathbf{u}_{\omega,j+1}^{h,(k)}, \boldsymbol{\lambda}_{j+\theta}^{h,(k)})\right\|^2,$$

because the contact constraints are the only nonlinearities of the problem. The resulting error decay is depicted on the right of Figure 8.11. Comparing the results with those in the left picture, one can observe that the number of total iterations until convergence has been reduced by about 10 steps and that also the computation with $\alpha = 0.8$ is able to detect the correct active set within 15 Newton iterations.

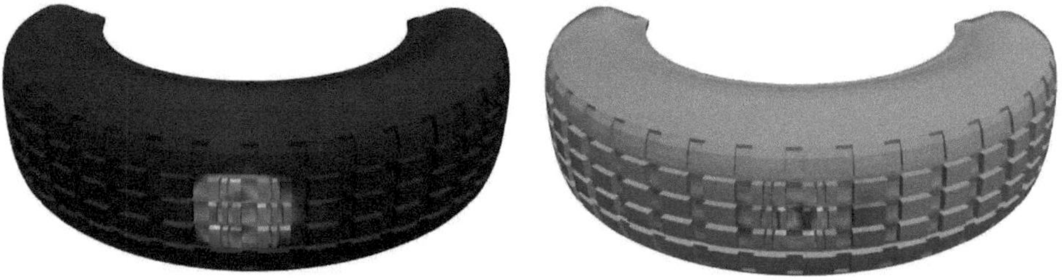

Figure 8.12: Effective stress of 3D tire problem at time t_{10}; left: disp = 0; right: disp = 2.

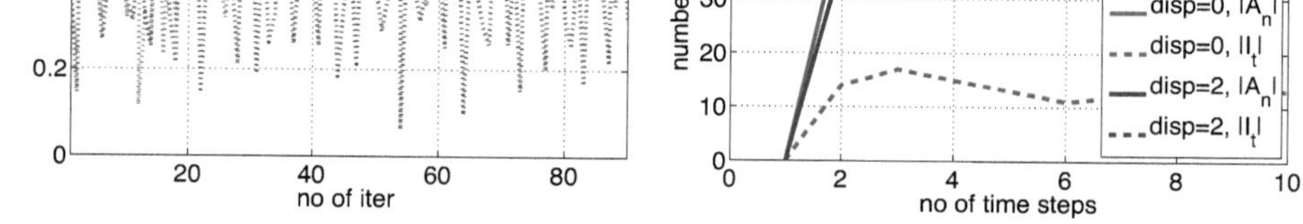

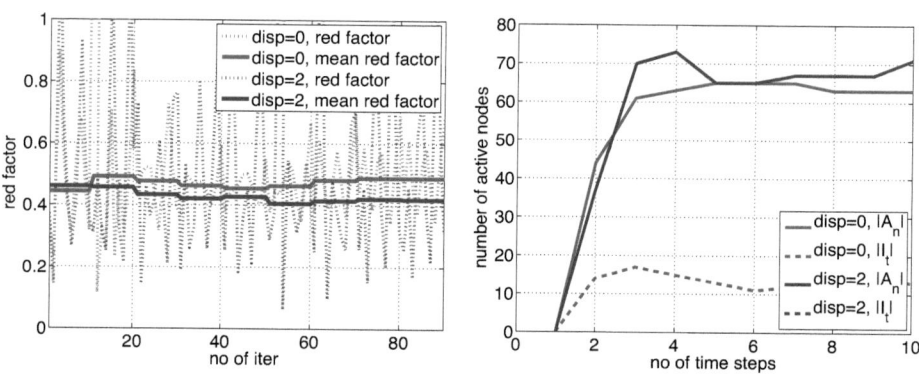

Figure 8.13: 3D nonlinear tire problem with $\alpha = 0.6$, $\tau_n = \tau_t = 0.15$; left: error reduction factor for first 10 iterations per time step; right: evolution of fine active sets.

Finally, we show a dynamic 3D example for the nonlinear Mooney–Rivlin material law (1.9). We choose the parameters $E_\Xi = E_\omega = 2.5 \cdot 10^7$, $\nu = 0.33$, $\varrho = 10^{-5}$ and $c_m = 0.5$, apply no volume forces and enforce frictional contact with the coefficient $\mathfrak{F} = 0.5$. We compute 10 time steps with a step size of $\Delta t = 10^{-5}$ and prescribe a time-dependent displacement of

$\frac{t}{10\Delta t}(0,\ 0,\ -0.75 + \text{sign}(x_1)\text{disp})$, disp $\in \{0, 2\}$, on the top, whereas all other boundaries are free. The effective stress of the corresponding solutions at time t_{10} is depicted in Figure 8.12.

The problem is solved using Algorithm 3 with $\alpha = 0.6$, $\tau_n = \tau_t = 0.15$ and $l_{\max} = 1$. On the left of Figure 8.13, the error reduction factor for the first 10 iterations of the time steps t_2 to t_{10} is plotted, together with the geometric mean for each time step. The right figure shows the evolution of the fine active sets $|\mathcal{A}_n^h|$ and $|\mathcal{I}_t^h|$ with respect to time. One can observe that the second test setting with disp $= 2$ yields no sticky nodes and thus gives a slightly better convergence rate than the problem with disp $= 0$ where sticky and slippy nodes appear. However, the mean error reduction factor is between 0.4 and 0.5 for both settings. These results illustrate the fact that Algorithm 3 is able to handle different types of nonlinearities within a single inexact Newton loop even on complex geometries.

9 Local time subcycling

The domain decomposition approach introduced in the previous two chapters has been motivated by the fact that one is interested in a detailed resolution of the solution within the fine patch. Hence, in addition to the different spatial mesh sizes, it is desirable to use different time scales for the coarse and the fine problems, which is the topic of the current chapter.

The main challenge in coupling different time step sizes lies in the construction of the interface constraints. A suboptimal choice of the coupling conditions can lead to numerical instability or to artificial dissipation at the interface, even if the time integrators in the subdomains are stable and energy-conserving [44, 45, 159]. In [42, 68], a time substepping scheme with linear interpolation of the velocity and the Lagrange multipliers has been proposed which is stable but dissipative at the interface. An improved energy-conserving scheme with linear interpolation of the multipliers has been analysed in [126, 141]. However, both of these methods are only formulated for matching grids at the interface and rely on the expensive exact solution of the resulting coupled system.

In this chapter, we present a time-discrete coupled system that uses different time step sizes in each subdomain, conserves the energy over a coarse time step and can efficiently be solved by the iterative algorithm considered in Chapter 7. Section 9.1 presents the theoretical background of the construction, namely, the so-called continuous Runge–Kutta and Nyström methods. In Section 9.2, we make use of such a continuous time-stepping scheme in order to define a mortar coupled solution with different time scales and analyse its asympotical convergence order for $\Delta t \to 0$. Afterwards, we sketch in Section 9.3 how this discrete coupled system can efficiently be solved using an iterative algorithm based on the overlapping domain decomposition. Finally, Section 9.4 contains several numerical tests.

9.1 Continuous output

From now on, we denote the time step associated with the coarse space $\mathbf{V}^H$ by ΔT and the time step on the patch ω by Δt. Furthermore, as we are interested in an energy-conserving time stepping scheme, we only consider the Newmark scheme used in (2.13) for the value $\theta = \frac{1}{2}$, i.e., the trapezoidal rule.

To derive suitable interface conditions for the case $\Delta t < \Delta T$, we need the values of the coarse solution not only at the discrete time steps $t_j = j\Delta T$ but also in between. Hence, we extend the trapezoidal rule to a Nyström method with dense output, also called a continuous Nyström method [128]. We refer to [75, 76] and the references therein for the definition of Nyström and continuous Runge–Kutta methods.

The first step of a s-stage Nyström method for the approximation of the first order system

$$\begin{pmatrix} \dot{\mathbf{u}} \\ \dot{\mathbf{v}} \end{pmatrix} = \begin{pmatrix} \mathbf{v} \\ \mathbf{f}(t, \mathbf{u}) \end{pmatrix}$$

with a given right hand side function $\mathbf{f}$ reads

$$\mathbf{k}_l = \mathbf{f}\left(t_0 + c_l \Delta T, \mathbf{u}_0 + c_l \Delta T \mathbf{v}_0 + \Delta T^2 \sum_{i=1}^{s} a_{li} \mathbf{k}_i \right), \quad l = 1, \dots, s, \tag{9.1a}$$

$$\mathbf{u}_1 = \mathbf{u}_0 + \Delta T \mathbf{v}_0 + \Delta T^2 \sum_{i=1}^{s} \bar{b}_i \mathbf{k}_i, \tag{9.1b}$$

$$\mathbf{v}_1 = \mathbf{v}_0 + \Delta T \sum_{i=1}^{s} b_i \mathbf{k}_i. \tag{9.1c}$$

The parameters c_l, a_{li}, b_i and $\bar{b}_i$ for $i, l \in \{1, \dots, s\}$ can concisely be written in form of a Butcher table:

$$\begin{array}{c|ccc} c_1 & a_{11} & \cdots & a_{1s} \\ \vdots & \vdots & & \vdots \\ c_s & a_{s1} & \cdots & a_{ss} \\ \hline & \bar{b}_1 & \cdots & \bar{b}_s \\ \hline & b_1 & \cdots & b_s \end{array}$$

For the trapezoidal rule, the Butcher table reads

$$\begin{array}{c|cc} 0 & 0 & 0 \\ 1 & \frac{1}{4} & \frac{1}{4} \\ \hline & \frac{1}{4} & \frac{1}{4} \\ \hline & \frac{1}{2} & \frac{1}{2} \end{array} \tag{9.2}$$

The method (9.1) is now extended to a continuous scheme which yields intermediate approximations $(\mathbf{u}_\vartheta, \mathbf{v}_\vartheta) \approx (\mathbf{u}(\vartheta \Delta T), \mathbf{v}(\vartheta \Delta T))$ for any $\vartheta \in [0, 1]$ without additional evaluations of $\mathbf{f}$. For this, the relations (9.1b), (9.1c) are changed to

$$\mathbf{u}_\vartheta = \mathbf{u}_0 + \vartheta \Delta T \mathbf{v}_0 + \Delta T^2 \sum_{i=1}^{s} \bar{b}_i(\vartheta) \mathbf{k}_i, \tag{9.3a}$$

$$\mathbf{v}_\vartheta = \mathbf{v}_0 + \Delta T \sum_{i=1}^{s} b_i(\vartheta) \mathbf{k}_i, \tag{9.3b}$$

where $\bar{b}_i(\vartheta)$, $b_i(\vartheta)$ are scalar functions satisfying $\bar{b}_i(0) = 0$, $\bar{b}_i(1) = \bar{b}_i$ and $b_i(0) = 0$, $b_i(1) = b_i$, respectively. The continuous method we propose uses the coefficients a_{li} and c_l of (9.2) combined with

$$\bar{b}_1(\vartheta) = \bar{b}_2(\vartheta) = \frac{\vartheta}{4}, \quad b_1(\vartheta) = b_2(\vartheta) = \frac{\vartheta}{2}. \tag{9.4}$$

By checking the corresponding order conditions [75], one can show that this method has a uniform local convergence order of $\mathcal{O}(\Delta T^2)$ for both the displacement $\mathbf{u}_\vartheta$ and the velocity $\mathbf{v}_\vartheta$. According to the definition of [128], the resulting continuous Nyström method has a uniform convergence order of one. However, as argued in the same paper, the error at time $t_{j+\vartheta} = (j+\vartheta)\Delta T$ can be split into the error of the discrete method at time t_j and the local error of the continuous extension. The former is of order $\mathcal{O}(\Delta T^2)$ for the second order scheme (9.2), whereas the latter is at least of second order. Hence, the continuous scheme yields a global error of $\mathcal{O}(\Delta T^2)$.

Moreover, one of the main reasons for the use of (9.4) is the following lemma in the spirit of [10]:

Lemma 9.1. *The interpolant* $(\mathbf{u}_\vartheta, \mathbf{v}_\vartheta)$ *defined by (9.3), (9.4) is stable, i.e., if the scheme is applied to the scalar model problem* $\ddot{u} = -\omega^2 u$, *there exists a constant* M *independent of* ω *and* ΔT *such that*

$$\max_{0\leq\vartheta\leq 1} \left(|\omega u(\vartheta\Delta T)|^2 + |v(\vartheta\Delta T)|^2\right) \leq M \tag{9.5}$$

holds for all $\omega, \Delta T > 0$.

Proof. Let $z := \omega\Delta T$ and $r = r(z) := \frac{z^2}{1+\frac{z^2}{4}}$. Substituting $f(t,u) = -\omega^2 u$ into (9.1a), we obtain from (9.2)

$$\begin{aligned} k_1 + k_2 &= -\omega^2\left(2u_0 + \Delta T v_0 + \frac{\Delta T^2}{4}(k_1+k_2)\right),\\ k_1 + k_2 &= -\frac{r(z)}{\Delta T^2}(2u_0 + \Delta T v_0). \end{aligned}$$

With this, the relations (9.3) yield

$$\begin{aligned} u_\vartheta &= u_0 + \vartheta\Delta T v_0 - \frac{\vartheta}{4} r(z)(2u_0 + \Delta T v_0),\\ v_\vartheta &= v_0 - \frac{\vartheta}{2\Delta T} r(z)(2u_0 + \Delta T v_0). \end{aligned}$$

The above relations can be written in the form

$$\begin{pmatrix} \omega u_\vartheta \\ v_\vartheta \end{pmatrix} = B_\vartheta(z) \begin{pmatrix} \omega u_0 \\ v_0 \end{pmatrix}, \tag{9.6}$$

with the amplification matrix

$$B_\vartheta(z) := \begin{pmatrix} 1 - \frac{\vartheta}{2} r & \vartheta z\left(1 - \frac{1}{4} r\right) \\ -\vartheta\frac{r}{z} & 1 - \frac{\vartheta}{2} r \end{pmatrix}. \tag{9.7}$$

The eigenvalues of (9.7) are given by the roots of

$$\begin{aligned} p(\lambda) &= \left(1 - \frac{\vartheta}{2} r - \lambda\right)^2 + \vartheta^2 r\left(1 - \frac{1}{4} r\right)\\ &= \lambda^2 + (\vartheta r - 2)\lambda + 1 + \vartheta(\vartheta - 1) r. \end{aligned}$$

This polynomial has two different complex roots, as the determinant

$$(\vartheta r-2)^2-4\left(1+\vartheta(\vartheta-1)r\right)=\vartheta^2 r(\underbrace{r}_{<4}-4)<0$$

is negative. Hence, $B_\vartheta(z)$ is diagonalizable, and the modulus of its eigenvalues is

$$|\lambda|^2=1+\underbrace{\vartheta(\vartheta-1)}_{\leq 0}r\leq 1.$$

This implies the stability result (9.5). □

Remark 9.2. The uniform convergence order of the scheme (9.3), (9.4) is not optimal. A continuous Nyström method of uniform order two can be obtained if the coefficients $b_i(\vartheta)$, $\bar{b}_i(\vartheta)$ are chosen as

$$\bar{b}_1(\vartheta)=\bar{b}_2(\vartheta)=\frac{\vartheta^2}{4},\quad b_1(\vartheta)=\vartheta-\frac{\vartheta^2}{2},\quad b_2(\vartheta)=\frac{\vartheta^2}{2}.$$

However, this scheme is unstable for the inner approximations, because the corresponding amplification matrix reads

$$B_\vartheta(z):=\begin{pmatrix}1-\dfrac{\vartheta^2}{2}r & \vartheta z\left(1-\dfrac{\vartheta}{4}r\right)\\ -\vartheta z\left(1-\dfrac{\vartheta}{4}r\right) & 1-\dfrac{\vartheta^2}{2}r\end{pmatrix}.$$

The eigenvalues of this matrix are given by the complex conjugate numbers

$$\lambda_{1/2}=1-\frac{\vartheta^2}{2}r\pm\vartheta z\left(1-\frac{\vartheta}{4}r\right)i,$$

whose modulus is bounded by one iff

$$\vartheta^2(1-\vartheta)^2r^2\leq 0.$$

Hence, the method is unconditionally unstable for $\vartheta<1$.

Using the relations (9.1bc) to eliminate the values $\mathbf{k}_1$, $\mathbf{k}_2$ from (9.3), we get

$$\mathbf{u}_\vartheta=(1-\vartheta)\mathbf{u}_0+\vartheta\mathbf{u}_1, \tag{9.8a}$$
$$\mathbf{v}_\vartheta=(1-\vartheta)\mathbf{v}_0+\vartheta\mathbf{v}_1, \tag{9.8b}$$

i.e., the continuous Nyström method (9.3), (9.4) can be interpreted as a linear interpolation of both the displacement and the velocity.

9.2 Time substepping

In this subsection, we use the continuous Nyström method described in Section 9.1 to define a finer time scale of step size Δt on the patch ω. We assume that the time step sizes have an integer ratio, i.e., $\frac{\Delta T}{\Delta t}=J$ holds for some $J\in\mathbb{N}$. For ease of notation, the time indices now refer to the fine time steps, such that the first macro time step lasts from t_0 to t_J. The continuity conditions on the interface Γ are derived from (9.8) at the values $\vartheta = \frac{j}{J}$, $j \in \{1, \ldots, J\}$, leading to a coupled system of equations consisting of one coarse problem on Ξ and J fine subproblems on ω.

First, we look at the solution of one fine time step at the patch ω, where we want to ensure the weak continuity of both the displacement and the velocity on Γ with the interpolated values from (9.8). In general, these quantities do not satisfy the relation (7.18d) at the interface, such that we need to enforce both continuity conditions separately. This is done by augmenting (7.18d) with an additional Lagrange multiplier $\boldsymbol{\xi}_\Gamma^h$, similar to the methods proposed in [15, 16, 125]:

$$\begin{pmatrix} \frac{1}{2}A_\omega^h & \frac{1}{\Delta t}M_\omega^h & (D^{hh})^T & \mathbf{0} \\ \frac{1}{\Delta t}M_\omega^h & -\frac{1}{2}M_\omega^h & \mathbf{0} & (D^{hh})^T \\ D^{hh} & \mathbf{0} & \mathbf{0} & \mathbf{0} \\ \mathbf{0} & D^{hh} & \mathbf{0} & \mathbf{0} \end{pmatrix} \begin{pmatrix} \mathbf{u}_j^h \\ \mathbf{v}_j^h \\ \boldsymbol{\zeta}_{\Gamma,j-1/2}^h \\ \boldsymbol{\xi}_{\Gamma,j-1/2}^h \end{pmatrix} = \begin{pmatrix} \boldsymbol{\rho}_{j-1}^h \\ \tilde{\boldsymbol{\rho}}_{j-1}^h \\ D^{hH}\boldsymbol{\nu}_j^H \\ D^{hH}\tilde{\boldsymbol{\nu}}_j^H \end{pmatrix}. \tag{9.9}$$

The residuals in (9.9) read

$$\begin{pmatrix} \boldsymbol{\rho}_{j-1}^h \\ \tilde{\boldsymbol{\rho}}_{j-1}^h \end{pmatrix} = \begin{pmatrix} \mathbf{f}_{j-1/2}^h \\ \mathbf{0} \end{pmatrix} + \begin{pmatrix} -\frac{1}{2}A_\omega^h & \frac{1}{\Delta t}M_\omega^h \\ \frac{1}{\Delta t}M_\omega^h & \frac{1}{2}M_\omega^h \end{pmatrix} \begin{pmatrix} \mathbf{u}_{j-1}^h \\ \mathbf{v}_{j-1}^h \end{pmatrix}, \tag{9.10a}$$

$$\begin{pmatrix} \boldsymbol{\nu}_j^H \\ \tilde{\boldsymbol{\nu}}_j^H \end{pmatrix} = \left(1 - \frac{j}{J}\right) \begin{pmatrix} \mathbf{u}_0^H \\ \mathbf{v}_0^H \end{pmatrix} + \frac{j}{J} \begin{pmatrix} \mathbf{u}_J^H \\ \mathbf{v}_J^H \end{pmatrix}. \tag{9.10b}$$

The feedback to the coarse grid equations (7.18ab) is carried out by the sum of the Lagrange multipliers, leading to

$$\begin{pmatrix} \frac{1}{2}A_\Xi^H & \frac{1}{\Delta T}M_\Xi^H \\ \frac{1}{\Delta T}M_\Xi^H & -\frac{1}{2}M_\Xi^H \end{pmatrix} \begin{pmatrix} \mathbf{u}_J^H \\ \mathbf{v}_J^H \end{pmatrix} = \begin{pmatrix} \boldsymbol{\rho}_0^H + \frac{1}{J}(D^{hH})^T \sum_{j=1}^J \boldsymbol{\zeta}_{\Gamma,j-1/2}^h \\ \tilde{\boldsymbol{\rho}}_0^H + \frac{1}{J}(D^{hH})^T \sum_{j=1}^J \boldsymbol{\xi}_{\Gamma,j-1/2}^h \end{pmatrix}, \tag{9.11}$$

with the residuals $\boldsymbol{\rho}_0^H$, $\tilde{\boldsymbol{\rho}}_0^H$ defined analogeously to (9.10a).

The reason for defining the backcoupling as stated in (9.11) is the following lemma:

Lemma 9.3. *For time-independent loads* $\mathbf{f}$, *the coupled system (9.9), (9.11) conserves the total energy (7.19) after one coarse time step.*

Proof. Using the notation $\Delta\mathbf{u}_j^h = \mathbf{u}_j^h - \mathbf{u}_{j-1}^h$, $\Delta\mathbf{u}_J^H = \mathbf{u}_J^H - \mathbf{u}_0^H$, we obtain

$$
\begin{aligned}
&\mathbb{E}_J^H - \mathbb{E}_0^H + \mathbb{E}_J^h - \mathbb{E}_0^h \\
&= \frac{1}{2}\Big((\Delta\mathbf{v}_J^H)^T M_\Xi^H \mathbf{v}_{J/2}^H + (\Delta\mathbf{u}_J^H)^T A_\Xi^H \mathbf{u}_{J/2}^H\Big) + (\Delta\mathbf{u}_J^H)^T\mathbf{f}^H \\
&\quad + \sum_{j=1}^{J}\left(\frac{1}{2}\Big((\Delta\mathbf{v}_j^h)^T M_\omega^h \mathbf{v}_{j-1/2}^h + (\Delta\mathbf{u}_j^h)^T A_\omega^h \mathbf{u}_{j-1/2}^h\Big) + (\Delta\mathbf{u}_j^h)^T\mathbf{f}^h\right) \\
&= \frac{1}{\Delta T}(\Delta\mathbf{v}_J^H)^T M_\Xi^H \Delta\mathbf{u}_J^H - \frac{1}{J}(\Delta\mathbf{v}_J^H)^T(D^{hH})^T\sum_{j=1}^{J}\boldsymbol{\xi}_{\Gamma,j-1/2}^h + \frac{1}{2}(\Delta\mathbf{u}_J^H)^T A_\Xi^H\mathbf{u}_{J/2}^H + (\Delta\mathbf{u}_J^H)^T\mathbf{f}^H \\
&\quad + \sum_{j=1}^{J}\left(\frac{1}{\Delta t}(\Delta\mathbf{v}_j^h)^T M_\omega^h(\Delta\mathbf{u}_j^h) + (\Delta\mathbf{v}_j^h)^T(D^{hh})^T\boldsymbol{\xi}_{\Gamma,j-1/2}^h + \frac{1}{2}(\Delta\mathbf{u}_j^h)^T A_\omega^h\mathbf{u}_{j-1/2}^h + (\Delta\mathbf{u}_j^h)^T\mathbf{f}^h\right) \\
&= \frac{1}{J}(\Delta\mathbf{u}_{\Gamma,J}^H)^T(D_\Gamma^{hH})^T\sum_{j=1}^{J}\boldsymbol{\varsigma}_{\Gamma,j-1/2}^h - \frac{1}{J}(\Delta\mathbf{v}_{\Gamma,J}^H)^T(D_\Gamma^{hH})^T\sum_{j=1}^{J}\boldsymbol{\xi}_{\Gamma,j-1/2}^h \\
&\quad + \sum_{j=1}^{J}\Big(-(\Delta\mathbf{u}_{\Gamma,j}^h)^T(D_\Gamma^{hh})^T\boldsymbol{\varsigma}_{\Gamma,j-1/2}^h + (\Delta\mathbf{v}_{\Gamma,j}^h)^T(D_\Gamma^{hh})^T\boldsymbol{\xi}_{\Gamma,j-1/2}^h\Big) \\
&= \frac{1}{J}(\Delta\mathbf{u}_{\Gamma,J}^H)^T(D_\Gamma^{hH})^T\sum_{j=1}^{J}\boldsymbol{\varsigma}_{\Gamma,j-1/2}^h - \frac{1}{J}(\Delta\mathbf{v}_{\Gamma,J}^H)^T(D_\Gamma^{hH})^T\sum_{j=1}^{J}\boldsymbol{\xi}_{\Gamma,j-1/2}^h \\
&\quad - \frac{1}{J}\sum_{j=1}^{J}\Big((\Pi^{hH}\Delta\mathbf{u}_{\Gamma,J}^h)^T(D_\Gamma^{hh})^T\boldsymbol{\varsigma}_{\Gamma,j-1/2}^h - (\Pi^{hH}\Delta\mathbf{v}_{\Gamma,J}^H)^T(D_\Gamma^{hh})^T\boldsymbol{\xi}_{\Gamma,j-1/2}^h\Big) = 0.
\end{aligned}
$$

□

As in Subsection 7.1.4, we next perform static condensation of the velocity in (9.9). Using the matrix $K_\omega^h = \frac{1}{2}A_\omega^h + \frac{2}{\Delta t^2}M_\omega^h$ from (2.22), this yields the following saddle point system for the fine displacement:

$$
\begin{pmatrix} K_\omega^h & (D^{hh})^T \\ D^{hh} & \mathbf{0} \end{pmatrix}
\begin{pmatrix} \mathbf{u}_j^h \\ \boldsymbol{\varsigma}_{\Gamma,j-1/2}^h + \frac{2}{\Delta t}\boldsymbol{\xi}_{\Gamma,j-1/2}^h \end{pmatrix}
= \begin{pmatrix} \boldsymbol{\rho}_{j-1}^h + \frac{2}{\Delta t}\widetilde{\boldsymbol{\rho}}_{j-1}^h \\ D^{hH}\boldsymbol{\nu}_j^H \end{pmatrix}. \tag{9.12}
$$

In order to obtain the correct velocity after (9.12) has been solved, we need to compute the solution of

$$
\begin{pmatrix} -\frac{1}{2}M_\omega^h & (D^{hh})^T \\ D^{hh} & \mathbf{0} \end{pmatrix}
\begin{pmatrix} \mathbf{v}_j^h \\ \boldsymbol{\xi}_{\Gamma,j-1/2}^h \end{pmatrix}
= \begin{pmatrix} \widetilde{\boldsymbol{\rho}}_{j-1}^h - \frac{1}{\Delta t}M_\omega^h\mathbf{u}_j^h \\ D^{hH}\widetilde{\boldsymbol{\nu}}_j^H \end{pmatrix}. \tag{9.13}
$$

If the global mass matrix is lumped (i.e., the submatrices $M_{\Gamma\Gamma}^h$, $M_{\Gamma\Gamma}^H$ are diagonal and $M_{\Gamma\omega}^h$, $M_{\omega\Gamma}^h$ vanish) or if the mass is removed from the interface Γ analogeously to Chapter 5 (i.e., $M_{\Gamma\Gamma}^h$, $M_{\Gamma\omega}^h$, $M_{\omega\Gamma}^h$ vanish), the solution of (9.13) is trivial. Furthermore, in these cases, the backcoupling term in $(9.11)_2$ vanishes because we have

$$
\begin{aligned}
&(D_\Gamma^{hh})^T \sum_{j=1}^{J} \boldsymbol{\xi}_{\Gamma,j-1/2}^h \\
&= \sum_{j=1}^{J} M_{\Gamma\Gamma}^h \left(\frac{1}{2} \left(\mathbf{v}_{\Gamma,j}^h + \mathbf{v}_{\Gamma,j-1}^h \right) - \frac{1}{\Delta t} \left(\mathbf{u}_{\Gamma,j}^h - \mathbf{u}_{\Gamma,j-1}^h \right) \right) \\
&= M_{\Gamma\Gamma}^h \Pi^{hH} \sum_{j=1}^{J} \left(\frac{1}{2} \left(\frac{2j-1}{J} \mathbf{v}_{\Gamma,J}^H + \frac{2J-2j+1}{J} \mathbf{v}_{\Gamma,0}^H \right) - \frac{1}{J\Delta t} \left(\mathbf{u}_{\Gamma,J}^H - \mathbf{u}_{\Gamma,0}^H \right) \right) \\
&= M_{\Gamma\Gamma}^h \Pi^{hH} \sum_{j=1}^{J} \left(\frac{2j-1}{2J} \mathbf{v}_{\Gamma,J}^H + \frac{2J-2j+1}{2J} \mathbf{v}_{\Gamma,0}^H \right) \\
&\quad - M_{\Gamma\Gamma}^h \Pi^{hH} \sum_{j=1}^{J} \left(\frac{1}{2} \left(\mathbf{v}_{\Gamma,J}^H + \mathbf{v}_{\Gamma,0}^H \right) + (M_{\Gamma\Gamma}^H)^{-1} (D_\Gamma^{hH})^T \boldsymbol{\xi}_{\Gamma,j-1/2}^h \right).
\end{aligned}
$$

Resolving the above equation for the sum of the multipliers, we obtain

$$
\begin{aligned}
&\left(\mathrm{Id} + M_{\Gamma\Gamma}^h \Pi^{hH} (M_{\Gamma\Gamma}^H)^{-1} (\Pi^{hH})^T \right) (D_\Gamma^{hh})^T \sum_{j=1}^{J} \boldsymbol{\xi}_{\Gamma,j-1/2}^h \\
&= M_{\Gamma\Gamma}^h \Pi^{hH} \sum_{j=1}^{J} \frac{2j-1-J}{2J} \left(\mathbf{v}_{\Gamma,J}^H - \mathbf{v}_{\Gamma,0}^H \right) = \mathbf{0}.
\end{aligned}
$$

Thus, from now on, we restrict ourselves to the case that $\sum_{j=1}^{J} \boldsymbol{\xi}_{\Gamma,j-1/2}^h = \mathbf{0}$, yielding the coarse problem

$$
\begin{pmatrix} \frac{1}{2} A_\Xi^H & \frac{1}{\Delta T} M_\Xi^H \\ \frac{1}{\Delta T} M_\Xi^H & -\frac{1}{2} M_\Xi^H \end{pmatrix} \begin{pmatrix} \mathbf{u}_J^H \\ \mathbf{v}_J^H \end{pmatrix} = \begin{pmatrix} \boldsymbol{\rho}_0^H + \frac{1}{J} (D^{hH})^T \sum_{j=1}^{J} \boldsymbol{\zeta}_{\Gamma,j-1/2}^h \\ \tilde{\boldsymbol{\rho}}_0^H \end{pmatrix}. \tag{9.14}
$$

Then, the restriction of (9.12) and (9.14) to the interface Γ yields the following coupled system of equations:

$$
\begin{aligned}
S_\Xi^H \mathbf{u}_{\Gamma,J}^H - \frac{1}{J} (D_\Gamma^{hH})^T \sum_{j=1}^{J} \boldsymbol{\zeta}_{\Gamma,j-1/2}^h - \bar{\boldsymbol{\rho}}_{\Xi,0}^H &= \mathbf{0}, \\
\tfrac{1}{\Delta T} \mathbf{u}_{\Gamma,J}^H - \tfrac{1}{2} \mathbf{v}_{\Gamma,J}^H - \tfrac{1}{\Delta T} \mathbf{u}_{\Gamma,0}^H - \tfrac{1}{2} \mathbf{v}_{\Gamma,0}^H &= \mathbf{0},
\end{aligned} \tag{9.15a}
$$

$$
\begin{aligned}
j = 1, \dots, J: \qquad S_\omega^h \mathbf{u}_{\Gamma,j}^h + (D_\Gamma^{hh})^T \boldsymbol{\zeta}_{\Gamma,j-1/2}^h + \frac{2}{\Delta t} (D_\Gamma^{hh})^T \boldsymbol{\xi}_{\Gamma,j-1/2}^h - \bar{\boldsymbol{\rho}}_{\omega,j-1}^h &= \mathbf{0}, \\
D_\Gamma^{hh} \mathbf{u}_{\Gamma,j}^h - D_\Gamma^{hH} \left(\left(1 - \frac{j}{J} \right) \mathbf{u}_{\Gamma,0}^H + \frac{j}{J} \mathbf{u}_{\Gamma,J}^H \right) &= \mathbf{0}, \\
(D_\Gamma^{hh})^T \boldsymbol{\xi}_{\Gamma,j-1/2}^h - \frac{2j-1-J}{2J} M_{\Gamma\Gamma}^h \Pi^{hH} \left(\mathbf{v}_{\Gamma,J}^H - \mathbf{v}_{\Gamma,0}^H \right) &= \mathbf{0}, \\
D_\Gamma^{hh} \mathbf{v}_{\Gamma,j}^h - D_\Gamma^{hH} \left(\left(1 - \frac{j}{J} \right) \mathbf{v}_{\Gamma,0}^H + \frac{j}{J} \mathbf{v}_{\Gamma,J}^H \right) &= \mathbf{0}.
\end{aligned} \tag{9.15b}
$$

The Schur matrices S_Ξ^H, S_ω^h used above have been defined in (7.24), and the residuals $\bar{\rho}_{\Xi,0}^H$, $\bar{\rho}_{\omega,j-1}^h$ are given by

$$\bar{\varrho}_{\Xi,0}^H := \rho_{\Gamma,0}^H + \frac{2}{\Delta T}\tilde{\rho}_{\Gamma,0}^H - K_{\Gamma\Xi}^H (K_{\Xi\Xi}^H)^{-1}\left(\rho_{\Xi,0}^H + \frac{2}{\Delta T}\tilde{\rho}_{\Xi,0}^H\right),$$
$$\bar{\rho}_{\omega,j-1}^h := \rho_{\Gamma,j-1}^h + \frac{2}{\Delta t}\tilde{\rho}_{\Gamma,j-1}^h - K_{\Gamma\omega}^h (K_{\omega\omega}^h)^{-1}\left(\rho_{\omega,j-1}^h + \frac{2}{\Delta t}\tilde{\rho}_{\omega,j-1}^h\right).$$

Besides the fact that the solution of (9.15) is energy-conserving, it permits optimal a priori estimates for the local error on the interior of ω. Details are given in Theorem 9.4 below. However, a direct solution of the coupled system (9.15) is very expensive, as it involves the simultaneous solution of one coarse problem on Ξ and J fine subproblems on ω. Here, we benefit from the iterative algorithm derived in Chapter 7 which provides an efficient way of solving the coupled system (9.15). The corresponding algorithm is summarized in Section 9.3.

In the rest of this section, we prove the following a priori estimates for the local error of the time stepping scheme (9.15) with respect to the time step sizes Δt, ΔT:

Theorem 9.4. *Let the initial conditions* $\left(\mathbf{u}_0^H, \mathbf{v}_0^H, \mathbf{u}_0^h, \mathbf{v}_0^h\right)$ *be given. Let*

$$\left(\mathbf{u}_J^H, \mathbf{v}_J^H, \mathbf{u}_1^h, \mathbf{v}_1^h, \boldsymbol{\zeta}_{1/2}^h, \ldots, \mathbf{u}_J^h, \mathbf{v}_J^h, \boldsymbol{\zeta}_{J-1/2}^h\right) \tag{9.16}$$

be the complete solution of (9.9), (9.14) with a mass matrix such that $M_{\Gamma\omega}^h = \mathbf{0}$, $M_{\omega\Gamma}^h = \mathbf{0}$, *and* $M_{\Gamma\Gamma}^h$ *is diagonal, and let*

$$\left(\hat{\mathbf{u}}^H(t), \hat{\mathbf{v}}^H(t), \hat{\mathbf{u}}^h(t), \hat{\mathbf{v}}^h(t), \hat{\boldsymbol{\zeta}}^h(t)\right) \tag{9.17}$$

be the time-continuous solution of the differential-algebraic equation (7.14) with the same mass matrix. Then, we obtain the following local error estimates:

$$\begin{aligned}
\|\mathbf{eu}_J^H\| &:= \|\mathbf{u}_J^H - \hat{\mathbf{u}}^H(t_J)\| && \le C\Delta T^3, && \\
\|\mathbf{ev}_J^H\| &:= \|\mathbf{v}_J^H - \hat{\mathbf{v}}^H(t_J)\| && \le C\Delta T^2, && \\
\|\mathbf{eu}_{\omega,j}^h\| &:= \|\mathbf{u}_{\omega,j}^h - \hat{\mathbf{u}}_\omega^h(t_j)\| && \le Cj\Delta t^3, && j \in \{1,\ldots,J\}, \\
\|\mathbf{eu}_{\Gamma,j}^h\| &:= \|\mathbf{u}_{\Gamma,j}^h - \hat{\mathbf{u}}_\Gamma^h(t_j)\| && \le C\tfrac{j(J-j)}{J^2}\Delta T^2 + \mathcal{O}(\Delta T^3), && j \in \{1,\ldots,J\}, \\
\|\mathbf{ev}_{\omega,j}^h\| &:= \|\mathbf{v}_{\omega,j}^h - \hat{\mathbf{v}}_\omega^h(t_j)\| && \le Cj\Delta t^3, && j \in \{1,\ldots,J\}, \\
\|\mathbf{ev}_{\Gamma,j}^h\| &:= \|\mathbf{v}_{\Gamma,j}^h - \hat{\mathbf{v}}_\Gamma^h(t_j)\| && \le C\tfrac{j(2J-j)}{J^2}\Delta T^2 + \mathcal{O}(\Delta T^3), && j \in \{1,\ldots,J\}, \\
\|\mathbf{e}\boldsymbol{\zeta}_{j-1/2}^h\| &:= \left\|\boldsymbol{\zeta}_{j-1/2}^h - \hat{\boldsymbol{\zeta}}^h(t_{j-1/2})\right\| && \le C\Delta T, && j \in \{1,\ldots,J\}.
\end{aligned}$$

The constants C of the leading order terms can depend on h, the material parameters and the time derivatives of (9.17) but are independent of J, j and ΔT.

Depending on the structure of the mass matrix, some of these estimates can be improved:

a) If the submatrices $M_{\Gamma\Xi}^H$, $M_{\Xi\Gamma}^H$ *vanish, we get*

$$\|\mathbf{ev}_{\Xi,J}^H\| \le C\Delta T^3.$$

b) If the interface mass matrix $M^h_{\Gamma\Gamma}$ vanishes, we obtain

$$\begin{aligned}\|\mathbf{ev}^H_J\| &\le C\Delta T^3, \\ \|\mathbf{ev}^h_{\Gamma,j}\| &\le C\tfrac{j(J-j)}{J^2}\Delta T^2 + \mathcal{O}(\Delta T^3), \quad && j \in \{1,\dots,J\}, \\ \|\mathbf{e}\boldsymbol{\zeta}^h_{j-1/2}\| &\le C\Delta t^2, && j \in \{1,\dots,J\}.\end{aligned}$$

Proof. The idea of the proof is similar to [125] where the time-continuous solution is considered as a perturbed solution of the discrete system, and the preturbation terms are estimated using Taylor expansion. To transfer this idea to the coupled system (9.15), we proceed in several steps. First, we compute the asymptotic expansion of the coarse Schur problem $(9.15\text{a})_1$, second, we recursively eliminate the inner degrees of freedom on ω to obtain a formulation involving only the fine interface displacements and multipliers, and finally, we link the results by means of the interface conditions $(9.15\text{b})_{2,4}$.

In the proof, we abbreviate $\hat{\mathbf{u}}^s_j := \hat{\mathbf{u}}^s(t_j)$, $s \in \{h, H\}$, and use the $\mathcal{O}$-notation with an explicit indication of the dependence on Δt, J or j.

a) Coarse problem on Ξ

Substitution of the exact solution into the discrete problem on Ξ gives

$$M^H_\Xi \frac{\hat{\mathbf{v}}^H_J - \hat{\mathbf{v}}^H_0}{\Delta T} + A^H_\Xi \frac{\hat{\mathbf{u}}^H_J + \hat{\mathbf{u}}^H_0}{2} - \frac{1}{J}\sum_{j=1}^{J}(D^{hH})^T\hat{\boldsymbol{\zeta}}^h_{j-1/2} - \frac{\mathbf{f}^H_J + \mathbf{f}^H_0}{2} = \boldsymbol{\delta}^H_1, \tag{9.18a}$$

$$\frac{\hat{\mathbf{v}}^H_J + \hat{\mathbf{v}}^H_0}{2} - \frac{\hat{\mathbf{u}}^H_J - \hat{\mathbf{u}}^H_0}{\Delta T} = \boldsymbol{\delta}^H_2, \tag{9.18b}$$

with suitable perturbation terms $\boldsymbol{\delta}^H_1$, $\boldsymbol{\delta}^H_2$ whose dependence on ΔT can be estimated using Taylor expansion:

$$\begin{aligned}\boldsymbol{\delta}^H_2 &= \hat{\mathbf{v}}^H_{J/2} - \dot{\hat{\mathbf{u}}}^H_{J/2} + \mathcal{O}(\Delta T^2) = \mathcal{O}(\Delta T^2), \\ \boldsymbol{\delta}^H_1 &= M^H_\Xi\dot{\hat{\mathbf{v}}}^H_{J/2} + A^H_\Xi\hat{\mathbf{u}}^H_{J/2} - \frac{1}{J}\sum_{j=1}^{J}(D^{hH})^T\left(\hat{\boldsymbol{\zeta}}^h_{J/2} + \frac{2j-1-J}{2J}\Delta T\dot{\hat{\boldsymbol{\zeta}}}^h_{J/2}\right) - \mathbf{f}^H_{J/2} + \mathcal{O}(\Delta T^2) \\ &= -\frac{\Delta T}{2J^2}(D^{hH})^T\dot{\hat{\boldsymbol{\zeta}}}^h_{J/2}\sum_{j=1}^{J}(2j-1-J) + \mathcal{O}(\Delta T^2) = \mathcal{O}(\Delta T^2).\end{aligned}$$

Subtracting the relations (9.18) from the discrete ones given by (9.15a), we obtain the following equations for the local error at time t_J:

$$M^H_\Xi \frac{\mathbf{ev}^H_J}{\Delta T} + A^H_\Xi \frac{\mathbf{eu}^H_J}{2} - \frac{1}{J}\sum_{j=1}^{J}(D^{hH})^T\mathbf{e}\boldsymbol{\zeta}^h_{j-1/2} = -\boldsymbol{\delta}^H_1, \tag{9.19a}$$

$$\frac{\mathbf{ev}^H_J}{2} - \frac{\mathbf{eu}^H_J}{\Delta T} = -\boldsymbol{\delta}^H_2. \tag{9.19b}$$

Substituting (9.19b) into (9.19a) gives

$$\left(\frac{2}{\Delta T^2}M^H_\Xi + \frac{1}{2}A^H_\Xi\right)\mathbf{eu}^H_J - \frac{2}{\Delta T}M^H_\Xi\boldsymbol{\delta}^H_2 - \frac{1}{J}\sum_{j=1}^{J}(D^{hH})^T\mathbf{e}\boldsymbol{\zeta}^h_{j-1/2} = -\boldsymbol{\delta}^H_1.$$

Using the Taylor expansion of the perturbation terms, the equations associated with the inner degrees of freedom yield

$$K^H_{\Xi\Xi}\mathbf{eu}^H_{\Xi,J} = -K^H_{\Xi\Gamma}\mathbf{eu}^H_{\Gamma,J} + \mathcal{O}(\Delta T), \tag{9.20}$$

with $K^H_{\Xi\Xi}$ defined as in (2.22):

$$K^H_{\Xi\Xi} = \frac{2}{\Delta T^2}M^H_{\Xi\Xi} + \frac{1}{2}A^H_{\Xi\Xi} = \frac{2}{\Delta T^2}M^H_{\Xi\Xi}\left(\mathrm{Id} + \frac{\Delta T^2}{4}(M^H_{\Xi\Xi})^{-1}A^H_{\Xi\Xi}\right). \tag{9.21}$$

For ΔT small enough, the inverse of (9.21) has the asymptotic expansion

$$\begin{aligned}(K^H_{\Xi\Xi})^{-1} &= \frac{\Delta T^2}{2}\left(\mathrm{Id} + \frac{\Delta T^2}{4}(M^H_{\Xi\Xi})^{-1}A^H_{\Xi\Xi}\right)^{-1}(M^H_{\Xi\Xi})^{-1}\\ &= \frac{\Delta T^2}{2}\sum_{k=0}^{\infty}\left(-\frac{\Delta T^2}{4}(M^H_{\Xi\Xi})^{-1}A^H_{\Xi\Xi}\right)^k(M^H_{\Xi\Xi})^{-1}\\ &= \frac{\Delta T^2}{2}(M^H_{\Xi\Xi})^{-1} + \mathcal{O}(\Delta T^4),\end{aligned} \tag{9.22}$$

leading to

$$\mathbf{eu}^H_{\Xi,J} = -\frac{\Delta T^2}{2}(M^H_{\Xi\Xi})^{-1}K^H_{\Xi\Gamma}\mathbf{eu}^H_{\Gamma,J} + \mathcal{O}(\Delta T^3). \tag{9.23}$$

For the boundary terms, we get

$$S^H_{\Xi}\mathbf{eu}^H_{\Gamma,J} - \frac{1}{J}\sum_{j=1}^{J}(D^{hH}_{\Gamma})^T\mathbf{e}\boldsymbol{\zeta}^h_{j-1/2} = \mathcal{O}(\Delta T), \tag{9.24}$$

with the Schur matrix S^H_{Ξ} defined in (7.24a). Hence, it only remains to estimate the error in the Lagrange multipliers, for which we need to look at the subproblems on the patch.

b) Fine subproblems on ω

We consider the fine time step (9.15b) from t_{j-1} to t_j, $j \in \{1, \dots, J\}$; substitution of the exact solution leads to

$$\begin{aligned} M^h_{\omega}\frac{\widehat{\mathbf{v}}^h_j - \widehat{\mathbf{v}}^h_{j-1}}{\Delta t} + A^h_{\omega}\frac{\widehat{\mathbf{u}}^h_j + \widehat{\mathbf{u}}^h_{j-1}}{2} + (D^{hh})^T\widehat{\boldsymbol{\zeta}}^h_{j-1/2} - \frac{\mathbf{f}^h_j + \mathbf{f}^h_{j-1}}{2} &= \boldsymbol{\delta}^h_{j1},\\ \frac{\widehat{\mathbf{v}}^h_{\omega,j} + \widehat{\mathbf{v}}^h_{\omega,j-1}}{2} - \frac{\widehat{\mathbf{u}}^h_{\omega,j} - \widehat{\mathbf{u}}^h_{\omega,j-1}}{\Delta t} &= \boldsymbol{\delta}^h_{\omega,j2},\\ \widehat{\mathbf{u}}^h_{\Gamma,j} - \Pi^{hH}\left(\left(1 - \frac{j}{J}\right)\widehat{\mathbf{u}}^H_{\Gamma,0} + \frac{j}{J}\widehat{\mathbf{u}}^H_{\Gamma,J}\right) &= \boldsymbol{\delta}^h_{\Gamma,j3},\\ \widehat{\mathbf{v}}^h_{\Gamma,j} - \Pi^{hH}\left(\left(1 - \frac{j}{J}\right)\widehat{\mathbf{v}}^H_{\Gamma,0} + \frac{j}{J}\widehat{\mathbf{v}}^H_{\Gamma,J}\right) &= \boldsymbol{\delta}^h_{\Gamma,j4}. \end{aligned}$$

As before, the first two perturbation terms give $\boldsymbol{\delta}^h_{j1} = \mathcal{O}(\Delta t^2)$, $\boldsymbol{\delta}^h_{\omega,j2} = \mathcal{O}(\Delta t^2)$. For the other two terms, Taylor expansion around t_j yields

$$\boldsymbol{\delta}^h_{\Gamma,j3} = -D^{hH}_{\Gamma}\left(\frac{j(J-j)}{J^2}\frac{\Delta T^2}{2}\ddot{\widehat{\mathbf{u}}}^H_{\Gamma,j} + \frac{j(J-j)(J-2j)}{J^3}\frac{\Delta T^3}{6}\dddot{\widehat{\mathbf{u}}}^H_{\Gamma,j} + \mathcal{O}(\Delta T^4)\right) \tag{9.25}$$

and a similar result for $\boldsymbol{\delta}^h_{\Gamma,j4}$. Next, we subtract the discrete from the perturbed equations to obtain

$$M^h_\omega \frac{\mathbf{ev}^h_j - \mathbf{ev}^h_{j-1}}{\Delta t} + A^h_\omega \frac{\mathbf{eu}^h_j + \mathbf{eu}^h_{j-1}}{2} + (D^{hh})^T \mathbf{e}\boldsymbol{\zeta}^h_{j-1/2} = -\boldsymbol{\delta}^h_{j1}, \tag{9.26a}$$

$$\frac{\mathbf{ev}^h_{\omega,j} + \mathbf{ev}^h_{\omega,j-1}}{2} - \frac{\mathbf{eu}^h_{\omega,j} - \mathbf{eu}^h_{\omega,j-1}}{\Delta t} = -\boldsymbol{\delta}^h_{\omega,j2}, \tag{9.26b}$$

$$\mathbf{eu}^h_{\Gamma,j} - \frac{j}{J}\Pi^{hH}\mathbf{eu}^H_{\Gamma,J} = -\boldsymbol{\delta}^h_{\Gamma,j3}, \tag{9.26c}$$

$$\mathbf{ev}^h_{\Gamma,j} - \frac{j}{J}\Pi^{hH}\mathbf{ev}^H_{\Gamma,J} = -\boldsymbol{\delta}^h_{\Gamma,j4}. \tag{9.26d}$$

Applying (9.26b) recursively, we obtain for the inner nodes

$$\mathbf{ev}^h_{\omega,j} = \frac{2}{\Delta t}\mathbf{eu}^h_{\omega,j} + \frac{4}{\Delta t}\sum_{i=1}^{j-1}(-1)^{j-i}\mathbf{eu}^h_{\omega,i} - 2\sum_{i=1}^{j}(-1)^{j-i}\boldsymbol{\delta}^h_{\omega,i2}. \tag{9.27}$$

Substituting this result into (9.26a) and remembering that the entries $M^h_{\omega\Gamma}$ of the mass matrix vanish, we get for the equations associated with the inner nodes

$$\mathbf{0} = \frac{1}{2}A^h_{\omega\omega}\left(\mathbf{eu}^h_{\omega,j} + \mathbf{eu}^h_{\omega,j-1}\right) + \frac{2}{\Delta t^2}M^h_{\omega\omega}\left(\mathbf{eu}^h_{\omega,j} - 3\mathbf{eu}^h_{\omega,j-1} + 4\sum_{i=1}^{j-2}(-1)^{j-i}\mathbf{eu}^h_{\omega,i}\right)$$
$$+ \frac{1}{2}A^h_{\omega\Gamma}\left(\mathbf{eu}^h_{\Gamma,j} + \mathbf{eu}^h_{\Gamma,j-1}\right) - \frac{2}{\Delta t}M^h_{\omega\omega}\left(\boldsymbol{\delta}^h_{\omega,j2} + 2\sum_{i=1}^{j-1}(-1)^{j-i}\boldsymbol{\delta}^h_{\omega,i2}\right) + \mathcal{O}(\Delta t^2).$$

Adding this equation and the one for the previous time step t_{j-1} gives

$$\mathbf{0} = \frac{1}{2}A^h_{\omega\omega}\left(\mathbf{eu}^h_{\omega,j} + 2\mathbf{eu}^h_{\omega,j-1} + \mathbf{eu}^h_{\omega,j-2}\right) + \frac{2}{\Delta t^2}M^h_{\omega\omega}\left(\mathbf{eu}^h_{\omega,j} - 2\mathbf{eu}^h_{\omega,j-1} + \mathbf{eu}^h_{\omega,j-2}\right)$$
$$+ \frac{1}{2}A^h_{\omega\Gamma}\left(\mathbf{eu}^h_{\Gamma,j} + 2\mathbf{u}^h_{\Gamma,j-1} + \mathbf{eu}^h_{\Gamma,j-2}\right) - \frac{2}{\Delta t}M^h_{\omega\omega}\left(\boldsymbol{\delta}^h_{\omega,j2} - \boldsymbol{\delta}^h_{\omega,(j-1)2}\right) + \mathcal{O}(\Delta t^2). \tag{9.28}$$

Next, we make use of (9.28) and the asymptotic expansion

$$(K^h_{\omega\omega})^{-1} = \frac{\Delta t^2}{2}(M^h_{\omega\omega})^{-1} - \frac{\Delta t^4}{8}(M^h_{\omega\omega})^{-1}A^h_{\omega\omega}(M^h_{\omega\omega})^{-1} + \mathcal{O}(\Delta t^6) \tag{9.29}$$

to get a recurrence relation for $\mathbf{eu}^h_{\omega,j}$ of the form

$$\mathbf{eu}^h_{\omega,j} = 2\left(2C^h_{\omega\omega} - \mathrm{Id}\right)\mathbf{eu}^h_{\omega,j-1} - \mathbf{eu}^h_{\omega,j-2} - \mathbf{a}^h_{\omega,j}. \tag{9.30}$$

In (9.30), we have used the definitions

$$C^h_{\omega\omega} := \frac{2}{\Delta t^2}(K^h_{\omega\omega})^{-1}M^h_{\omega\omega} = \mathrm{Id} - \frac{\Delta t^2}{4}(M^h_{\omega\omega})^{-1}A^h_{\omega\omega} + \mathcal{O}(\Delta t^4), \tag{9.31}$$

$$\begin{aligned}\mathbf{a}^h_{\omega,j} &:= \frac{1}{2}(K^h_{\omega\omega})^{-1}A^h_{\omega\Gamma}\left(\mathbf{eu}^h_{\Gamma,j} + 2\mathbf{eu}^h_{\Gamma,j-1} + \mathbf{eu}^h_{\Gamma,j-2}\right) - \Delta t C^h_{\omega\omega}\left(\boldsymbol{\delta}^h_{\omega,j2} - \boldsymbol{\delta}^h_{\omega,(j-1)2}\right) + \mathcal{O}(\Delta t^4)\\ &= \frac{\Delta t^2}{4}(M^h_{\omega\omega})^{-1}A^h_{\omega\Gamma}\left(\mathbf{eu}^h_{\Gamma,j} + 2\mathbf{eu}^h_{\Gamma,j-1} + \mathbf{eu}^h_{\Gamma,j-2}\right) - \Delta t\left(\boldsymbol{\delta}^h_{\omega,j2} - \boldsymbol{\delta}^h_{\omega,(j-1)2}\right) + \mathcal{O}(\Delta t^4),\end{aligned} \tag{9.32}$$

and set all terms with time index ≤ 0 to zero. The solution of (9.30) is given by

$$\mathbf{eu}^h_{\omega,j} = -\sum_{i=1}^{j} B^h_{\omega,j-i}\mathbf{a}^h_{\omega,i}, \tag{9.33}$$

with $B^h_{\omega,i}$ being recursively defined by

$$\begin{aligned} B^h_{\omega,0} &= \mathrm{Id}, \quad B^h_{\omega,1} = 2\left(2C^h_{\omega\omega} - \mathrm{Id}\right), \\ B^h_{\omega,k+2} &= 2\left(2C^h_{\omega\omega} - \mathrm{Id}\right) B^h_{\omega,k+1} - B^h_{\omega,k}. \end{aligned} \tag{9.34}$$

Equations (9.33), (9.34) can easily be checked by substitution into (9.30).

By induction, it can be shown that the matrix $B^h_{\omega,k}$ has the explicit representation

$$B^h_{\omega,k} = (-1)^k \sum_{i=0}^{k} \binom{k+i+1}{2i+1} \left(-4C^h_{\omega\omega}\right)^i. \tag{9.35}$$

With Maple and (9.31), the formula (9.35) becomes asymptotically

$$B^h_{\omega,k} = (-1)^k \sum_{i=0}^{k} \binom{k+i+1}{2i+1} (-4)^i \left(\mathrm{Id} - \mathcal{O}(i\Delta t^2)\right) = (k+1)\mathrm{Id} + \mathcal{O}(k^3\Delta t^2), \tag{9.36}$$

such that we obtain with (9.32), (9.33), (9.36) and $\boldsymbol{\delta}^h_{\omega,i2} = \mathcal{O}(\Delta t^2)$

$$\begin{aligned} \mathbf{eu}^h_{\omega,j} &= -\frac{\Delta t^2}{4}(M^h_{\omega\omega})^{-1}A^h_{\omega\Gamma}\sum_{i=1}^{j}(j-i+1)\left(\mathbf{eu}^h_{\Gamma,i} + 2\mathbf{eu}^h_{\Gamma,i-1} + \mathbf{eu}^h_{\Gamma,i-2}\right) \\ &\quad + \Delta t \sum_{i=1}^{j}(j-i+1)\left(\boldsymbol{\delta}^h_{\omega,i2} - \boldsymbol{\delta}^h_{\omega,(i-1)2}\right) + \mathcal{O}(j^4\Delta t^4) \\ &= -\frac{\Delta t^2}{4}(M^h_{\omega\omega})^{-1}A^h_{\omega\Gamma}\left(\mathbf{eu}^h_{\Gamma,j} + \sum_{i=1}^{j-1}\Big((j-i+1) + 2(j-i) + (j-i+1)\Big)\mathbf{eu}^h_{\Gamma,i}\right. \\ &\quad + \Delta t \sum_{i=1}^{j} \boldsymbol{\delta}^h_{\omega,i2} + \mathcal{O}(j^4\Delta t^4) \\ &= -\frac{\Delta t^2}{4}(M^h_{\omega\omega})^{-1}A^h_{\omega\Gamma}\left(\mathbf{eu}^h_{\Gamma,j} + \sum_{i=1}^{j-1} 4(j-i)\mathbf{eu}^h_{\Gamma,i}\right) + \mathcal{O}(j\Delta t^3). \end{aligned} \tag{9.37}$$

c) Interface conditions

Next, we consider the interface equations of (9.26a), where we use (9.26d) and (9.37) to obtain asymptotically

$$\begin{aligned} &-(D^{hh}_\Gamma)^T\mathbf{e}\boldsymbol{\zeta}^h_{j-1/2} \\ &= \frac{1}{\Delta t}M^h_{\Gamma\Gamma}\left(\mathbf{ev}^h_{\Gamma,j} - \mathbf{ev}^h_{\Gamma,j-1}\right) + \frac{1}{2}A^h_{\Gamma\Gamma}\left(\mathbf{eu}^h_{\Gamma,j} + \mathbf{eu}^h_{\Gamma,j-1}\right) + \frac{1}{2}A^h_{\Gamma\omega}\left(\mathbf{eu}^h_{\omega,j} + \mathbf{eu}^h_{\omega,j-1}\right) + \boldsymbol{\delta}^h_{\Gamma,j1} \\ &= \frac{1}{J\Delta t}M^h_{\Gamma\Gamma}\Pi^{hH}\mathbf{ev}^H_{\Gamma,J} + \frac{1}{\Delta t}M^h_{\Gamma\Gamma}\left(\boldsymbol{\delta}^h_{(j-1)4} - \boldsymbol{\delta}^h_{j4}\right) + \frac{1}{2}A^h_{\Gamma\Gamma}\left(\mathbf{eu}^h_{\Gamma,j} + \mathbf{eu}^h_{\Gamma,j-1}\right) + \mathcal{O}(\Delta t^2) \\ &\quad - \frac{\Delta t^2}{8}A^h_{\Gamma\omega}(M^h_{\omega\omega})^{-1}A^h_{\omega\Gamma}\left(\mathbf{eu}^h_{\Gamma,j} + \mathbf{eu}^h_{\Gamma,j-1} + 4\sum_{i=1}^{j-1}(2j-2i-1)\mathbf{eu}^h_{\Gamma,i}\right). \end{aligned} \tag{9.38}$$

Summation over j and (9.26c) lead to

$$
\begin{aligned}
(D_\Gamma^{hh})^T \sum_{j=1}^{J} \mathbf{e}\boldsymbol{\zeta}_{j-1/2}^h &= -\frac{1}{\Delta t} M_{\Gamma\Gamma}^h \Pi^{hH} \mathbf{ev}_{\Gamma,J}^H - \frac{1}{\Delta t} M_{\Gamma\Gamma}^h \sum_{j=1}^{J} (\boldsymbol{\delta}_{(j-1)4}^h - \boldsymbol{\delta}_{j4}^h) + \mathcal{O}(J\Delta t^2) \\
&\quad - \frac{1}{2} A_{\Gamma\Gamma}^h \sum_{j=1}^{J} \frac{2j-1}{J} \Pi^{hH} \mathbf{eu}_{\Gamma,J}^H + \frac{\Delta t^2}{8} A_{\Gamma\omega}^h (M_{\omega\omega}^h)^{-1} A_{\omega\Gamma}^h \\
&\quad \sum_{j=1}^{J} \left(\frac{2j-1}{J} \Pi^{hH} \mathbf{eu}_{\Gamma,J}^H + 4 \sum_{i=1}^{j-1} (2j-2i-1) \frac{i}{J} \Pi^{hH} \mathbf{eu}_{\Gamma,J}^H \right) \\
&= -\frac{1}{\Delta t} M_{\Gamma\Gamma}^h \Pi^{hH} \mathbf{ev}_{\Gamma,J}^H - \frac{1}{\Delta t} M_{\Gamma\Gamma}^h \sum_{j=1}^{J} (\boldsymbol{\delta}_{(j-1)4}^h - \boldsymbol{\delta}_{j4}^h) + \mathcal{O}(J\Delta t^2) \\
&\quad - \frac{J}{2} A_{\Gamma\Gamma}^h \Pi^{hH} \mathbf{eu}_{\Gamma,J}^H + \frac{\Delta t^2}{8} A_{\Gamma\omega}^h (M_{\omega\omega}^h)^{-1} A_{\omega\Gamma}^h \frac{J(J^2+2)}{3} \Pi^{hH} \mathbf{eu}_{\Gamma,J}^H.
\end{aligned}
$$

With (9.25) and Taylor expansion around $t_{J/2}$, we get

$$
\begin{aligned}
\sum_{j=1}^{J} (\boldsymbol{\delta}_{(j-1)4}^h - \boldsymbol{\delta}_{j4}^h) &= \frac{\Delta T^2}{2} \sum_{j=1}^{J} D_\Gamma^{hH} \left(\frac{j(J-j)}{J^2} \left(\ddot{\mathbf{v}}_{\Gamma,J/2}^H + \frac{2j-J}{6J} \Delta T \dddot{\mathbf{v}}_{\Gamma,J/2}^H \right) \right. \\
&\quad \left. - \frac{(j-1)(J-j+1)}{J^2} \left(\ddot{\mathbf{v}}_{\Gamma,J/2}^H + \frac{2j-2-J}{6J} \Delta T \dddot{\mathbf{v}}_{\Gamma,J/2}^H \right) \right) + \mathcal{O}(\Delta T^4) \\
&= \frac{\Delta T^2}{2J^2} \sum_{j=1}^{J} (J-2j+1) D_\Gamma^{hH} \ddot{\mathbf{v}}_{\Gamma,J/2}^H + \mathcal{O}(\Delta T^4) \\
&\quad + \frac{\Delta T^3}{12J^3} \sum_{j=1}^{J} (-6j^2 + 6j(J+1) - (J^2+3J+2)) \dddot{\mathbf{v}}_{\Gamma,J/2}^H = \mathcal{O}(\Delta T^4).
\end{aligned}
$$

With (9.19b), this leads to

$$
\begin{aligned}
\frac{1}{J} (D_\Gamma^{hh})^T \sum_{j=1}^{J} \mathbf{e}\boldsymbol{\zeta}_{j-1/2}^h &= -\frac{1}{\Delta T} M_{\Gamma\Gamma}^h \Pi^{hH} \mathbf{ev}_{\Gamma,J}^H - \frac{1}{2} A_{\Gamma\Gamma}^h \Pi^{hH} \mathbf{eu}_{\Gamma,J}^H + \mathcal{O}(\Delta T^2) \\
&= -K_{\Gamma\Gamma}^h \Pi^{hH} \mathbf{eu}_{\Gamma,J}^H + \frac{2}{\Delta T} M_{\Gamma\Gamma}^h \Pi^{hH} \boldsymbol{\delta}_{\Gamma,2}^H + \mathcal{O}(\Delta T^2).
\end{aligned}
\tag{9.39}
$$

Substituting (9.39) into (9.24), we obtain

$$
\left(S_\Xi^H + (\Pi^{hH})^T K_{\Gamma\Gamma}^h \Pi^{hH} \right) \mathbf{eu}_{\Gamma,J}^H = \mathcal{O}(\Delta T).
$$

The asymptotic expansion

$$
\begin{aligned}
& S_\Xi^H + (\Pi^{hH})^T K_{\Gamma\Gamma}^h \Pi^{hH} \\
&= K_{\Gamma\Gamma}^H + K_{\Gamma\Xi}^H (K_{\Xi\Xi}^H)^{-1} K_{\Xi\Gamma}^H + (\Pi^{hH})^T \left(\frac{1}{2} A_{\Gamma\Gamma}^h + \frac{2J^2}{\Delta T^2} M_{\Gamma\Gamma}^h \right) \Pi^{hH} \\
&= \frac{2}{\Delta T^2} \left(M_{\Gamma\Gamma}^H + M_{\Gamma\Xi}^H (M_{\Xi\Xi}^H)^{-1} M_{\Xi\Gamma}^H + J^2 (\Pi^{hH})^T M_{\Gamma\Gamma}^h \Pi^{hH} \right) \left(\mathrm{Id} + \mathcal{O}(\Delta T^2) \right)
\end{aligned}
$$

yields $\left(S_{\Xi}^{H}+(\Pi^{hH})^{T}K_{\Gamma\Gamma}^{h}\Pi^{hH}\right)^{-1}=\mathcal{O}(\Delta T^{2})$ and thus the error estimate

$$\mathbf{eu}_{\Gamma,J}^{H}=\mathcal{O}(\Delta T^{3}).$$

From (9.19b), we obtain

$$\mathbf{ev}_{\Gamma,J}^{H}=\mathcal{O}(\Delta T^{2})$$

for the velocity at the interface, and the error in the sum of the Lagrange multipliers follows from (9.39) to be

$$\sum_{j=1}^{J}(D_{\Gamma}^{hH})^{T}\mathbf{e}\boldsymbol{\zeta}_{j-1/2}^{h}=\mathcal{O}(J\Delta T).$$

Substituting these results into (9.23) and (9.19b) yields

$$\begin{aligned}\mathbf{eu}_{\Xi,J}^{H}&=\mathcal{O}(\Delta T^{3}),\\ \mathbf{ev}_{\Xi,J}^{H}&=\mathcal{O}(\Delta T^{2}).\end{aligned}\tag{9.40}$$

With (9.25), (9.26c), (9.26d), this leads to

$$\begin{aligned}\mathbf{eu}_{\Gamma,j}^{h}&=\mathcal{O}\left(\frac{(J-j)j}{J^{2}}\Delta T^{2}\right)+\mathcal{O}(\Delta T^{3}),\\ \mathbf{ev}_{\Gamma,j}^{h}&=\mathcal{O}\left(\frac{(J-j)j}{J^{2}}\Delta T^{2}\right)+\mathcal{O}\left(\frac{j}{J}\Delta T^{2}\right)+\mathcal{O}(\Delta T^{3}).\end{aligned}\tag{9.41}$$

The error for the inner degrees of freedom follows from (9.37) and the inner equations of (9.26a)

$$\begin{aligned}\mathbf{eu}_{\omega,j}^{h}&=\mathcal{O}(j\Delta t^{3}),\\ \mathbf{ev}_{\omega,j}^{h}&=\mathcal{O}(j\Delta t^{3}).\end{aligned}$$

The error of a single Lagrange multiplier can be obtained from (9.38):

$$\mathbf{e}\boldsymbol{\zeta}_{j-1/2}^{h}=\mathcal{O}(\Delta T).\tag{9.42}$$

Some of the above estimates can be improved if additional assumptions are made on the structure of the mass matrix.

- On the one hand, if the coarse mass matrix is lumped, the estimate (9.40) can be improved to

  $$\mathbf{ev}_{\Xi,J}^{H}=\mathcal{O}(\Delta T^{3})$$

  from the inner equations of (9.19a).

- On the other hand, if the modified mass matrix is used such that $M_{\Gamma\Gamma}^{h}$ vanishes, the bound (9.42) for the Lagrange multiplier can be ameliorated to

  $$\mathbf{e}\boldsymbol{\zeta}_{j-1/2}^{h}=\mathcal{O}(\Delta t^{2})$$

from (9.38). With this, we can improve the estimate for the complete coarse velocity from (9.19a) to

$$\mathbf{ev}_J^H = \mathcal{O}(\Delta T^3),$$

and in addition obtain a better bound for the fine velocity at the interface

$$\mathbf{ev}_{\Gamma,j}^h = \mathcal{O}\left(\frac{(J-j)j}{J^2}\Delta T^2\right) + \mathcal{O}(\Delta T^3).$$

□

We remark that there is a difference between the time-continuous solutions of the systems with consistent and lumped mass matrix or with and without mass modification. The corresponding discretization error has been analysed in Chapter 6 for the modified and in, e.g., [170] for the lumped mass matrix.

Theorem 9.4 implies that if we modify the mass matrix M_ω^h near the interface Γ, the local errors at time t_J, i.e., after one large time step, are of order $\mathcal{O}(\Delta T^3)$ for both displacement and velocity, and thus the time stepping scheme is globally second order convergent (cf. page 153). Further, we obtain optimal local estimates for the fine grid quantities in the interior of ω.

9.3 Approximate solution scheme

As stated in the previous section, the solution of the coupled problem (9.15) is very expensive. However, it can easily be treated with an iterative solution scheme similar to Algorithm 1. One iteration consists of the solution of a coarse problem on $\mathbf{V}^H$ with the macro time step size ΔT, followed by J intermediate fine scale problems with step size Δt and Dirichlet boundary conditions on Γ. The backcoupling for the next iteration step is performed via the difference in the coarse and fine boundary stresses at Γ. The corresponding numerical scheme for the iterative computation of

$$\hat{\mathbf{z}}_J := \left(\mathbf{u}_{\Gamma,J}^H,\ \boldsymbol{\mu}_{\Gamma,J/2}^H,\ \mathbf{u}_{\Gamma,1}^h,\ \boldsymbol{\zeta}_{\Gamma,1/2}^h\ \ldots,\ \mathbf{u}_{\Gamma,J}^h,\ \boldsymbol{\zeta}_{\Gamma,J-1/2}^h\right), \tag{9.43}$$

given all values at time t_0, is stated in Algorithm 4.

Remark 9.5. For $l \geq 1$, the residuals (9.47b) to (9.47d) vanish.

An appropriate stopping criterion for Algorithm 4 can be derived as in Section 7.2.4 based on the residual

$$\mathbf{r}_\Xi^{H,(l)} = \frac{1}{J}(D_\Gamma^{hH})^T \sum_{J-1}^{J} \delta\boldsymbol{\zeta}_{\Gamma,j-1/2}^{h,(l-1)} - (D_\Gamma^{HH})^T \delta\boldsymbol{\mu}_{\Gamma,J/2}^{H,(l-1)} = (S_\Xi^H + S_\omega^H)\delta\mathbf{u}_{\Gamma,J}^{H,(l)}.$$

This leads to the following relative algebraic error estimator for $l \geq 1$:

$$\left(\eta_{J,\text{alg}}^{(l)}\right)^2 := \frac{\left(\delta\mathbf{u}_{\Gamma,J}^{H,(l)},\mathbf{r}_\Xi^{H,(l)}\right)}{\left(\mathbf{u}_{\Gamma,J}^{H,(l)},(S_\Xi^H + S_\omega^H)\mathbf{u}_{\Gamma,J}^{H,(l)}\right)}. \tag{9.48}$$

Algorithm 4 Two-way coupling scheme with time subcycling (time step $t_0 \to t_J$)

Starting from some initial guess $\widehat{\mathbf{z}}_J^{(0)}$, compute sequentially for $l = 0, 1, \ldots, (l_{\max} - 1)$

(*i*) Solve problem on coarse space $\mathbf{V}^H$ with interface load on Γ inherited from fine computation on ω:

$$\begin{pmatrix} (S_\Xi^H + S_\omega^H) & \mathbf{0} \\ S_\omega^H & (D_\Gamma^{HH})^T \end{pmatrix} \begin{pmatrix} \delta \mathbf{u}_{\Gamma,J}^{H,(l)} \\ \delta \boldsymbol{\mu}_{\Gamma,J/2}^{H,(l)} \end{pmatrix} = \begin{pmatrix} \mathbf{r}_\Xi^{H,(l)} + \mathbf{r}_\omega^{H,(l)} \\ \mathbf{r}_\omega^{H,(l)} \end{pmatrix}. \tag{9.44}$$

(*ii*) Loop on time substeps: $j = 1, \ldots, J$:

Solve problem on fine space $\mathbf{V}^h$ with weakly imposed trace on Γ inherited from coarse computation on Ω:

$$\begin{pmatrix} S_\omega^h & (D_\Gamma^{hh})^T \\ D_\Gamma^{hh} & \mathbf{0} \end{pmatrix} \begin{pmatrix} \delta \mathbf{u}_{\Gamma,j}^{h,(l)} \\ \delta \boldsymbol{\zeta}_{\Gamma,j-1/2}^{h,(l)} \end{pmatrix} = \begin{pmatrix} \mathbf{r}_{\omega,j}^{h,(l)} \\ \boldsymbol{\nu}_{\Gamma,j}^{h,(l)} \end{pmatrix} + \begin{pmatrix} \frac{2(J-2j+1)}{\Delta T^2} M_{\Gamma\Gamma}^h \Pi^{hH} \delta \mathbf{u}_{\Gamma,J}^{H,(l)} \\ \frac{j}{J} D_\Gamma^{hH} \delta \mathbf{u}_{\Gamma,J}^{H,(l)} \end{pmatrix}. \tag{9.45}$$

(*iii*) Update the solution vector:

$$\widehat{\mathbf{z}}_J^{(l+1)} := \widehat{\mathbf{z}}_J^{(l)} + \delta \widehat{\mathbf{z}}_J^{(l)}. \tag{9.46}$$

The residuals of (9.44), (9.45) are given by

$$\mathbf{r}_\Xi^{H,(l)} = \bar{\boldsymbol{\rho}}_{\Xi,0}^H - S_\Xi^H \mathbf{u}_{\Gamma,J}^{H,(l)} + \frac{1}{J} \sum_{j=1}^{J} (D_\Gamma^{hH})^T \boldsymbol{\zeta}_{\Gamma,j-1/2}^{h,(l)}, \tag{9.47a}$$

$$\mathbf{r}_\omega^{H,(l)} = \bar{\boldsymbol{\rho}}_{\omega,0}^H - S_\omega^H \mathbf{u}_{\Gamma,J}^{H,(l)} - (D_\Gamma^{HH})^T \boldsymbol{\mu}_{\Gamma,J/2}^{H,(l)}, \tag{9.47b}$$

$$\mathbf{r}_{\omega,j}^{h,(l)} = \bar{\boldsymbol{\rho}}_{\omega,j-1}^{h,(l)} - S_\omega^h \mathbf{u}_{\Gamma,j}^{h,(l)} - (D_\Gamma^{hh})^T \boldsymbol{\zeta}_{\Gamma,j-1/2}^{h,(l)} + \frac{2(J-2j+1)}{\Delta T^2} M_{\Gamma\Gamma}^h \Pi^{hH} \left(\mathbf{u}_{\Gamma,J}^{H,(l)} - \mathbf{u}_{\Gamma,J}^{H,(0)} - \Delta T \mathbf{v}_{\Gamma,J}^{H,(0)} \right), \tag{9.47c}$$

$$\boldsymbol{\nu}_{\Gamma,j}^{h,(l)} = D_\Gamma^{hH} \mathbf{u}_{\Gamma,0}^H + \frac{j}{J} D_\Gamma^{hH} \left(\mathbf{u}_{\Gamma,J}^{H,(l)} - \mathbf{u}_{\Gamma,0}^H \right) - D_\Gamma^{hh} \mathbf{u}_{\Gamma,j}^{h,(l)}. \tag{9.47d}$$

9.4 Numerical results

9.4.1 Discretization error of time subcycled system

First, we numerically investigate the time discretization error of the coupled system (9.15). For this, we consider a simple 2D test setting with two blocks on top of each other such that $\Omega = [0,1] \times [0,3]$ and $\omega = [0,1] \times [0,1.5]$. We impose homogeneous Dirichlet boundary conditions on the top, the time-dependent Neumann forces

$$\mathbf{g}_N(t) = \begin{cases} -\frac{7}{8} \cdot 10^9 \cdot \max(0, 0.5 - x_1)\, t & t \le 8 \cdot 10^{-5}, \\ -\frac{7}{8} \cdot 10^9 \cdot \max(0, 0.5 - x_1)\, (1.6 \cdot 10^{-4} - t) & 8 \cdot 10^{-5} < t \le 1.6 \cdot 10^{-4}, \\ 0, & \text{else} \end{cases}$$

on the bottom and homogeneous Neumann conditions elsewhere. The volume load as well as the initial velocity are set to $\mathbf{0}$. In order to measure only the error due to time discretization, we use a uniform quadrilateral grid of mesh size $H = h = 2^{-3}$. The material parameters are chosen according to $\nu_\Xi = \nu_\omega = 0.33$, $E_\Xi = E_\omega = 10^5$ and a discontinuous density of $\rho_\Xi = 1$, $\rho_\omega = 0.01$.

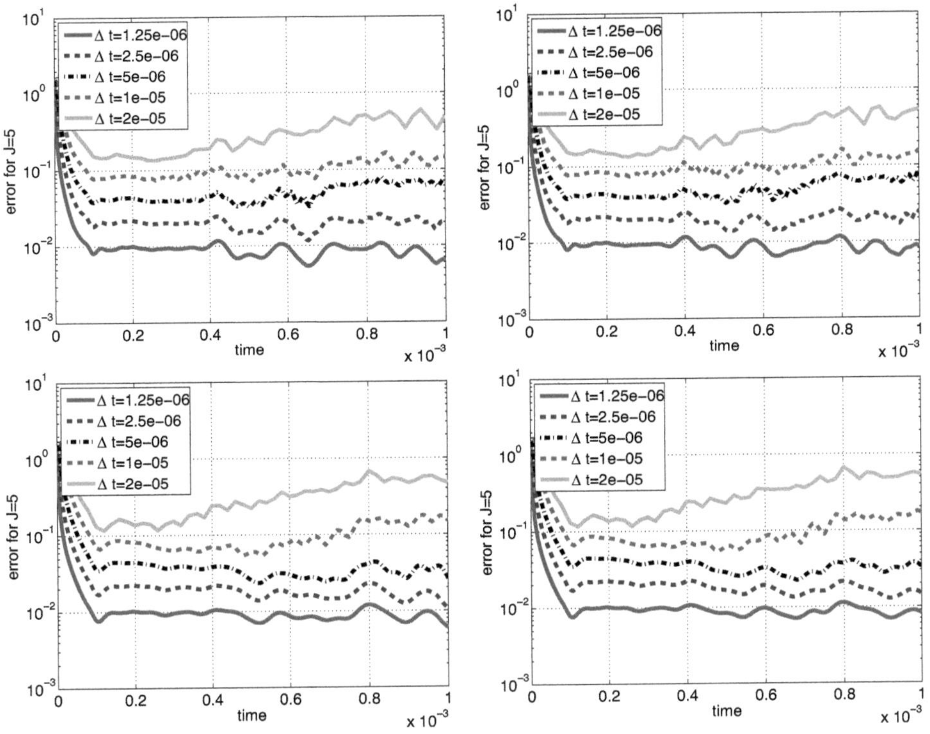

Figure 9.1: Evolution of global relative error e_{time} for time subcycling with $J = 5$; left: without mass modification; right: mass modification near Γ; upper row: consistent mass matrix; lower row: lumped mass matrix.

In the following, we consider four different types of mass matrix, namely either consistent or row-sum lumped, as well as either with or without a modification of M_ω^h near Γ. For each type, the results for the computations with different values of Δt and J are compared with the "reference solution" $\hat{\mathbf{u}}$ computed with an overall fine time step of size $2.5 \cdot 10^{-7}$.

In Figure 9.1, the evolution of the global relative energy error given by

$$e_{\text{time}}^2 := \frac{(\mathbf{u}^H - \hat{\mathbf{u}}^H)^T A_\Xi^H (\mathbf{u}^H - \hat{\mathbf{u}}^H) + (\mathbf{u}^h - \hat{\mathbf{u}}^h)^T A_\omega^h (\mathbf{u}^h - \hat{\mathbf{u}}^h)}{(\hat{\mathbf{u}}^H)^T A_\Xi^H \hat{\mathbf{u}}^H + (\hat{\mathbf{u}}^h)^T A_\omega^h \hat{\mathbf{u}}^h}, \tag{9.49}$$

ΔT	cons	mod,cons	lump	mod,lump
$2.0 \cdot 10^{-5}$	0.1437	0.1557	0.1488	0.1690
$1.0 \cdot 10^{-5}$	0.0708	0.0832	0.0259	0.0332
$5.0 \cdot 10^{-6}$	0.0179	0.0234	0.0106	0.0146
$2.5 \cdot 10^{-6}$	0.0066	0.0084	0.0059	0.0081
mean red. rate	3.1336	2.9629	3.7952	3.5158

Table 9.1: Global rel. error at time $T = 10^{-3}$ for $J = 5$ and different mass matrices.

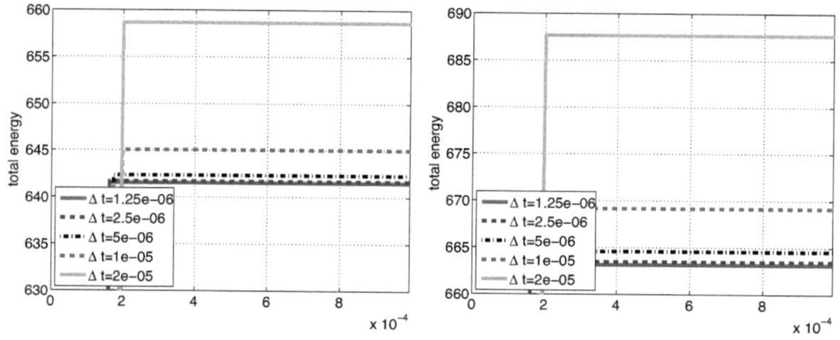

R_H	en. diff.
1	22.2746
2	5.0064
3	1.4162
4	0.3172

Figure 9.2: Evolution of total energy for time subcycling with $J = 5$; left: consistent mass matrix; middle: lumped mass matrix; right: difference in total energy between lumped and consistent mass for $H = 2^{-2-R_H}$ and $\Delta t = 5 \cdot 10^{-6}$.

with $(\mathbf{u}^H, \mathbf{u}^h)$ denoting the solution of (9.15), is displayed for time subcycling with $J = 5$. As the results for $J \in \{1, 2, 10\}$ are very similar to Figure 9.1, they are not shown separately. One can observe that the global time discretization error decreases for the coupled system with $J > 1$ in the same way as for the standard version with $J = 1$. Furthermore, the error reduction for the consistent mass matrix behaves similarly to the modified / lumped versions, although this case is not covered by Theorem 9.4.

In Table 9.1, the global error at time $T = 10^{-3}$ is compared for $J = 5$ and different values of ΔT. Again, the values for $J \in \{1, 2, 10\}$ are similar and thus omitted. One can see that the computations with a lumped mass matrix seem to have a better convergence rate than with a consistent mass matrix, and that the mean decay rate is a litte less than the expected order of $2^2 = 4$. However, as the results for $J = 1$ and $J > 1$ show the same behaviour, the latter is likely to be due to the definition of the problem.

Figure 9.2 illustrates the evolution of the total energy (7.19) after each coarse time step for the consistent as well as for the lumped mass matrix. One can observe that after the first excitation on the bottom of the body, the energy is constant with respect to time. The different values of the energy in the left and right picture are caused by the different types of mass matrices; however, for $H \to 0$, the total energies converge to the same value, as indicated by the table on the right of Figure 9.2.

Next, we fix $\Delta T = 4 \cdot 10^{-5}$ and investigate the evolution of the relative energy error on the patch ω given by

$$e^2_{\omega,\text{time}} := \frac{(\mathbf{u}^h - \hat{\mathbf{u}}^h)^T A^h_\omega (\mathbf{u}^h - \hat{\mathbf{u}}^h)}{(\hat{\mathbf{u}}^h)^T A^h_\omega \hat{\mathbf{u}}^h}, \tag{9.50}$$

for different values of J. The corresponding results for the different kinds of mass matrix are shown in Figure 9.3 and Table 9.2. Although the convergence rates are somewhat below the theoretically optimal second order convergence, one can observe that the time subcycling decreases the local energy error in all cases, and that the error reduction rates are in the same range as those for the global error summarized in Table 9.1.

Finally, the pictures in Figure 9.4 display the evolution of the relative approximation error introduced by the time subcycling, i.e., we compute (9.50) with a "reference solution" $\hat{\mathbf{u}}^h$ with the global time step Δt. As the results for the modified and the standard mass matrix are very similar, we only show the results for the latter case. Comparing the values in Figure 9.4 with those of Figure 9.3, one can see that the approximation error is dominated by the discretization error for $J \leq 8$.

J	cons	mod,cons	lump	mod,lump
1	0.4989	0.5506	0.4881	0.5530
2	0.1448	0.1624	0.1529	0.1751
4	0.0741	0.0879	0.0267	0.0346
8	0.0182	0.0240	0.0107	0.0146
16	0.0054	0.0078	0.0058	0.0078
mean red. rate	3.2180	2.9909	3.3188	3.1146

Table 9.2: Rel. error on ω at time $T = 10^{-3}$ for $\Delta T = 4 \cdot 10^{-5}$ and different mass matrices.

9.4.2 Algebraic error of Algorithm 4

In this subsection, we test the error reduction properties of Algorithm 4 for the approximate solution of the coupled system (9.15). We consider the same test setting as in the previous subsection and fix $\Delta t = 10^{-3}$, $E_\Xi = 10^2$ and $\rho_\Xi = 0.01$. The material parameters on the patch as well as the mesh size are varied to $\rho_\omega = 10^{\text{par}_\varrho} \rho_\Xi$, $E_\omega = 10^{\text{par}_E} E_\Xi$ and $h = 2^{-L} H$.

In order to get a measure of the convergence rate of Algorithm 4, we look at the asymptotic value of the ratio $\eta^{(l+1)}_{J,\text{alg}} / \eta^{(l)}_{J,\text{alg}}$ of the error measures defined in (9.48). The corresponding values for the lumped mass matrix without modification, $0 \leq L \leq 3$, $J = 100$ and different values of the material parameters are depicted in Figure 9.5. The left picture shows the error reduction factor for $\text{par}_\rho = \text{par}_E$, $\text{par}_E \in \{0, 2, 4, 6\}$, whereas the right picture plots the values for $\text{par}_\varrho = \text{par}_E - 2$. One can observe that the error quotient is bounded from above independently of L as well as of the ratio E_ω / E_Ξ. The results for the modified mass matrix are very similar and are thus omitted. Likewise, the results for $2 \leq J \leq 100$ are not shown separately as the computed convergence rate is basically independent of J, at least for $\text{par}_E - \text{par}_\varrho \leq 2$. However, this does not hold if the stiffness term is dominant, i.e., if $E_\omega \gg \varrho_\omega$, as can be seen from the results in Figure 9.6

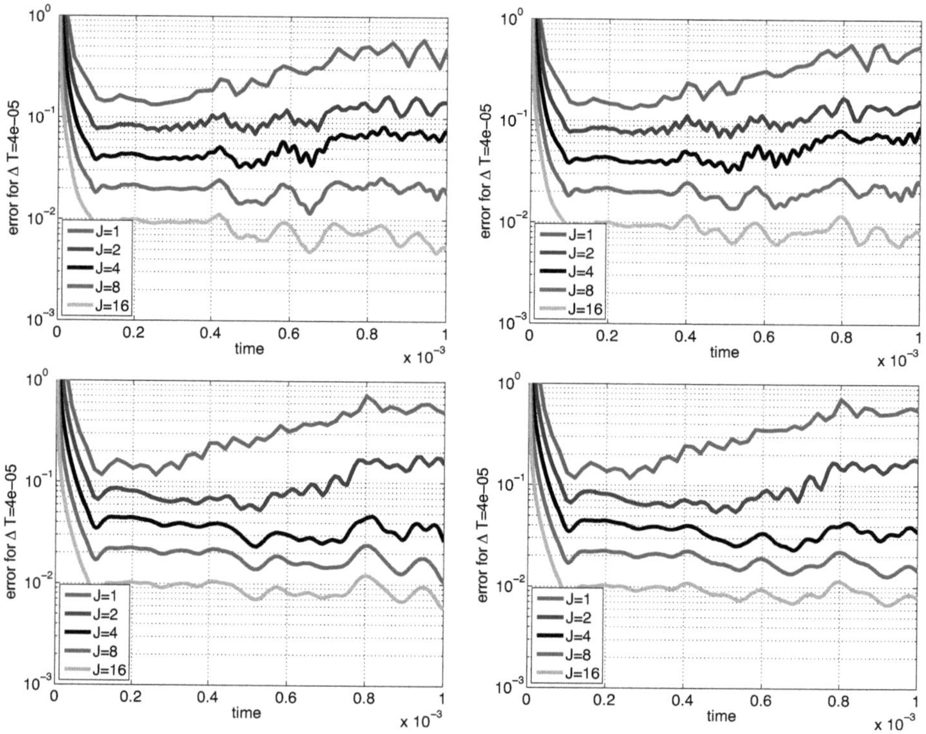

Figure 9.3: Evolution of local rel. error $e_{\text{time},\omega}$ for time subcycling with $\Delta T = 4 \cdot 10^{-5}$, $J \in \{1, 2, 4, 8, 16\}$; left: without mass modification; right: mass modification near Γ; upper row: consistent mass matrix; lower row: lumped mass matrix.

showing the error quotient with respect to J for $\text{par}_\varrho = 0$ and $\text{par}_E = 4, 6$, respectively. One can observe that for $\text{par}_E = 6$, the size of the quotient is strongly varying with respect to J, with peaks for $J \in \{2, 25, 40, 55, 70, 95\}$. However, the error quotient is bounded from above away from 1 in all cases, indicating that Algorithm 4 yields a suitable solver or preconditioner for the energy-conserving discrete problem (9.15).

9.4.3 Tire application in 2D

Finally, we present a more complex example using the nonconforming geometry presented in Subsection 8.3.3. We set $L = 2$, $\Delta t = 5 \cdot 10^{-6}$, $\theta = 1$ and use the Mooney–Rivlin material law (1.9) with the parameters $E_\Xi = E_\omega = 4.4 \cdot 10^6$, $\nu = 0.33$ and $c_m = 0$. The density of the inner ring of the tire is given by $\varrho_\Xi = 1$, whereas the salients have a density of 10^{-3}. We remark that due to the geometry of the patches (cf. Figures 7.1 and 8.6), the density is now discontinuous in ω. The initial and boundary conditions are chosen as in Subsection 8.3.3, and we consider

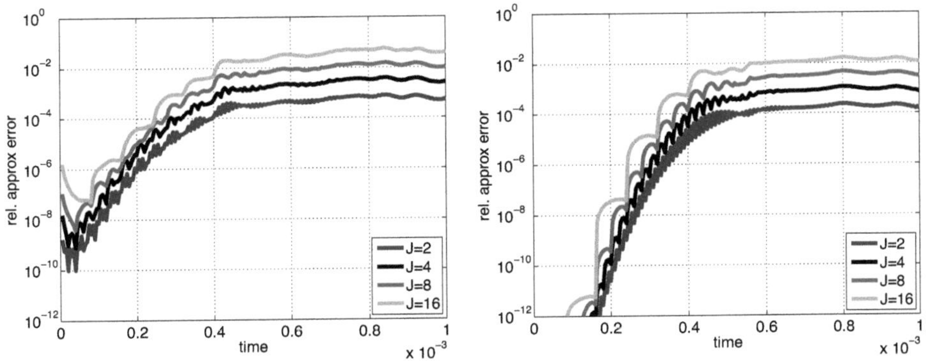

Figure 9.4: Evolution of the relative approximation error for time subcycling with fixed value of $\Delta T = 4 \cdot 10^{-5}$, $J \in \{2, 4, 8, 16\}$ without mass modification; left: consistent mass matrix; right: lumped mass matrix.

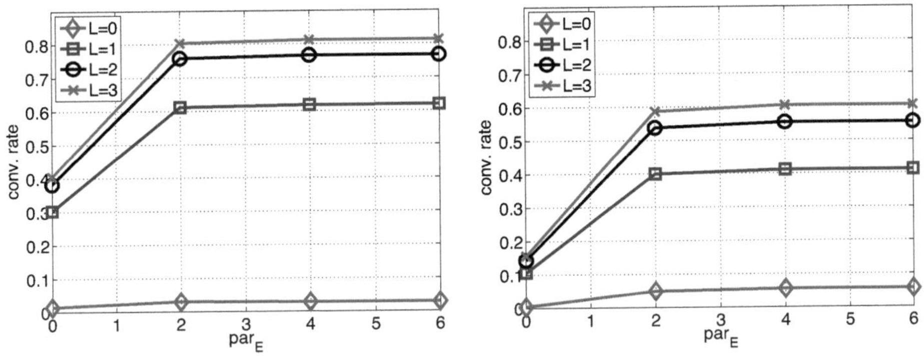

Figure 9.5: Asympt. error red. rates for $L \in \{0, 1, 2, 3\}$, $J = 100$, $\varrho_\omega = 10^{\mathrm{par}_\rho} \varrho_\Xi$, $E_\omega = 10^{\mathrm{par}_E} E_\Xi$ and lumped mass matrix; left: $\mathrm{par}_\rho = \mathrm{par}_E$; right: $\mathrm{par}_\rho = \mathrm{par}_E - 2$.

frictionless contact with the flat obstacle given by $x_2 = -2.04$. We employ lumped mass matrices as well as a redistribution of the mass near the contact boundary Γ_C. The tangential matrices are updated in each inner iteration ($l_{\max} = 1$), and the resulting quasi-Newton method is solved in each time step subject to $\|\mathbf{R}^{(k)}\| \leq 10^{-5}$.

In Figure 9.7, the results for the computations with $J = 1$ and $J = 10$ are compared at different time instances, whereas the evolution of the active set is shown on the left of Figure 9.8. Although one can observe slight differences, the overall quality of the results with time subcycling is evident. We remark that for this example, we have not used any active coarse nodes for the coarse approximate problem.

The total number of Newton iterations as well as the asymptotic error reduction factor are

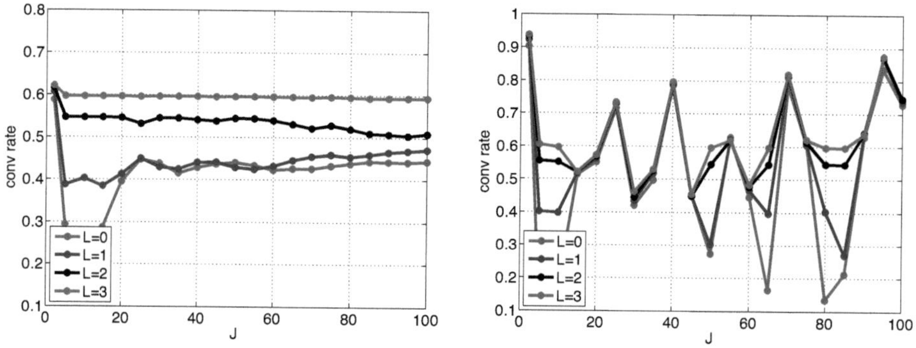

Figure 9.6: Asymptotic error reduction rates for $L \in \{0, 1, 2, 3\}$, $\varrho_\omega = \varrho_\Xi$, $E_\omega = 10^{\text{par}_E} E_\Xi$ and lumped mass matrix with respect to J; left: $\text{par}_E = 4$; right: $\text{par}_E = 6$.

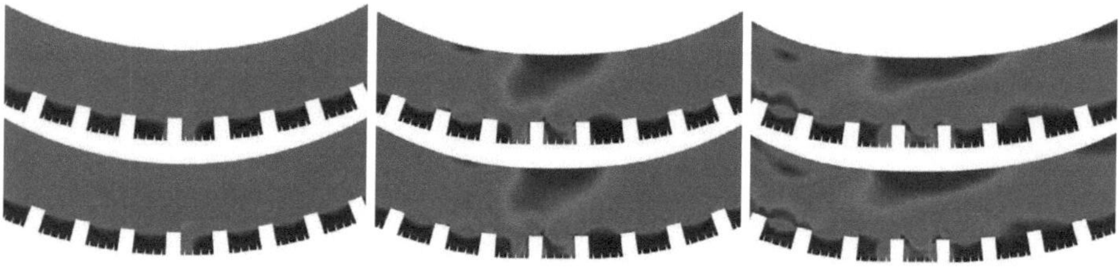

Figure 9.7: 2D tire example with $L = 2$ and $\Delta t = 5 \cdot 10^{-6}$; upper row: $J = 1$; lower row: $J = 10$; from left to right: effective stress at times t_{50}, t_{150} and t_{250}.

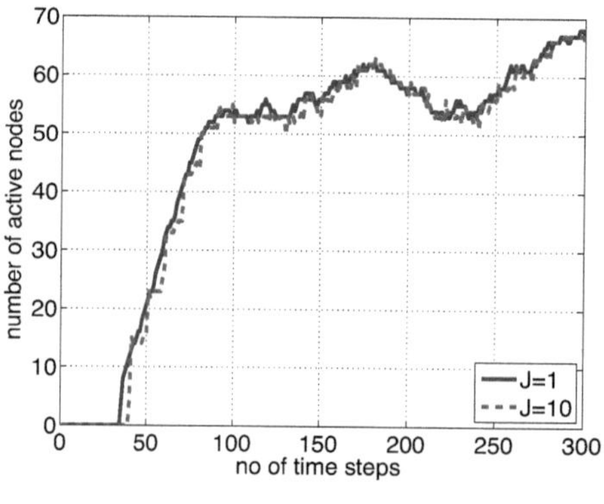

J	1	10
total no. of NI	1843	252
max no. of NI per ΔT	6	8
asy. err. red. factor	0.3403	0.4178

Figure 9.8: 2D tire problem with time subcycling; left: evolution of active sets; right: convergence statistics.

summarized on the right of Figure 9.8. We emphasize that each Newton iteration consists of the solution of one global coarse and J fine local systems. In comparison with $J = 1$, the number of fine solves has increased by 37% for the computation with $J = 10$, but the number of coarse global problems has decreased by 86%. Hence, the subcycling scheme is extremely useful if the coarse global problems are much more expensive than the local solves.

10 Concluding remarks

This thesis is devoted to the construction and analysis of robust numerical algorithms for the solution of dynamic frictional contact problems, taking into account plastic and multiscale behaviour. The main challenges of such a scheme have separately been addressed in the four parts of this work.

The first introductory part has presented the governing equations for the elastoplastic contact problem and its discretization in space and time. The inequality conditions for both plasticity and frictional contact have been formulated using additional variables representing the contact stress or the plastic strain, respectively. The discrete spaces for these quantities have been chosen as the span of biorthogonal basis functions, such that the constraints decouple and can be enforced locally. To be able to treat the inequalities within the same Newton loop as the material nonlinearities, the local KKT conditions have been transformed into a set of nondifferentiable but semismooth functions. The resulting semismooth Newton scheme is extremely efficient due to its locally superlinear convergence as well as to the avoidance of nested iterations for the different nonlinearities.

In the second part, the influence of the choice of the NCP function on the robustness of the semismooth Newton iteration has been investigated. It has been demonstrated that the NCP function used for, e.g., the radial return method yields a suboptimal convergence radius which can considerably be improved if the original function is multiplied with a nonzero factor. Due to the similarity of the flow rules for plasticity and friction, the same modification of the NCP function can be used for both types of nonlinearities, yielding an improvement in either case. Furthermore, it has been shown that the locality of the plastic and contact constraints allows for a simple static condensation of the Newton updates for the dual variables, such that only a system of the size of the displacement has to be solved in each Newton step. The condensed systems for both plasticity and frictional contact have explicitly been stated, and their solvability has theoretically been investigated. Several complex numerical examples have illustrated the increased robustness and the flexibility of the proposed algorithms.

The third part has dealt with a regularization of the spatially discrete dynamic contact problem by means of a modification of the mass matrix near the potential contact boundary. The benefit of such a mass redistribution has been motivated from a DAE point of view as well as by means of an oscillating system with two degrees of freedom. Afterwards, we have presented several possibilities to construct the modified mass matrix by using non-standard quadrature rules near the contact boundary. For sufficiently exact quadrature formulas, the modified mass matrix preserves the moments of the original system up to second order. In addition, we have presented a different interpretation of the modified mass matrix in terms of stable interpolation operators which allowed us to prove optimal a priori error estimates for the semi-discrete as well as for the fully discrete system.

Motivated by industrial applications where a detailed solution is desired in a small part of the domain, the final part of this work has dealt with an iterative solution scheme based on a decomposition of the total domain into overlapping subdomains with independent grids. Observing that in the linear case, the iterative process can be interpreted as a Gauß–Seidel scheme applied to an augmented system, the convergence rate as well as the spectral properties of the method have been analysed with respect to the discretization and material parameters. Afterwards, we have extended the scheme to the nonlinear and the geometrical nonconforming case and shown several numerical examples illustrating its convergence properties. Finally, the scheme has been generalized to the use of different time step sizes in the subdomains. We have derived a coupled energy-conserving system, analysed its asymptotic convergence rate, and presented an efficient iterative solver. A complex numerical example has illustrated that the method can indeed obtain qualitatively good results with a largely reduced number of global linear systems to be solved.

In summary, a proper combination of the techniques presented in this work yields a robust and efficient numerical algorithm that is able to solve threedimensional dynamic frictional contact problems with a small contact zone and still give a detailed resolution of the local effects.

However, there are several interesting extensions and open questions that can be addressed in the future. Firstly, the basic assumption of small displacements and stresses is not valid in many important applications, such that one can investigate how the presented approaches can be transferred to the case of finite deformation and finite plasticity. Next, the concept of the two-scale iterative algorithm described in the fourth part can be extended to a multi-scale scheme that possibly includes an adaptive choice or enlargement of the fine-scale contact zone. Another issue is the theoretical analysis of the modification of the algorithm discussed on page 122. In addition, the use of the algorithm as a preconditioner for a Krylov subspace method has not yet been investigated numerically.

Bibliography

[1] J. Ahrens, B. Geveci, and C. Law. ParaView: An end-user tool for large data visualization. In C. D. Hansen and C. R. Johnson, editors, *The Visualization Handbook*, pages 717–732. Elsevier, 2005. www.paraview.org. viii

[2] P. Alart and A. Curnier. A mixed formulation for frictional contact problems prone to Newton like solution methods. *Comput. Methods Appl. Mech. Engrg.*, 92:353–375, 1991. 26, 27

[3] J. Alberty, C. Carstensen, and D. Zarrabi. Adaptive numerical analysis in primal elastoplasticity with hardening. *Comput. Methods Appl. Mech. Engrg.*, 171:175–204, 1999. 23

[4] F. Armero and E. Petöcz. A new dissipative time-stepping algorithm for frictional contact problems: formulation and analysis. *Comput. Methods Appl. Mech. Engrg.*, 179:151–178, 1999. 25

[5] V. I. Arnold. *Mathematical methods of classical mechanics.* Springer, New York, 2 edition, 1989. 11

[6] I. Babuška and M. Suri. Locking effects in the finite element approximation of elasticity problems. *Numer. Math.*, 62:439–463, 1992. 13

[7] G. Baker and V. Dougalis. The effect of quadrature errors on finite element approximations for second order hyperbolic equations. *SIAM J. Numer. Anal.*, 13(4):577–598, 1976. 92, 97, 99, 101

[8] P. Ballard. The dynamics of discrete mechanical systems with perfect unilateral constraints. *Arch. Rational Mech. Anal.*, 154:199–274, 2000. 68

[9] P. Bastian, K. Birken, K. Johannsen, S. Lang, N. Neuß, H. Rentz–Reichert, and C. Wieners. UG – a flexible software toolbox for solving partial differential equations. *Comput. Vis. Sci.*, 1:27–40, 1997. viii

[10] A. Bellen and M. Zennaro. Stability properties of interpolants for Runge–Kutta methods. *SIAM J. Numer. Anal.*, 25(2):411–432, 1988. 153

[11] F. Ben Belgacem. The mortar finite element method with Lagrange multipliers. *Numer. Math.*, 84:173–197, 1999. 105

[12] C. Bernardi, C. Canuto, and Y. Maday. Generalized inf-sup conditions for Chebyshev spectral approximation of the Stokes problem. *SIAM J. Numer. Anal.*, 25(6):1237–1271, 1988. 56

[13] C. Bernardi, Y. Maday, and A. T. Patera. A new nonconforming approach to domain decomposition: The mortar element method. In H. Brezis and J.-L. Lions, editors, *Nonlinear partial differential equations and their applications*, volume XI of *Collège de France Seminar*, pages 13–51. Pitman, 1994. 105

[14] P. Betsch and P. Steinmann. Conservation properties of a time FE method – part II: Time-stepping schemes for non-linear elastodynamics. *Internat. J. Numer. Methods Engrg.*, 50:1931–1955, 2001. 25

[15] P. Betsch and P. Steinmann. Conservation properties of a time FE method – part III: Mechanical systems with holonomic constraints. *Internat. J. Numer. Methods Engrg.*, 53:2271–2304, 2002. 155

[16] P. Betsch and P. Steinmann. A DAE approach to flexible multibody dynamics. *Multibody Syst. Dyn.*, 8:367–391, 2002. 155

[17] D. Bigoni and D. Zaccaria. Loss of strong ellipticity in non-associative elastoplasticity. *J. Mech. Phys. Solids*, 40(6):1313–1331, 1992. 15

[18] P. E. Bjørstad and O. B. Widlund. Iterative methods for the solution of elliptic problems on regions partitioned into substructures. *SIAM J. Numer. Anal.*, 23(6):1097–1120, 1986. 105, 116, 127

[19] M. Borri, C. Bottasso, and L. Trainelli. Integration of elastic multibody systems by invariant conserving/dissipating algorithms. part II: Numerical schemes and applications. *Comput. Methods Appl. Mech. Engrg.*, 190:3701–3733, 2001. 67

[20] D. Braess. Enhanced assumed strain elements and locking in membrane problems. *Comput. Methods Appl. Mech. Engrg.*, 165:155–174, 1998. 106

[21] D. Braess. *Finite Elemente.* Springer, Berlin, 3rd edition, 2003. 56, 95, 96, 106, 110

[22] D. Braess, C. Carstensen, and B. D. Reddy. Uniform convergence and a posteriori error estimators for the enhanced strain finite element method. *Numer. Math.*, 96:461–479, 2004. 13, 106

[23] J. H. Bramble and J. E. Pasciak. A domain decomposition technique for Stokes problems. *Appl. Numer. Math.*, 6:251–261, 1989. 106

[24] J. H. Bramble, J. E. Pasciak, and A. H. Schatz. An iterative method for elliptic problems on regions partitioned into substructures. *Math. Comp.*, 46(174):361–369, 1986. 127

[25] S. C. Brenner. Korn's inequalities for piecewise H^1 vector fields. *Math. Comp.*, 73(247):1067–1087, 2003. 108, 112

[26] S. C. Brenner and L. Scott. *The Mathematical Theory of Finite Element Methods.* Springer, New York, 1994. 13, 21, 108

[27] F. Brezzi and M. Fortin. *Mixed and hybrid finite element methods.* Springer, New York, 1991. 108

[28] S. Brunßen, C. Hager, B. I. Wohlmuth, and F. Schmid. Simulation of elastoplastic forming processes using overlapping domain decomposition and inexact Newton methods. In B. D. Reddy, editor, *IUTAM Symposium on theoretical, computational and modelling aspects of inelastic media*, pages 155–164. Springer Science+Business media, 2008. 33

[29] S. Brunßen, F. Schmid, M. Schäfer, and B. I. Wohlmuth. A fast and robust method for contact problems by combining a primal-dual active set strategy and algebraic multigrid. *Internat. J. Numer. Methods Engrg.*, 69:524–543, 2007. 41

[30] C. Carstensen. Numerical analysis of the primal problem of elastoplasticity with hardening. *Numer. Math.*, 82:577–597, 1999. 20, 23

[31] T. Chan, B. Smith, and J. Zou. Overlapping Schwarz methods on unstructured meshes using non-matching coarse grids. *Num. Math.*, 73:149–167, 1996. 105

[32] R. W. Chaney. Piecewise C^k functions in nonsmooth analysis. *Nonlinear Anal.*, 15:649–660, 1990. 35

[33] K. S. Chavan, B. P. Lamichhane, and B. I. Wohlmuth. Locking-free finite element methods for linear and nonlinear elasticity in 2D and 3D. *Comput. Meth. Appl. Mech. Engrg.*, 196:4075–4086, 2007. 106, 132

[34] V. Chawla and T. A. Laursen. Energy consistent algorithms for frictional contact problems. *Internat. J. Numer. Methods Engrg.*, 42:799–827, 1998. 17, 25, 68

[35] P. W. Christensen. A nonsmooth Newton method for elastoplastic problems. *Comput. Methods Appl. Mech. Engrg.*, 191:1189–1219, 2002. 25, 26, 27, 34, 36, 41, 46

[36] P. W. Christensen. A semi-smooth Newton method for elasto-plastic contact problems. *Internat. J. Solids Structures*, 39:2323–2341, 2002. 26

[37] P. W. Christensen, A. Klarbring, J. S. Pang, and N. Strömberg. Formulation and comparison of algorithms for frictional contact problems. *Internat. J. Numer. Methods Engrg.*, 42:145–173, 1998. 26

[38] P. W. Christensen and J. S. Pang. Frictional contact algorithms based on semismooth Newton methods. In M. Fukushima and L. Qi, editors, *Reformulation: Nonsmooth, piecewise smooth, semismooth and smoothing methods*, pages 81–116. Kluwer Academic Publishers, 1999. 41

[39] J. Chung and G. M. Hulbert. A time integration algorithm for structural dynamics with improved numerical dissipation: the generalized α-method. *J. Appl. Mech.*, 60(2):371–375, 1993. 25

[40] P. G. Ciarlet. *Mathematical Elasticity*, volume I. North Holland, 1998. 11, 92

[41] F. Clarke. *Optimization and nonsmooth analysis.* Wiley, New York, 1983. 35

[42] A. Combescure and A. Gravouil. A numerical scheme to couple subdomains with different time-steps for predominantly linear transient analysis. *Comput. Methods Appl. Mech. Engrg.*, 191:1129–1157, 2002. 151

[43] H.-H. Dai. Model equations for nonlinear dispersive waves in a compressible Mooney–Rivlin rod. *Acta Mechanica*, 127:193–207, 1998. 13

[44] W. J. T. Daniel. A study of the stability of subcycling algorithms in structural dynamics. *Comput. Methods Appl. Mech. Engrg.*, 156:1–13, 1998. 151

[45] W. J. T. Daniel. Subcycling first- and second-order generalizations of the trapezoidal rule. *Internat. J. Numer. Methods Engrg.*, 42:1091–1119, 1998. 151

[46] R. Dautray and J.-L. Lions. *Mathematical analysis and numerical methods for science and technology – Evolution problems*, volume 5. Springer, Berlin Heidelberg, 1992. 92

[47] G. De Saxce and Z. Q. Feng. New inequality and functional for contact with friction: The implicit standard material approach. *Mech. Struct. Mach.*, 19:301–325, 1991. 26

[48] R. S. Dembo, S. C. Eisenstat, and T. Steinhaug. Inexact Newton methods. *SIAM J. Numer. Anal.*, 19(2):400–408, 1982. 36, 105, 136

[49] P. Deuflhard. *Newton methods for nonlinear problems.* Springer, Berlin Heidelberg, 2004. 33, 36, 105

[50] P. Deuflhard, R. Krause, and S. Ertel. A contact-stabilized Newmark method for dynamical contact problems. *Internat. J. Numer. Methods Engrg.*, 73(9):1274–1290, 2008. 67, 87

[51] Z. Dostál, D. Horák, R. Kučera, V. Vondrák, J. Haslinger, J. Dobiaš, and S. Pták. FETI based algorithms for contact problems: scalability, large displacements and 3D Coulomb friction. *Comp. Methods Appl. Mech. Engrg.*, 194:395–409, 2005. 105

[52] G. Duvaut and J.-L. Lions. *Inequalities in mechanics and physics.* Springer, Berlin, 1976. 20

[53] C. Eck, J. Jarušek, and M. Krbec. *Unilateral contact problems. Variational methods and existence theorems.* CRC Press Taylor & Francis Group, Boca Raton, 2005. 11, 21

[54] S. C. Eisenstat and H. F. Walker. Choosing the forcing terms in an inexact Newton method. *SIAM J. Sci. Comput.*, 17(1):16–32, 1996. 36, 105, 141

[55] L. C. Evans. *Partial differential equations.* American Mathematical Society, Providence, 1998. 91

[56] F. Facchinei, A. Fischer, and C. Kanzow. Inexact Newton methods for semismooth equations with applications to variational inequality problems. In G. Di Pillo and F. Giannessi, editors, *Nonlinear Optimization and Applications*, pages 125–139. Internat. Center for Numer. Methods in Engrg., Plenum Press: New York, 1996. 36

[57] F. Facchinei and J.-S. Pang. *Finite-dimensional variational inequalities and complementary problems*. Springer, New York, 2003. 28, 33, 34, 35

[58] A. Fischer. Solution of monotone complementarity problems with locally Lipschitzian functions. *Math. Prog.*, 76:513–532, 1997. 35

[59] B. Flemisch and B. I. Wohlmuth. Stable Lagrange multipliers for quadrilateral meshes of curved interfaces in 3D. *Comput. Methods Appl. Mech. Engrg.*, 196:1589–1602, 2007. 22, 57

[60] C. Geiger and C. Kanzow. *Theorie und Numerik restringierter Optimierungsaufgaben*. Springer, Berlin Heidelberg, 2002. 26

[61] R. Glowinski, J. He, A. Lozinski, J. Rappaz, and J. Wagner. Finite element approximation of multi-scale elliptic problems using patches of elements. *Numer. Math.*, 101:663–687, 2005. 105

[62] R. Glowinski and P. Le Tallec. *Augmented Lagrangian and operator splitting methods in nonlinear mechanics*. SIAM Studies in Applied Mathematics 9, Philadelphia, 1989. 26

[63] R. Glowinski, J. L. Lions, and R. Trémolières. *Numerical analysis of variational inequalities*. North-Holland, Amsterdam, 1981. 20

[64] P. Goldfeld, L. F. Pavarino, and O. B. Widlund. Balancing Neumann-Neumann preconditioners for mixed approximations of heterogeneous problems in linear elasticity. *Numer. Math.*, 95:283–324, 2003. 105

[65] H. Goldstein. *Classical mechanics*. Addison-Wesley, Reading, MA, 2 edition, 1980. 11

[66] M. Gonzales, B. Schmidt, and M. Ortiz. Energy-stepping integrators in Lagrangian mechanics. *Internat. J. Numer. Methods Engrg.*, 82(2):205–241, 2010. 25

[67] O. Gonzalez. Exact energy and momentum conserving algorithms for general models in nonlinear elasticity. *Comp. Methods Appl. Mech. Engrg.*, 190:1763–1783, 2000. 25, 145

[68] A. Gravouil and A. Combescure. Multi-time-step explicit-implicit method for non-linear structural dynamics. *Internat. J. Numer. Methods Engrg.*, 50:199–225, 2001. 151

[69] C. Hager, P. Hauret, P. Le Tallec, and B. I. Wohlmuth. Overlapping domain decomposition for multiscale dynamic contact problems. IANS Preprint 2010/007, Universität Stuttgart, 2010. 13, 106, 107, 109, 123, 131, 132, 138

[70] C. Hager, S. Hüeber, and B. I. Wohlmuth. A stable energy conserving approach for frictional contact problems based on quadrature formulas. *Internat. J. Numer. Methods Engrg.*, 73:205–225, 2008. 68, 78

[71] C. Hager, S. Hüeber, and B. I. Wohlmuth. Numerical techniques for the valuation of basket options and its Greeks. *J. Comput. Fin.*, 13(4):1–31, 2010. 33

[72] C. Hager and B. Wohlmuth. Semismooth Newton methods for variational problems with inequality constraints. *GAMM-Mitt.*, 33(1):8–24, 2010. 33

[73] C. Hager and B. I. Wohlmuth. Analysis of a space-time discretization for dynamic elasticity problems based on mass-free surface elements. *SIAM J. Numer. Anal.*, 47(3):1863–1885, 2009. 68, 101

[74] C. Hager and B. I. Wohlmuth. Nonlinear complementarity functions for plasticity problems with frictional contact. *Comput. Methods Appl. Mech. Engrg.*, 198:3411–3427, 2009. 37

[75] E. Hairer, S. P. Nørsett, and G. Wanner. *Solving ordinary differential equations I: Nonstiff problems.* Springer, Berlin Heidelberg, 1987. 151, 153

[76] E. Hairer and G. Wanner. *Solving ordinary differential equations II: Stiff and differential-algebraic problems.* Springer, Berlin Heidelberg, 1991. 68, 108, 151

[77] W. Han and B. D. Reddy. Computational plasticity: the variational basis and numerical analysis. *Comput. Mech. Advances*, 2:283–400, 1995. 11, 14, 15, 20, 23

[78] W. Han and B. D. Reddy. *Plasticity - Mathematical theory and analysis.* Springer, 1999. 11, 20

[79] P. T. Harker and J.-S. Pang. Finite-dimensional variational inequalitiy and nonlinear complementarity problems: A survey of theory, algorithms and applications. *Math. Progr.*, 48:161–220, 1990. 20

[80] J. Haslinger, I. Hlaváček, J. Nečas, and J. Lovíšek. *Solution of variational inequalities in mechanics.* Springer, New York, 1988. 20

[81] P. Hauret and P. Le Tallec. Energy-controlling time integration methods for nonlinear elastodynamics and low-velocity impact. *Comput. Methods Appl. Mech. Engrg.*, 195:4890–4916, 2006. 16, 67

[82] P. Hauret and P. Le Tallec. Two-scale Dirichlet–Neumann preconditioners for elastic problems with boundary refinements. *Comput. Methods Appl. Mech. Engrg.*, 196:1574–1588, 2007. 114

[83] P. Hauret, J. Salomon, A. Weiß, and B. I. Wohlmuth. Energy consistent co-rotational schemes for frictional contact problems. *SIAM J. Sci. Comput.*, 30:2488–2511, 2008. 145

[84] S. Hayashi, N. Yamashita, and M. Fukushima. A combined smoothing and regularization method for monotone second-order cone complementarity problems. *SIAM J. Optim.*, 15(2):593–615, 2005. 35

[85] C. Hesch and P. Betsch. A mortar method for energy-momentum conserving schemes in frictionless dynamic contact problems. *Internat. J. Numer. Methods Engrg.*, 77:1468–1500, 2009. 25

[86] H. M. Hilber, T. J. R. Hughes, and R. L. Taylor. Improved numerical dissipation for time integration algorithms in structural dynamics. *Earthquake Engrg. Struct. Dyn.*, 5(3):283–292, 1977. 25

[87] M. Hintermüller, K. Ito, and K. Kunisch. The primal-dual active set strategy as a semi–smooth Newton method. *SIAM J. Optim.*, 13(3):865–888, 2003. 28, 33, 41

[88] I. Hlaváček, J. Haslinger, J. Nečas, and J. Lovíšek. *Numerical solution of variational inequalities.* Springer Series in Applied Mathematical Sciences. Springer, New York, 1988. 20

[89] G. Hofstetter and R. L. Taylor. Non-associative Drucker–Prager plasticity at finite strains. *Commun. Appl. Numer. Methods*, 6:583–589, 1990. 15

[90] S. Hüeber, A. Matei, and B. I. Wohlmuth. Efficient algorithms for problems with friction. *SIAM J. Sci. Comput.*, 29(1):70–92, 2007. 23

[91] S. Hüeber, G. Stadler, and B. I. Wohlmuth. A primal-dual active set algorithm for three-dimensional contact problems with Coulomb friction. *SIAM J. Sci. Comput.*, 30(2):572–596, 2008. 25, 41, 53, 55, 57

[92] S. Hüeber and B. I. Wohlmuth. An optimal a priori error estimate for non-linear multibody contact problems. *SIAM J. Numer. Anal.*, 43(1):157–173, 2005. 23

[93] S. Hüeber and B. I. Wohlmuth. A primal-dual active set strategy for non–linear multibody contact problems. *Comput. Methods Appl. Mech. Engrg.*, 194:3147–3166, 2005. 16, 23

[94] T. J. R. Hughes. *The finite element method: Linear, static and dynamic finite element analysis.* Prentice-Hall, 1987. 13, 24

[95] A. Ibrahimbegovic. *Nonlinear solid mechanics: Theoretical formulations and finite element solution methods.* Springer, Dordrecht, 2009. 12

[96] K. Ito and K. Kunisch. Semi-smooth Newton methods for variational inequalities of the first kind. *M2AN, Math. Model. Numer. Anal.*, 37:41–62, 2003. 41

[97] K. Ito and K. Kunisch. The primal-dual active set method for nonlinear optimal control problems with bilateral constraints. *SIAM J. Control. Optim.*, 43(1):357–376, 2004. 41

[98] C. Kane, J. E. Marsden, M. Ortiz, and M. West. Variational integrators and the Newmark algorithm for conservative and dissipative mechanical systems. *Internat. J. Numer. Methods Engrg.*, 49:1295–1325, 2000. 25

[99] C. Kane, E. A. Repetto, M. Ortiz, and J. E. Marsden. Finite element analysis of nonsmooth contact. *Comput. Methods Appl. Mech. Engrg.*, 180:1–26, 1999. 87

[100] E. Kasper and R. Taylor. A mixed-enhanced strain method. part I: Geometrically linear problems. *Comput. Struct.*, 75:237–250, 2000. 106, 132

[101] E. Kasper and R. Taylor. A mixed-enhanced strain method. part II: Geometrically non-linear problems. *Comput. Struct.*, 75:251–260, 2000. 106, 132

[102] H. B. Khenous, P. Laborde, and Y. Renard. On the discretization of contact problems in elastodynamics. In P. Wriggers and U. Nackenhorst, editors, *Analysis and Simulation of Contact Problems*, volume 27 of *Lecture Notes in Applied and Computational Mechanics*, pages 31–38. Springer, 2006. 67, 69, 73

[103] H. B. Khenous, P. Laborde, and Y. Renard. Mass redistribution method for finite element contact problems in elastodynamics. *Europ. J. Mech. A/Solids*, 27:918–932, 2008. 67, 68, 69, 73

[104] N. Kikuchi and J. T. Oden. *Contact problems in elasticity: A study of variational inequalities and finite element methods.* SIAM Studies in Applied Mathematics 8, Philadelphia, 1988. 11, 16, 20

[105] D. Kinderlehrer and G. Stampacchia. *An introduction to variational inequalities and their applications.* SIAM, 2000. 20

[106] C. Klapproth, A. Schiela, and P. Deuflhard. Consistency results on Newmark methods for dynamical contact problems. *Numer. Math.*, 116:65–94, 2010. 67, 87

[107] A. Klawonn and O. Rheinbach. Robust FETI-DP methods for heterogeneous three dimensional elasticity problems. *Comput. Methods Appl. Mech. Engrg.*, 196:1400–1414, 2007. 105

[108] A. Klawonn and O. B. Widlund. FETI and Neumann–Neumann iterative substructuring methods: Connections and new results. *Comm. on Pure Appl. Math.*, 54:57–90, 2001. 105

[109] A. Klawonn, O. B. Widlund, and M. Dryja. Dual-primal FETI methods for three-dimensional elliptic problems with heterogeneous coefficients. *SIAM J. Numer. Anal.*, 40(1):159–179, 2003. 105

[110] R. Kornhuber, R. Krause, O. Sander, P. Deuflhard, and S. Ertel. A monotone multigrid solver for two body contact problems in biomechanics. *Computing Vis. Sc.*, 11(1):3–15, 2007. 67, 87, 88

[111] T. Koziara and N. Bicanic. Semismooth Newton method for frictional contact between pseudo-rigid bodies. *Comput. Methods Appl. Mech. Engrg.*, 197:2763–2777, 2008. 26

[112] J. Kruis. *Domain decomposition methods for distributed computing.* Saxe-Coburg Publications, Stirling, 2006. 105

[113] D. Kuhl and M. A. Crisfield. Energy–conserving and decaying algorithms in non-linear structural dynamics. *Internat. J. Numer. Methods Engrg.*, 45:569–599, 1999. 25

[114] K. Kunisch and A. Rösch. Primal-dual active set strategy for a general class of constrained optimal control problems. *SIAM J. Optim.*, 13(2):321–334, 2002. 41

[115] B. P. Lamichhane, B. D. Reddy, and B. I. Wohlmuth. Convergence in the incompressible limit of finite element approximations based on the Hu–Washizu formulation. *Numer. Math.*, 104:151–175, 2006. 106, 132

[116] T. A. Laursen. *Computational contact and impact mechanics.* Springer, Berlin Heidelberg, 2002. 11, 16, 17, 23, 25, 26, 69

[117] T. A. Laursen and V. Chawla. Design of energy conserving algorithms for frictionless dynamic contact problems. *Internat. J. Numer. Methods Engrg.*, 40:836–886, 1997. 25, 68

[118] T. A. Laursen and G. R. Love. Improved implicit integrators for transient impact problems – geometric admissibility within the conserving framework. *Internat. J. Numer. Methods Engrg.*, 53:245–274, 2002. 25, 68

[119] T. A. Laursen and X. N. Meng. A new solution procedure for application of energy-conserving algorithms to general constitutive models in nonlinear elastodynamics. *Comput. Methods Appl. Mech. Engrg.*, 190:6309–6322, 2001. 25

[120] T. A. Laursen and J. Simo. A continuum-based finite element formulation for the implicit solution of multibody, large deformation frictional contact problems. *Internat. J. Numer. Methods Engrg.*, 36:3451–3485, 1993. 16, 17

[121] A. Lauser, C. Hager, R. Helmig, and B. I. Wohlmuth. A new approach for phase transitions in miscible multi-phase flow in porous media. SimTech-Preprint 2010-34, Universität Stuttgart, submitted to Adv. Water Resour., 2010. 33

[122] P. Le Tallec. Domain decomposition methods in computational mechanics. *Comput. Mech. Adv.*, 1:121–220, 1994. 105

[123] P. Le Tallec. Numerical methods for nonlinear three-dimensional elasticity. volume 3 of *Handbook of Numerical Analysis*, pages 465–622. North Holland, 1994. 11

[124] A. Lew, J. E. Marsden, M. Ortiz, and M. West. Variational time integrators. *Internat. J. Numer. Methods Engrg.*, 60:153–212, 2004. 11

[125] C. Lunk and B. Simeon. Solving constrained mechanical systems by the family of Newmark and α-methods. *Z. Angew. Math. Mech.*, 86(10):772–784, 2006. 155, 159

[126] N. Mahjoubi and S. Krenk. Multi-time-step domain coupling method with energy control. *Internat. J. Numer. Methods Engrg.*, doi: 10.1002/nme.2878, 2010. 151

[127] L. Marini and A. Quarteroni. A relaxation procedure for domain decomposition methods using finite elements. *Numer. Math.*, 55:575–598, 1989. 127

[128] A. Marthinsen. Continuous extensions to Nyström methods for second order initial value problems. *BIT*, 36(2):309–332, 1996. 151, 153

[129] T. W. McDewitt and T. A. Laursen. A mortar-finite element formulation for frictional contact problems. *Internat. J. Numer. Methods Engrg.*, 48(10):1525–1547, 2000. 22

[130] R. Mifflin. Semismooth and semiconvex functions in constrained optimization. *SIAM J. Control Optim.*, 15:959–972, 1977. 33

[131] P. M. Naghdi. A critical review of the state of finite plasticity. *Z. Angew. Math. Phys.*, 41(3):315–394, 1990. 14

[132] P. Neff and C. Wieners. Comparison of models for finite plasticity: A numerical study. *Comput. Visual. Sci.*, 6:23–35, 2003. 14

[133] S. Nemat-Nasser. On finite deformation elasto-plasticity. *Internat. J. Sol. Struct.*, 18:857–872, 1982. 14

[134] R. A. Nicolaides. Existence, uniqueness and approximation for generalized saddle point problems. *SIAM J. Numer. Anal.*, 19(2):349–357, 1982. 56

[135] R. W. Ogden. *Non-linear elastic deformations.* Ellis Harwood Ltd., Chichester, and Halsted Press/John Wiley & Sons, New York, 1984. 11, 12, 13

[136] M. Ortiz and E. P. Popov. Accuracy and stability of integration algorithms for elasto-plastic constitutive relations. *Internat. J. Numer. Methods Engrg.*, 21:1561–1576, 1985. 25

[137] A. Pandolfi, C. Kane, J. E. Marsden, and M. Ortiz. Time-discretized variational formulation of non-smooth frictional contact. *Internat. J. Numer. Methods Engrg.*, 53:1801–1829, 2002. 25

[138] J.-S. Pang. Newton's method for B-differentiable equations. *Math. Oper. Res.*, 15(2):311–341, 1990. 36

[139] J.-S. Pang and S. A. Gabriel. NE/SQP: A robust algorithm for the nonlinear complementarity problem. *Math. Progr.*, 60:295–337, 1993. 26

[140] A. Popp, M. W. Gee, and W. A. Wall. A finite deformation mortar contact formulation using a primal–dual active set strategy. *Internat. J. Numer. Methods Engrg.*, 79(11):1354–1391, 2009. 16

[141] A. Prakash and K. D. Hjelmstad. A FETI-based multi-time-step coupling method for Newmark schemes in structural dynamics. *Internat. J. Numer. Methods Engrg.*, 61:2183–2204, 2004. 151

[142] M. Puso and T. A. Laursen. A mortar segment-to-segment contact method for large deformation solid mechanics. *Comput. Methods Appl. Mech. Engrg.*, 193(6-8):601–629, 2004. 16

[143] L. Qi. Convergence analysis of some algorithms for solving nonsmooth equations. *Math. Oper. Res.*, 18:227–244, 1993. 36

[144] L. Qi and J. Sun. A nonsmooth version of Newton's method. *Math. Prog.*, 58:353–367, 1993. 33, 34, 35

[145] A. Quarteroni and A. Valli. *Domain decomposition methods for partial differential equations.* Oxford University Press, 1999. 105, 113, 127

[146] B. D. Reddy and J. C. Simo. Stability and convergence of a class of enhanced strain methods. *SIAM J. Numer. Anal.*, 32:1705–1728, 1995. 13, 106

[147] Y. Renard. A uniqueness criterion for the Signorini problem with Coulomb friction. *SIAM J. Math. Ana.*, 38(2):452–467, 2006. 16, 17

[148] Y. Renard. The singular dynamic method for constrained second order hyperbolic equations: Application to dynamic contact problems. *J. Comput. Appl. Math.*, 234(1):906–923, 2010. 67

[149] Y. Saad. *Iterative methods for sparse linear systems.* SIAM, Philadelphia, 2 edition, 2003. 116

[150] O. Schenk and K. Gärtner. Solving unsymmetric sparse systems of linear equations with pardiso. *J. Future Generation Computer Systems*, 20(3):475–487, 2004. viii

[151] O. Schenk and K. Gärtner. On fast factorization pivoting methods for symmetric indefinite systems. *Elec. Trans. Numer. Anal.*, 23:158–179, 2006. viii

[152] C. Schwab. *p- and hp-finite element methods. Theory and applications in solid and fluid mechanics.* Oxford University Press, New York, 1998. 21

[153] L. R. Scott and S. Zhang. Finite element interpolation of nonsmooth functions satisfying boundary conditions. *Math. Comput.*, 54(190):483–493, 1990. 92, 93, 117

[154] B. Simeon. On Lagrange multipliers in flexible multibody dynamics. *Comput. Methods Appl. Mech. Engrg.*, 195:6993–7005, 2006. 69

[155] J. Simo and T. A. Laursen. Augmented Lagrangian treatment of contact problems involving friction. *Comput. Struct.*, 42:97–116, 1992. 26

[156] J. Simo and N. Tarnow. The discrete energy-momentum method. Conserving algorithms for nonlinear elastodynamics. *Z. Angew. Math. Phys.*, 43(5):757–792, 1992. 25, 81

[157] J. C. Simo and T. J. R. Hughes. *Computational inelasticity.* Springer, Berlin Heidelberg, 1998. 11, 14, 15, 16, 21, 26, 28, 41

[158] B. Smith, P. Bjørstad, and W. Gropp. *Domain decomposition. Parallel multilevel methods for elliptic partial differential equations.* Cambridge Univ. Press, 1996. 105

[159] P. Smolinski, S. Sleigh, and T. Belytschko. Stability of an explicit multi-time step integration algorithm for linear structural dynamics equations. *Comput. Mech.*, 18:236–244, 1996. 151

[160] V. Thomée. *Galerkin finite element methods for parabolic problems.* Springer, Berlin Heidelberg, 1997. 92

[161] A. Toselli and O. B. Widlund. *Domain decomposition methods – algorithms and theory.* Springer Berlin Heidelberg, 2005. 105, 116, 120

[162] K. C. Valanis and J. F. Peters. Ill-posedness of the initial and boundary value problems in non-associative plasticity. *Acta Mechanica*, 114:1–25, 1996. 15

[163] C. Wieners. Nonlinear solution methods for infinitesimal perfect plasiticity. *Z. Angew. Math. Mech.*, 87(8):643–660, 2007. 21, 26

[164] B. I. Wohlmuth. A mortar finite element method using dual spaces for the Lagrange multiplier. *SIAM J. Numer. Anal.*, 38:989–1012, 2000. 22, 105

[165] B. I. Wohlmuth. *Discretization methods and iterative solvers based on domain decomposition.* Springer, Berlin, 2001. 22, 109

[166] B. I. Wohlmuth and R. Krause. Monotone methods on non-matching grids for non linear contact problems. *SIAM J. Sci. Comput.*, 25:324–347, 2003. 16, 23

[167] P. Wriggers. *Nichtlineare Finite-Element-Methoden.* Springer, Berlin Heidelberg, 2001. 11, 16

[168] P. Wriggers. *Computational contact mechanics.* J. Wiley & Sons Ltd, 2002. 11

[169] S. J. Wright. *Primal-dual interior point methods.* SIAM, Philadelphia, 1997. 26

[170] S. R. Wu. Lumped mass matrix in explicit finite element method for transient dynamics of elasticity. *Comput. Methods Appl. Mech. Engrg.*, 195:5983–5994, 2006. 165

[171] X. Xu and L. Qin. Spectral analysis of Dirichlet–Neumann operators and optimized Schwarz methods with Robin transmission conditions. *SIAM J. Numer. Anal.*, 47(6):4540–4568, 2010. 119, 120

[172] B. Yang, T. A. Laursen, and X. N. Meng. Two dimensional mortar contact methods for large deformation frictional sliding. *Internat. J. Numer. Methods Engrg.*, 62:1183–1225, 2005. 16

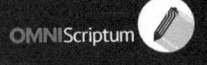

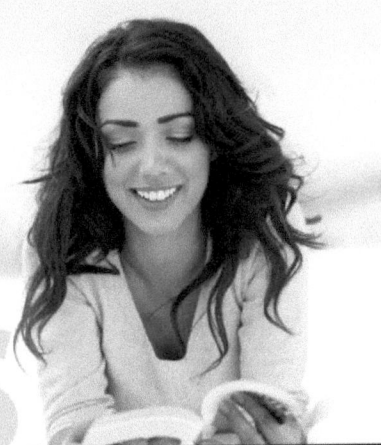

MIX
Papier aus verantwortungsvollen Quellen
Paper from responsible sources
FSC® C105338

Printed by Books on Demand GmbH, Norderstedt / Germany